FOR ALL THOSE WHO YET LOVE AND DREAM

TABLE OF CONTENTS

CHAPTER 1

LUMINOUS WORLD: BARON KARL VON REICHENBACH

ACADEMICIAN

One chapter in forgotten science history introduces one of the greatest researchers of all time, whose investigation of basic life-related energies stands paramount in the history of qualitative science. His name forgotten and ignored by modernists, the life and work of Baron Karl von Reichenbach stands as a monument. He is a true scientific legend, a giant, a reminder that the world is more marvelous than we are led to believe by those who misalign our perceptions and misdirect our views. It is for this reason that I have chosen to begin the LOST SCIENCE series with his biography.

Our story begins in the Kingdom of Wurttemberg. Born in Stuttgart (1788), Karl von Reichenbach became a laudable personage of great scientific stature. Known for his humility and deep sensitivity, the enormous scientific contributions made by him in European industry and research are legendary. His father, the Court Librarian, was able to supply Karl with a rich reserve of arcane treasures. Books of a most wonderful kind flooded his young life with the stimulating and refreshing visions of a hundred forgotten naturalists.

After a stormy youth as a chief conspirator against the Napoleonic occupation in Germany, Karl emerged as a scholar of high merit. Earning his doctorate in natural sciences and theology, he became a knowledgeable and enthusiastic contributor in chemical, geological, metallurgical, and meteorological sciences.

Very gradually distinguishing himself as an exemplary industrial engineer, he began establishing ironworks (Villengen, Baden), charcoal furnaces (Hausach, Baden), metallurgical and chemical works (Blansko, Moravia), steelworks (Turnitz, Austria), and blast-furnaces (Gaya, Moravia). His wealth increasing beyond all reckoning, he purchased lands literally from the Danube to the Rhine. His fame and reputation as an industrialist and research scientist spread across Europe. In short, he was an exemplary scientist-mogul of legendary proportion.

Reichenbach discovered paraffin in 1830, one practical result of his own research with coal tar and coal tar derivatives. He did not stop making chemical discoveries of commercial impact however. From coal tar he extracted the antiseptic Eupion (1831), the preservative and therapeutic agent Creosote (1832), the indigo dye Pittical (1833) and Cidreret (a red dyestuff), Picamar (a perfume base), as well as Kapnomor, and Assamar. The successful commercial development of these organic substances brought him into greater wealth. Reichenbach's discoveries founded the huge dye and chemical industries by which Germany made legendary fortunes, which few but German chemists remember.

The Baron engaged the first exacting geological survey of Moravia. He loved all things natural, especially things which were considered extraordinary or rare. To this end he collected things such as meteorites, a collection which was famous in his day. While most academes ridiculed the notion of sky-falling stones ("aeroliths"), he published several notable treatises on the subject.

An avid observer of all anomalous natural phenomena, the various exotic forms of lightning and auxiliary atmospheric phenomena comprised another of his beloved scientific domains. His numerous and scholarly scientific descriptions of rare lightning forms and other strange natural occurrences flooded the periodicals of his time, making him an early enthusiast of what later would be termed "Fortean Phenomena".

Possessing the unlimited resources of both the very finest scientific materials and vast wealth, Baron Reichenbach ventured into scientific domains which few have successfully engaged. His pursuit of rare and erudite natural phenomena proceeded without limit. His fascination with the unknown became much more than a passionate devotion to an idle curiosity.

After completing his national industrial marvels, his devotion to these fascinations became a research endeavor of enormous thematic depth. Reichenbach discovered a glowing energy form which totally revolutionized his own world view, as well as those who earnestly followed his marvelous publications. Until his death in 1869, he maintained that nature was suffused with a mysterious luminous energy from which it derived its vivifying powers. How this great discovery was made begins the controversial period of Reichenbach's life, that period when he dared academic prejudice and plunged into the unknown.

SLEEPWALKERS

Scientific curiosity drew Baron von Reichenbach into a serious study of illnesses termed "neurasthenias". He was perhaps first to address these "psychosomatic" illnesses. Somnambulism, night cramp, night fears, and emotional hysteria were remarkably incomprehensible maladies. Each such illness was utterly fascinating to him. They seemed to affect only certain "sensitive" or "nervous" individuals. The mystical nature of these ailments, especially that of "sleepwalking", provoked fear among all classes of people during this time period. No class, ethnic, or religious group lacked victims of the conditions, which seemed to carelessly select its helpless victims. But beneath the surface of these extraordinary maladies Reichenbach suspected the extraordinary.

Most physicians and other professionals were as helpless before these strange maladies as their poor victims. There was no working theory by which to penetrate the mystery and discover, if fortunate, the cause and the cure. Many fell away to the common superstitions surrounding the conditions, fearful of venturing into its lairs. But Reichenbach was not one given to superstitious fear or fantasy. Though he suspected the extraordinary, he also expected to discover a new force at work: an undiscovered natural cause. Therefore he walked boldly into the study with no preconceptions.

The symptoms of "sleepwalking" was somewhat well known and greatly feared by the ordinary villagers. Having a monthly regularity, usually appearing with the full moon, he attempted to scientifically address the phenomenon. "Somnambulism", the technical term, is a condition in which sleeping individuals suddenly rise (yet asleep) and walk for long time periods until awakened. When in the grips of this strange seizure, the somnambulist walks out across precarious ledges and rooftops. In a complete state of trance, somnambulists remain absolutely unaware of their endangered states. Unaware of the often frightening heights to which their sleepwalking brought them, many somnambulists died (and yet die) through tragic falls.

Most victims of the condition were seen by their frightened observers, walking with eyes opened. Sometimes these persons spoke aloud in gibberish, moving their hands about as if conversing in a state of full consciousness. They could not be waken when in this condition. It was as if they had slipped into another world, within which they led other lives. When under the strange spell, no manner of arousal could break their trance-like state. Prisoners to forces beyond the human understanding of the time, few would escape the cruel grip of their illness until death. Lives wasted by the malady which none dared mention, they lived out their time in quiet fear and obscurity.

Dreaded by parents of young children, the outward first signs of this catatonic grip began as severe and sudden muscle cramps. The illness progressively worsened with age, children absorbed into the somnambulistic world with frightful speed. Ultimately these victims would die in some horrid and freakish accident during a sleepwalking episode. Their bodies in a strange state of muscular catatonia, it was possible for these victims to sustain deep gashing wounds entirely without pain until awakened. Widely separated sleepwalking cases seemed unified on specific nights of the month, a bizarre coalition.

The condition seemed especially aggravated during nights of the full moon, arms reaching out toward that celestial body as if signalling mysterious spiritualistic messages. This was the source of superstitious fears surrounding the phenomenon, the almost paganistic movement which these persons displayed in seeking out the moon. It was during these opened displays that whole villages might know the presence of a somnambulist. This is why parents were so careful to lock in their afflicted children, regardless of age.

Often the most unsuspecting stimulations would arouse them from the seizure after a certain time had passed, where sharp pinpricks otherwise could not elicit even a vague conscious response. A sudden swoon, and the victim would "come to their senses", often with hysterical fear and shock the result. Imagine innocently going to sleep, and then awaking with a start atop a precarious ledge or rooftop alone! Many victims of the sleepwalking illness had to be locked into their bed chambers during the night by caring parents, some of whom had prematurely aged with the strain. Most victims who were severely afflicted could never hold steady employment or perform the simple duties of married life. Most withered away behind walls. Victims. Unknown and unfulfilled lives.

There were others who suffered from "night fears" and emotional "hysterias", often provoked into episodes by the approach of sunset and the full moon. Thought to be allied with madness and spiritism, "night phobics" and "somnambulists" were feared as persons influenced entirely by occult forces. Most townsfolk feared that the condition was a contagious evil. Those with sleepwalkers in their families were often shunned by all others. Called "lunaticks" by most country folks, the conditions were considered a curse, a plague, a mark of evil, the opened cause of some horrid unconfessed deed. Many families having these afflicted victims were barred from religious attendance. Gradually separated from social mainstreams, these families eventually perished in forced obscurity.

Judging from the symptomology and the equally strange "lunar attractions", Reichenbach believed the illnesses were a response to more fundamental natural forces. Other colleagues were not willing to risk their reputations by making any statements on the issue. Because of a long standing prejudicial poise, academes were not willing to study these specific illnesses or so-called "occult" forces. Too great a change of scientific foundations would be required. Furthermore, they challenged his data gathering methods, declaring that no strict quantitative measurements could ever be made in the study of "hysterias". In the absence of such kinds of data, his study would fall apart.

It was clear that influences such as these could never be accurately assessed without the human agent as subjective observer. The human subject was viewed by Reichenbach to be a laboratory, a world in which perceptual energies operate. There was no other means for studying such phenomena. Until new and organismic meters could be developed, the human agent was the laboratory. This new scientific poise, a shift from quantitative to qualitative, attracted the critical attention of his colleagues.

A new qualitative view of natural phenomena would gradually reveal a forgotten world where permeating energies were discovered everywhere. Many academes viewed this as a dangerous "return to superstition and ignorance", but the Baron would later state that nature was fundamentally composed of experience permeating energies. Their influence, he insisted, so deeply suffused observers that quantitative methods could not sufficiently reveal their presence.

Only the human organism, as laboratory and detector, could best serve as sensitive indicator of otherwise unrecognized "mystery forces". Psychic forces could not yet be directly measured by laboratory instruments. He fully anticipated that later scientific developments would provide some kind of material detector for these mysterious powers, meters which imitated organismic response. Several such devices were later developed and implemented as interactions between materials and human energies were accidentally discovered (Torr, Joire, Bose, Pavlita, Meinke, Hanks).

Determined to discover the true natural cause of somnambulism and its allied emotional hysterias, he gathered literally hundreds of case histories from the surrounding countryside. Most were afraid to speak of the condition. Baron Reichenbach made the very first venture into a new scientific territory when once he observed the phenomenon for himself. The task would first entail a sociological profile, filled with new philosophical insights and new phenomena. Data itself would require philosophical re-interpretation until satisfying models for the problem could be developed. Only a penetrating mind could see the implications which innumerable case studies would soon reveal. Furthermore, the acquisition of necessary data would entail visiting and consulting with hundreds, possibly thousands of families before any definitive statements could begin.

The compassion stirring within him had scientific weaponry as its advantage. He would seek out the strange cause behind the terrifying effect. Ultimately, this research might lead to some kind of a cure. Few would reject his powerful, confident, and benevolent presence. It is doubtful that many other investigators could have found such ease in communicating the true motives of his search. Establishing trust with his "sensitives" was the first real step in securing data of greater content. Parents, however aged, were very quick to tell the Baron just when and where the first occasion of sleepwalking began in their own children.

In order to truly comprehend more of the attributes associated with these maladies, the Baron probed victims with deep and personal questions. In this, he preceded Sigmund Freud's talking-cure method. Despite his lengthy and confidential discussions with sleepwalkers, he noted that cures did not result. Talking did not remove the symptoms by helping the victims to "face their fears". No, he continued to believe that this peculiar class of maladies had a deep, unrecognized natural cause.

As case studies became less informational and more human, he realized the gravity of what later were termed "emotional illnesses". The Baron recognized that, despite the emotions conjured by the illness, emotion itself was not the root of the condition. After compiling and studying thousands of such records, Baron von Reichenbach discovered certain curious features which always accompanied those who were afflicted with night hysteria and somnambulism. As he went about carefully seeking his cases studies the prevalence of these phenomena truly shocked him. There were cases everywhere.

Parents told that night cramps, night fears, and sleepwalking appeared when their children were yet very young. In most cases, the conditions gradually disappeared with increasing age. Night fear, night cramps, and somnambulism always followed the appearance of specific lunar phases, reaching maximum expressions at full moon. Afflicted individuals were not all older in years. Very little children were also afflicted. These little ones were too young to be actively aware of moon-related superstitions or frightening pagan fantasies. Their particular form of hysteria or somnambulism was not a response to family atmospheres of fright. It was a natural response to an external natural influence.

The Baron, observant and sharp-witted, noted that most of the visited families were not particularly superstitious people to begin with. Neither were these people excessively religious or religiously fearful of "lunatick" influences. Though some may have resorted to old folk-arts of exorcism and talismanic magick, most had given up the search for an immediate relief to their plight. The parents of sleepwalking children were care-worn, silently suffering individuals.

SENSITIVES

So, there was an overwhelming number of cases where very young children began manifesting night fears and sleepwalking without any pre-established "provoking atmosphere" of religious dread. These were spontaneous conditions, manifesting so early in childhood that they could not be the result of suggestion. Were the condition an hereditary weakness, then more family members would suffer from it. But this was not the case.

In addition, the cases with which the Baron was principally intrigued were all widely isolated cases. His case studies revealed a wide regionally dispersed incidence of the malady. The Baron gradually expanded his inquiry concerning nocturnal phobia and somnambulism across Europe over a wider population cross-section. There were so many cases to chronicle. Each told of the same pattern of symptoms. The age when this malady first manifested often commenced with the child's ability to walk. The question was disturbing. Why would healthy young children suddenly exhibit the varieties of sleepwalking symptoms?

His great range of cases now taught that the distribution and occurrence of night fears and somnambulism had no geographic preference. Specific bordering European nations did not exhibit variations in the incidence of these conditions. Family structure did not influence their dread appearance. Otherwise dysfunctional families often did not produce case studies. Religious persuasion had no discernible effect either. He voiced the opinion that certain land regions might be devoid of sleepwalkers, being revealed only through more refined examinations of social groupings. There were no preferences with respect to sexuality. Male and female victims showed equal representation, although women were more frequently cited by professionals as "more susceptible to hysteria and night fears".

All of these peculiar maladies were horrid secrets, kept well within the family, never openly mentioned. Social taboo maintained the wall of secrecy behind which afflicted persons maintained their own safety. Since sleepwalking and night fears were each traditionally associated with lunacy, to admit being a sleepwalker or having uncontrolled emotional reactions at sunset could be a life-threatening affair. There were more ancient times when whole families, having a single such sleepwalking member, were burned at the stake. Many noble families and persons of wealth were found by him to have had hysterical or somnambulistic family members.

Public exposure by hostile neighbors could be the prelude to "institutionalization". Since most cases kept the malady a grave secret, few ever discussed the problem enough to "share symptoms". This was another major recognition. Many afflicted individuals suspected that others were inflicting "the evil eye" upon them. Ritual exorcisms, common in the folk-religions of the Mediterranean and Eastern Europe, were summoned with great caution. Fear of exposure by authority figures of dubious alignment prevented both the cry for help and the sought cure.

Those afflicted with the symptoms of sleepwalking maintained these symptoms throughout adulthood, only occasionally relieved by greatly misunderstood alleviations. Furthermore, the Baron found that spells of sleepwalking were usually preceded by curious prickling sensations, "cramps", and muscle "spasms". This muscle "tetanus" warned of the impending crisis which signalled imminent sleepwalking episodes to concerned family members. Nightfall threw those having nocturnal phobia into paroxysms of crying and trembling with apparently no reason at all.

The Baron isolated the principle signals of the condition's temporal onset. Varying in intensity with subjects and seasons, piercing muscular pains afflicted his sensitives in various parts of the torso: a sign that sleepwalking would soon commence. Sleepwalking was similar to cataleptic seizures, the victim losing complete consciousness during the attack. Parents recognized the early symptoms, preparing each month for sleepless nights. The condition is far more prevalent today than is commonly assumed or admitted.

Watching little children being seized with these terrible and uncontrolled behaviors broke the hearts of many decent and perplexed parents. Physicians, often called for persistent "nightmares" or "fevers" recognized the signs of sleepwalking. Vain was the help of physicians, whose herbal preparations offered no real cure. Sleepwalkers were never themselves sure whether their dreams were real excursions or fantasies, lacking all sense of reality. Certain individuals interviewed by the Baron remarked that their hands and arms became stiff, painfully twitching uncontrollably just as the full moon phase was approaching.

Other cases, deemed "hysterical", remarked that these cramps could be painfully felt throughout the entire torso during this lunar peak season. Such body-permeating tetanus blocked breathing, stiffening the torso as in death. In many cases, this muscular tetanus resulted in complete immobilization and partial paralysis throughout the week prior to uncontrollable sleepwalking episodes in certain cases. Little could be done to give them aid. The body stiffening, fright the result, uncontrolled shaking, those who watched were utterly helpless.

Through his very extensive collection of case studies, the Baron was able to predict the profile of persons most susceptible to this strange malady with an equally curious precision. He quickly discovered that such "sensitivity" was not at all uncommon. In fact, he was thoroughly surprised to find that such sensitivity permeated all classes and nationalities. It was easy to find subjects. Chaperons always present, the experiments were conducted in all dignity and scientific rigor. The Baron was meticulous and patient, recording everything that he observed with a special penetrating clarity which became his own unique trademark.

This peculiar neuro-sensitivity lay at the very bottom of their equally peculiar condition. This neuro-sensitivity was an organismic state with which they were each born. The first group comprised individuals of a very nervous and sickly nature whose extreme neuro-sensitivity commenced only with ill-health. These were termed his "sick sensitives". The second group were comprised of vibrantly healthy individuals having extreme sensitivity to all stimuli: "healthy sensitives". When sensitive states were examined, the Baron found a series of amazing and unsuspected correlations.

Sensitive individuals evidenced special neurological states: heightened states where feeling was easily stimulated and prolonged. Sensitive persons felt and sensed more of the world than most other persons could. In this sense they were indeed "special and distinct". The Baron registered a large population of sensitives, first from his own districts and provinces. These could be summoned to the Baron's estate for an exhaustive series of qualitative observations.

He now had lists of many hundred reliable and true sensitives from all classes and European nations. This, for the time period, was a remarkable feat. Moreover, he had the financial means to maintain his controversial work, besides transporting and housing his sensitives throughout the lengthy examination periods required by his rigorous and thorough qualitative methods.

The Castle Reisenberg could comfortably house his guests under supervision and safety, servants supplying all required needs throughout the many days of research. He was best suited for such a task, having the time, wealth, and academic position to engage the undertaking. He carefully arranged observation of such sleepwalkers with an aim toward dispelling the problem completely.

Reichenbach's "sensitives" were the "extremely susceptible persons" treated by Franz Anton Mesmer. Persons prone to "hysteria" and "neurasthenia" were those later examined and treated by Sigmund Freud. What is generally not well known is the historically strong connection among these three personages.

ANIMAL MAGNETISM

Vilified and outcast by the European medical guild of the 1700's, the name of Franz Anton Mesmer remains as mysterious today as it was in his own day. Mesmer's earliest work centers around the development of a strange battery-like accumulator by which his earliest and most famous cures were wrought. Since medical practitioners accused him of exclusively using hypnotic spells and suggestions, the historical reference to this battery remained shrouded in forgotten archives, a true mystery. Where, however, do we find its most complete description? Nowhere else but in the writings of Baron von Reichenbach! No doubt the result of his father's library, this lost information was fortunately preserved. According to Reichenbach's own reference, hypnotism is not what Mesmer employed in his work at all.

The battery was developed slowly, the result of an attempt to mimic conditions found at certain "sacred spots" of the Austrian countryside. Mesmer constructed the battery to imitate natural configurations. It has a decidedly organic aspect to its internal structure. A grounded device, the wooden tub housed several thick layers of wet vegetable matter and iron slag. A single iron rod ran through the entire composition, closed at the top with a circular wooden barrier. While working with the development and application of his special battery, Mesmer himself received a distinct impulse when he touched the single iron pole. Electrostatic shock was not unknown. Familiar with these, he declared that this energy was completely different in nature, having a more body permeating and "thrilling" aspect when experienced. But, none of Mesmer's subjects reported that the effect was identical with common electrostatic shocks.

These shocks were thrilling, vivifying, exciting, and definitely curative. Those who touched the exposed rod experienced a sudden tingling rush which permeated their bodies, bringing delighted shrieks or sighs, but most often stimulating a sudden unconsciousness. Many seemed to faint to the ground, although on awakening none reported pain or spasm at the onset of the "fainting". Mesmer's attendants caught "fainting" patients. Many of the upper class came only for the diversion, but found themselves leaving the experience in some way relieved of unsuspected emotional blocks. When these individuals awoke, they were apparently relieved of inaccessible, life-distorting emotional blocks. Mesmer simply watched cures taking place. Patients were simply directed to grasp the free terminal of the large tub-shaped battery. The nobles treasured his science, seeing him as a modern alchemist. Their desire was to keep him near the Court.

In later years, critics failed to consider the Mesmer battery (the "baguet") while verbally slandering Mesmer's character. Nonetheless, he managed several notable cures among the upper class. These remained his loyal patrons until death. Upon examination, electrostatic energy could not have been developed by the Mesmer battery. The arrangement represented a electrical short-circuit. In addition, the monopole could not have produced adequate electrical voltage to achieve such permeating physiological effects. Furthermore, no low voltage or high amperage current could have been developed in this structure.

Last, electrostatic shocks are not vivifying. They do not increase life-potential. They can kill. They are not thrilling, they hurt. They do not bring relief, they produce tension. Mesmer had discovered a distinct form of energy which few academicians and other professionals refused to acknowledge. The energy with which Mesmer dealt was classed among those peculiar vivifying energies termed "vitalistic".

The name Mesmer remains significant in the forgotten science which bridges the medieval scientific arts with the science of the early Victorian Epoch. The vilification of Mesmer entailed greater cause than most suspect. Obvious is the danger which Mesmeric medical practice posed to ordinary physicians. This is why most professionals do not like the association of Freud with either Mesmer or Reichenbach. Yet, like Mesmer himself, the theme and association is an indelible historical fact. Baron von Reichenbach had carefully studied all the historical references concerning Mesmer while yet in his youth. Familiarity with the widest possible range of different scientific topic areas was a lesson graciously learned from his father, the Court Librarian.

Sigmund Freud, unable to help certain "hysterical" patients, travelled to France in order to learn from a neurologist who used both Mesmerism and hypnotism, Dr. Jean Martin Charcot. His first work began with an absolute reliance on hypnotic methods. Later awareness of 'subconscious" symbols and emotional associations caused him to deviate from Freud's original means for treating neurasthenia and hysteria.

"Hysteria" and "neurasthenia" typified those whose temperament seemed highly strung and nervous. They also were chronically fatigued and indifferent to life. In addition, such persons were noticeably unable to experience the normal intensity of their senses. Thus separated from the world at large, neurasthenics and those prone to hysteria often progressed into deeper states of alienation: the journey from neurosis to psychosis. This collective title branded their victims with such completely negative associations that none dared enter the study which could possibly lead to a "cure". Persons designated as neurasthenics and somnambulists were not treated in the same manner as others. With such secrecy, ignorance was given its freedom.

Any other researcher besides Baron von Reichenbach would never have achieved such a great depth of accumulated case studies. But local people and other of his countrymen were quick to respond to his kind and compassionate queries. Years later, studying Reichenbach's work, Freud rejected causes of neurasthenia which involved "external influences". Citing the power of dreams and symbols as a distinctly permeating energy, he delved into supposed suppressed memories and painful "traumatic" life episodes. This change of direction did not adequately and effectively provide a treatment which changed patients in a short time period.

Freudian "talk-cure" required expensive, extensive, and intensely personal interviews between patient and physician. In many cases, significant cures were not effected at all. "Hysterical" patients did not find complete alleviation of symptoms after these supposed subconscious excursions. Reichenbach, however, looked for purely physical causes of the malady. If an external energy was influencing a person's physiology, then every illness termed "neurasthenic" could be cured.

Studying his case histories, the Baron realized that these individuals were not "raving mad lunaticks" . These were not persons enslaved to "inaccessible memories". They were ill, but their malady did not proceed from emotional or mental cause at all. In fact, far more persons evidenced the condition in mild form than most suspected. The Baron did not believe that dreams or suppressed memories were the real roots of sleepwalking at all. Neither did he therefore apply any of the "talking tools" later exclusively implemented by Freud to free his patients.

Negative dreams, images, phobias, and associations, the so called subconscious motivators, always seemed to follow rather than precede episodes of sleepwalking. Children, far too young to have formed any such associations, were some of the noteworthy victims. Reichenbach believed that subconscious "inductions" followed more mysterious natural energies. Negative thoughts, emotions, and imagery polarized around the entrance of such natural energies. Sleepwalking symptoms would first appear when these mysterious currents entered a person's physiology. All the foul and negative associations would follow much later. Reichenbach believed that the sleepwalking malady was a consequence of external, body permeating forces. He expressed the belief that a force, a new and yet unmeasured force, was the cause of all these case histories.

An early investigator of qualitative phenomena, the Baron was well aware of the phenomena which often attend such research. There were those who criticized the use of human agents as measuring tools. These colleagues protested that human subjects were often easily influenced by all kinds of suggestions and other effects, and were therefore completely unreliable. Reichenbach agreed that verbal suggestion was a problem. He had learned, at the very onset, not to "lead the subject on" with excessive questions. But as to their sensitivity to "all kinds of effects", well...that was the point of using them! Only human agents could experience the very effects and influences which he was trying to detect!

Treating the phenomenon of suggestion against true perception, he often created ecstatic tensions in a room to test the honesty and reliability of his sensitives. Designed to evoke suggestions in such subjects, the Baron became an experimental adept in these regards. He selected out only those sensitives who were adamant concerning their perceptions, eliminating those highly suggestible persons who could easily fault his stringent scientific requirements with their imaginations.

The Baron was well aware that suggestive questions could falsify all of his accurate data. Truth, after all, was what he sought. Each of his numerous sensitives corroborated their experiences without provocations. He later formally reported the clear distinctions between actual sensitivity and mere suggestion, a daring but necessary disclosure. The Baron decided to utilize sensitives from every social class and nationality where possible. No other academician would dare touch the issue for fear of losing title and position. Just as had been done with Franz Anton Mesmer, many feared reprisals for the mere association with "vitalistic" research. Most researchers of high rank were thus eliminated from the most exciting and astounding research venue of the early Victorian Era. Later Victorian academes broke the conservative tradition and plunged into the study of vitalism, replete with its references to animal magnetism and, or course, Mesmer (Crookes, Lodge, White, Tesla, Lahkovsky).

Reichenbach insisted that Mesmerism had nothing to do with hypnotism. Furthermore, he found that hypnotism had no curative effect on somnambulists. No manner of suggestion successfully intervened with the sleepwalking activity. He therefore placed no confidence in the purely psychological cause of somnambulism. He did not equate Mesmer with hypnotism, knowing and practicing both hypnotic suggestion and "Mesmeric passes".

He next intensely studied Mesmer's "animal magnetism". Reichenbach discovered that the application of this force in no way involved the use of hypnotic suggestion, a verbally applied means. The ill-famed "Mesmerism" involves the passing of hands over persons who are afflicted with diseases in the hopes of effecting alleviations. He mastered the method with great proficiency. In Mesmeric passes of the hand one could distinctly sense the movement of a mysterious "influence", which proceeded from its administrator to the recipient. It was after all a simple exchange of an unknown energy which mesmer termed "animal magnetism". Subjects, in several instances, were attracted to the hand of the administrator, hence the term "magnetism". Reichenbach found that these "magnetic passes" of the hand over somnambulists could temporarily reduce their symptomologies of cramp or muscular tetanus.

OCCULT FORCE

He now realized that the entire physiological organization of these unfortunate sensitives was somehow being influenced by an aggravating agency which would be found in the external natural world. Just as his passing hand could bring relief, so too a mysterious "passing energy" brought them into misery. The root and cause of all emotional ailments had to be an invasive external force. This agency had to be a force, a radiance, or a current which acted as an allergen to sensitive persons. This, he insisted was the cause of all these bizarre symptoms. He therefore tried to isolate this "occult force". But, where would he begin? How would he find an energy which had been attended by so many centuries of fear and mystification?

Energies have sources. Energies manifest as radiances and currents. What was this fundamental "occult" energy? Was it electricity? Was it magnetism? What was animal magnetism? Was it a combination of known forces, or something completely distinct? The questions outweighed the answers.

He had found that equally basic environmental states were required before somnambulism would be triggered. It was also obvious to him that lunar influences were, of course, the "forbidden" causative agency. At first he dodged this issue completely. Reichenbach explored the possibility that some accepted, though previously unrecognized force combination, might be the "irritant", the true natural cause of sleepwalking. Reichenbach did not first grope for the improbable, proceeding from the known to the unknown.

Was the effect a chemical one? Were certain strange aerial agencies the cause of sleepwalking? Could sleepwalking be an allergic reaction to some wind-spread dust? It was obvious that not all persons were plagued with the yearly onset of "hayfever", despite its wide manifestation during late summer. The pollen of trees and flowers did not produce the allergic symptoms in all people. There were a few individuals who manifested specific allergic reactions to roses or gardenias, oak trees or dogwoods, goldenrod or hay. In a similar way could not this sleepwalking not be an allergic reaction? But, what allergen continued to exist during the snowy winters?

In a series of very basic hypothetical assumptions, he cited electrostatic energy. This seemed the first likely choice. Who has not known a sleepless night? A permeating natural condition might directly impact the delicate nerves of sensitives. If some mysterious physical force was directly influencing these persons, then perhaps most people were basically "insensitive" to its pervasive influence. Somnambulistic muscle tetanus bore an unmistakable likeness to electrical shock responses over a long period of time. The Baron at first believed that sleepwalking might be caused by some kind of greatly sustained regional electrification.

If the somnambulistic condition was an irritable response to electrifications, similar to allergic response, then it would be possible to measure neighborhood electrostatic strengths against the sleepwalking response. If the "permeating force" hypothesis was going to work at all, it certainly required a more complete and practical analysis now. Qualitative experimentation would be the necessary route toward ascertaining this truth, since only sensitives could reveal the effects which he sought. The sensitives were his "detectors". But he could balance their response against a quantitative measure.

Neurosensitives might be more susceptible to such regional electrical irritations precisely because their neural apparatus is so different. Perhaps their myelin neurosheaths were thinner than normal. Perhaps their inter-synaptic spacings were closer. Perhaps their neurochemistry produced enhanced and prolonged neural firing. Then, any electrostatic environment would send them into convulsive fits and spasms for a sustained period of time.

Yes. Perhaps an invisible electrostatic condition was activating the primary tetanus response in certain "sensitive" individuals. Forces and irritability. His thesis was beginning to take a more scientific form now. He formally postulated that these episodes of muscle cramping, painful twitching, irritability, and finally sleepwalking was actually the result of a special sensitivity to natural electricity. The response resembled a prolonged and compounded "electro-tetanus": a physiological intolerance to a weak, though progressive regional electrical state. But this is not what he found.

His own familiarity with strange forms of lightning led him to believe that invisible irritating electrostatic shocks, a quiet kind of "heat lightning", might be the first triggering mechanism of somnambulism. Invisible electrostatic shocks were measured across large areas of ground. Like "heat lightning" these covered very large areas of ground, sending almost imperceptible electrical shocks throughout grounds, buildings, animals, and people. Were not animals observably disturbed long prior to lightning storms? Sensitive meters measured sudden electrostatic impulses when ground-connected. Such measurements indicated that large area "invisible" lightning discharges were shuddering through the ground incessantly. Such shocks would definitely be perceived by the body as an irritant.

The central problem with his electrostatic hypothesis was that many sleepwalkers did not commence their trance-like behavior during these surges at all! Neither did they respond to thunderstorms. In fact, though houses were riddled with many thousands of surging electrostatic volts during such storms, these individuals did not show any of the muscle tetanus symptoms. Not so much as a single muscular spasm, the usual onset of the sleepwalking episodes, was observed during severe thunderstorms.

If not electrostatic, then what? Magnetic influence? Perhaps the body was sensitive to sudden fluctuations in the terrestrial magnetic field which stimulated the seizures. Such forces had been measured during solar eruptions, visibly rocking and shaking compass needles. How much more would supersensitive human neurology respond to such pervasive magnetic influences? Applications of bar magnets to sensitive individuals already proved to produce muscle tetanus reactions. In fact, on several occasions, the Baron stimulated a trance-state in some sensitives merely by passing a bar-magnet over them. Painful cramping and trance were each repeatedly induced among several different sensitives until the Baron was sure that the response was real.

It was only necessary to perform several experiments now to measure the regional magnetic fluctuation as sleepwalking symptoms began to appear. Only then would the correlation be sealed and proven. Measurable magnetic surges occurred throughout the day and night for weeks. Surprisingly however, there were few correlations between these magnetic surges and the somnambulistic symptoms. How could this be? The hand-held magnets produced both defined tetanus and trance states, while terrestrial magnetic surges did not. Here was a true mystery. Bar-magnets were several orders more powerful than the regional influence. But the regional influence was supposedly the cause of sleepwalking symptoms.

Here was a true conundrum. How were permanent magnets and terrestrial magnetism different? They quantitatively measured out to be the same force. Master of the scientific method, Reichenbach had exhausted the existing registry of academically acknowledged forces. With the exception of the bar-magnet activity, no correlation could be shown to exist between terrestrial magnetism and sleepwalking. Was there an unrecognized force then? What force was projected by bar-magnets which was not projected by terrestrial magnetism?

By now, he seemed to have exhausted the known forces and their combinations. It was a noble effort, an apologist's elegant attempt. He pondered long on these questions. Though he originally dodged the issue, he was ready to try the last resort. There was that one common factor in all somnambulistic case studies. Only one. And that common element had more mythology associated with it than most academicians cared to recount. It was...the moon. Something about the radiations from the moon.

MOONLIGHT

The thought that somnambulism might brought on by some aspect of moonlight was a strange and "non-academic" one. Nevertheless, the presence of full moonlight always produced the most dramatic sleepwalking episodes. The Baron began analyzing his own findings, now correlating "common features" among all his many thousands of case studies. It paid well to do such basic research, acquiring data with no predetermined schema in mind. There were several scientifically plausible connections to his data in this hypothesis.

Certain lunar phases were always marked by the parents as signs of the impending sleepwalking episodes. This line of thought brought on a revolution in his scientific approach which led to a startling discovery. If sensitive neurophysiologies responded to mysterious "permeating" regional influences, then these influences were completely unrecognized by academic science. The new force which he originally proposed.

Baron von Reichenbach produced a most remarkable series of experiments whose single aim was to discover the now obvious connection which existed between the lunar radiance and somnambulism. Most of his colleagues, esteemed academicians, scoffed at such a very simple and obviously superstitious-laden hypothesis. The age-old association of mental illness and lunar phase could not be taken seriously!

Reichenbach was not dissuaded from his straight course now. He would test moonlight on his sensitives, one by one. His method began with a simple series of tests in more controlled environments. One by one, sensitives were permitted to rest within a completely darkened room. The curtains drawn, the lunar light completely absent, he observed a small alleviation of their muscular symptoms. This first discovery revealed the curious and sometimes "spontaneous cure" which these persons often experienced when remaining completely indoors during these lunar phases.

Into a light-sealed room, the Baron arranged for a thin ray of moonlight to impinge on particular parts of the face, arms, and hands of sensitives at rest. The first sensation which sensitives reported was a disagreeable warmth, an uncomfortable irritation which flooded their being. A claustrophobic sensation permeated their bodies and they become restless. Here then was the first symptom of the sleepwalker in action.

With longer exposure, the cramping and muscular twitching gradually began. Uncomfortable heat and muscular tetanus began manifesting in their bodies. The Baron found it amazing that the removal of moon rays revealed a long-lingering effect. Sensitives maintained their greatly irritated states by exposures of only a minute or two! Though this effect very gradually faded away, it offered evidence of the allergenic reaction which he had previously hypothesized. Mirror-reflected moonlight gave weaker, but similar effects.

Remarkable! A mysterious and previously unknown force was here, in the moonlight itself. In addition to these pain-inducing effects, the Baron observed that sensitives were strongly attracted into the moonlight. They each displayed a desire to touch and be drawn more into the moonlight. Could this physiological attraction explain why they so often, quite unconsciously, were led outdoors during their trance-states?

Here then was a great discovery. Moonlight did indeed produce "allergic" irritations in certain sensitive persons. An unexpected discovery of enormous import. He published these early findings, only after confirming these findings in several hundred other cases. An allergic reactivity to lunar spectra existed among these strange neurosensitives perhaps because its spectrum contained certain elemental irritants. This hypothesis was very easy to test. The Baron placed a large glass prism in the moonbeam, splitting the lunar light into its own distinct rainbow. The lunar spectrum contained the sleepwalker's irritants in distinct colors. Lunar red produced the irritating heat, lunar green actually induced cramping on contact! Longer exposures to moonlight induced partial paralysis, amounting to a peculiar loss of consciousness. Thereafter, partial sleepwalking episodes were actually induced. Here then was the real cause of somnambulism and cramp. Once thought to be an occult or spiritistic phenomenon.

One by one, during individual sessions, the Baron gave a rod of glass to his sensitives and asked each to touch a shaft of moonlight which passed, well-insulated, through the room. Thrusting the rod into the lightbeam produced sickness, sometimes vomiting. Most certainly the glass was conducting something more than light. He next gave a plate of metal to his subjects, requesting that they introduce the metal into the shaft of moonlight. Movement of the metal plate into the light shaft produced the cramping response. Lunar radiations were being conducted by the metal plates directly into the body of the neurosensitive. How could ordinary moonlight perform such extreme responses? What exactly was it about moonlight which caused such powerful reactions in the musculature of neurosensitives? And how was moonlight conducted along metal plates? This was not light which entered their bodies.

Moonlight was stimulating a new conductivity throughout the metal. This was communicated into the sensitives by conduction, provoking the somnambulistic symptoms! Was this the mysterious energy he had previously hypothesized? A new series of experiments marked a clear division between his former apologetic and latter revolutionary research. He began devising novel apparatus specifically for making precise qualitative observations. Positioning large metal plates on outer window ledges which faced the moon, he designed special conductive apparatus to which sensitives were exposed. Thick braided wires, each being brought into a chamber through the window, were held by each sensitive during individual examinations.

With the plate under full moonlight, he observed remarkable projective effects. When he simply approached each of his isolated subjects with the braided wire, they each began sensing the onset of severe muscular cramps through several feet of space. It was clear that an unknown energy was actually radiating from the termination! This energy began in light rays, was absorbed and conducted through metals, and then could discharge from conductors like light! Fantastic! The fact that human sensation alone could experience the effects validated the qualitative nature of Reichenbach's work.

More tests revealed that, for each sensitive, painful muscle cramping began within a specified distance from the braid end. Here was an "objective" measure of human sensitivity. He was now able to measurably distinguish among his sensitives. Those who felt the discharge from furthest distances were true "highest sensitives". Those who required contact with the braid were "lowest sensitives". Direct contact with the braid always gave the most severe and painful cramps. This contact always evoked prolonged reactions. A one minute touch often brought a one hour spasm.

This experimental arrangement was prepared and conducted thousands of times with hundreds of different sensitives. Always the same results were achieved. Sensitives experientially corroborated each of these findings with great accuracy. No device, no measuring instrument could achieve an equivalent energetic detection. In this first simple demonstration, a world of new forces and their interaction with matter was being revealed.

Asking each of his separate subjects to describe the currents which they painfully could feel while holding the wire braid, each independently offered identical statements. The contact seemed "hot ... irritating ... uncomfortable". But this was just what they reported that direct moonlight produced! Since the very same effects could be communicated through a wire braid, the energy had little to do with the light at all. It was obvious that a special energy, radiating from the moon, was merely conducted along light rays. Now he had to isolate and understand this species of energy with a determined effort.

He attempted measuring the electrical charge condition of the discharges. The most sensitive electroscopes showed absolutely no deflections when connected to the braid. This was therefore not an electrostatic manifestation which had been overlooked by scientists of the seventeenth century. In the same manner, extremely fine compass needles were not moved by the mysterious current. The energy was therefore not magnetic in species. It was just as he had determined previously. What then was it? Would other celestial bodies produce the same kinds of effects?

The Baron performed the identical experiment with solar light. Thrusting glass rods and metal rods into an isolated solar beam, sensitives reported an anomalous "cool" sensation. They actually preferred this energetic effect to that of moonlight since it was wonderfully refreshing. Using the large glass prism, Reichenbach discovered that sunlight also possessed specific spectral components in which the mysterious energy seemed most concentrated. A suffusing and irritating "heat" was reported in red solar light. This heat provoked a "stuffy claustrophobic" feeling but no muscular spasms.

A wonderful vivifying force was discovered in the violet spectral end of solar light. Sensitives felt stronger and more alive when touching wire exposed to solar violet light. In addition, sensitives were able to discern the "violet excitations" and the "red irritations" in metal objects which had been merely exposed to solar light for several minutes! This significant discovery opened a new door.

OD ENERGY

New knowledge! First, sensitives could actually detect and report the penetrating effects of a new energy species in moonlight. Second, this conducted energy was not itself light. Had it been light, its effects could not have provoked spasms by conductive contact alone. Not light, but carried along lightbeams. A curious paradox! Third, this conducted energy produced defined sensations when physiologically contacted and conducted. Fourth, this unknown energy was capable of being both absorbed and conducted along metal wires. Fifth, it was neither electrical nor magnetic energy. Sixth, it became radiant when discharged from points across space. Seven, after brief exposures, matter could store the energy for prolonged periods. Eight, mirrors could reflect the currents. Nine, all of his sensitives gave closely identical reports during their independent sessions.

Clearly this was a completely unrecognized force having its own laws and properties. It was an identity which evoked identical reports in the greatest majority of sensitives. This was indeed the mystery energy which caused somnambulism and its frightful attendant maladies. How this mysterious energy permeated whole regions of ground was now clear. Light energy saturated homes with the currents, and sensitives responded to the currents. Specific areas of land could probably absorb more of this energy than others. These places would show higher incidence of sleepwalkers. Similarly, there were probably places which were absolutely free of sleepwalkers.

Sensitives were again asked to rest in a completely darkened room and report their sensations. They were not told what to expect. In the dark, the Baron introduced the wire braid. He had placed it into various portions of solar or lunar spectra, waiting for their honest responses. Without prompting or coaching, each independently reported the very same sensations and impressions. As the braid came within a set distance, each immediately felt the strange sensations which radiated from the end. Moonlight always produced hot irritations and cramps. Sunlight violet always produced cool pleasantries, the reversal of night fears. Here was a more scientific distinction which differentiated the historical preferences for sunlight or moonlight in different individuals.

Neither of the lunar or solar current effects, however strong, registered on sensitive thermometers. An entirely distinct, and previously unsuspected world of forces was at work! Here was an opened door, through which all of the academically ridiculed energies found entrance. Human physiology was the invaluable tool by which it was detected. Here now was where the reviled ancient sciences found their vindication, evidencing the qualitative sensitivity of ancient naturalists who spoke of "the radiant world".

There were instances when sleepwalkers were haphazardly insulated from lunar light in the normal course of their family lives. When moonlight was prevented from reaching the interior of their homes, reports always mentioned a lower incidence of sleepwalking. Certain families had learned that their children could be "cured" from the sleepwalking affliction by simply putting them in a more interior room of the house. This effectively, and most fortunately, insulated them from lunar light. So, the Baron decided to maximize the conditions of this "insulation", and thereby cure all his cases.

Reports also mentioned that misty or cloud-laden skies blocked expected sleepwalking episodes, to the great relief of parents. Whenever full moon days were accompanied by overcast skies there was no sleepwalking. Reichenbach sought the development of an insulator to help these infirm persons, a means by which living quarters could be isolated and "sterilized" from all photo-contaminations. Although thick barriers could not block out the strange currents completely, proper insulation was not without its curative effects.

It was also found that extreme neurosensitives could feel the effects of both moonlight and sunlight right through walls and ceilings! The Baron's persistent and repeated experiments found that woolen cloth, especially in heavily woven layers, actually blocked the mystery currents. It was possible now to help those suffering from sleepwalking now!

Drawing out heavy curtains over windows and bedsteads would resist and block the strange energies. Plenty of sunshine would actually be therapeutic for their "nervous" physiological states. Extreme such sensitivities would require that heavy tapestries be placed around sensitives against lunar light. It was apparent that these mystery currents were powerful allergic agents: photo-allergens. Without the insidious implications of madness and dysfunction now, the victims of somnambulism had found their help. The Baron established rooms for their recuperation. Those suffering from night hysteria and night cramps took sunlight and found that their symptoms were disappearing in time. The cures were miraculous and mystifying.

The causative energy was itself a true mystery, requiring a name. Looking into Germanic mythology, he sought some term which could describe the permeating nature of this strange current. "Odos" in Ancient Greek meant "roadway". "Voda" in Old Norse means "I go quickly...I stream forth". "Odylle", "Ode", and "Od", the names which Reichenbach gave to this singularly fundamental energy also referred to Wodin, the "all-transcending one".

The name was the first of a new technical lexicon which Reichenbach would build throughout the next few decades, despite final and loud academic protests. Od energy represented a Victorian revelation, an opening of ancient knowledge. Od energy was far more than an ordinary inertial force. Od was an energy which somehow linked sensation and the world, a personal energy which connected individuals directly with the very core of natural reality.

Not fully able to comprehend the entirety of Od and what it represented, Reichenbach began studying the primary attributes of Od energy. He first wished to find out the propagation speed of the Od currents in wire conductors. Metals apparently conducted the currents with special strength. Taking a large length of braided wire, the Baron asked his subjects to hold their end of the braid. In an isolated chamber, he then touched the termination to a heavy metal plate which sat under pure sunshine. He timed the effect. When the sensitive reported the sensation, the Baron could calculate the actual conductive velocity along the braid.

Repeating this experiment several hundred times, he satisfied himself that the Od currents were extremely slow ones. Surprisingly, they travelled some 1.5 yards per second at best, saturating and creeping through the conductive lines as a vegetative flux. Od intensities grew with time, until conductor saturations were observed. Time was always required between the connective application and the sensitive perception of the energy in his subjects. Some 30 seconds was often required before any manifestation could be experienced after an initial application of light at the other end.

It was found that Od saturated matter in a fluidic manner, propagating organismically in distinct thready penetrations. Saturated objects "spilled over" with Od currents. Steel objects and given volumes of water each held their maximum Od charge for 10 minutes or more. Od was sensibly conducted along electrical insulators such as silk threads, cotton threads, glass rods of great length, wooden dowels, and long resinous strands. Also noteworthy in these regards was the way in which the energy would "load" this conductive matter, being stored for several minutes after the connective wire was removed. In some cases, the energy would remain in certain materials for up to one hour's time before leaking away! The healing Od violet energy of sunlight could be stored in Leyden Jars for a very long time. Od entered the body-interior of materials which it traversed. This mode of propagation differed completely from electrical charging, where charges traverse the conductive surface only.

"Od" was an apt name for a power which traversed all matter. Od, the permeating current, was now studied with attention to the details of its behavior. Od could permeate a great length of wire, while the temperature currents of heat and cold measurably could not; a distinguishing feature. Despite the great length of these wires in some cases, new sensitives continued to accurately and independently distinguish the "heat" of moonlight, and the "cold" of sunlight.

Solid metals, "continuous metals", were the best Od conductors. Loosely woven matter, like cloths, were highly resistive to the flow. New measurements with the most sensitive thermometers could neither reveal the heat nor the cold which sensitives felt when touching Od charged wires. Here were highly consistent experiential states, effects which could not be mechanistically resolved. Od was not heat or cold, not temperature, yet it was able to be perceived as such in sensitives. A more thorough and exhaustive scientific approach was now obviously necessitated.

DARKROOMS

A chance observation plunged Reichenbach into a new research avenue which demanded a total change of his methods. During an examination, the Baron arranged his solar light experiment and introduced his familiar braided line to a select group of sensitives. The room was excessively darkened, the sunshine saturating the light-receiving plate outside. Sensitives each began reporting a visible flame of white light which projected vertically from the braid end. The Baron noted these reports with mounting excitement.

In completely darkened chambers his sensitives began visualizing luminosities among all the items which had been exposed to both the sun and the moon. Metals objects gave a steady, flame-like radiance which absolutely fascinated his medium sensitives. Reichenbach found that his high sensitives each had personal familiarity with the phenomenon, assuming that everyone could see the lights. It was found that each gained this ability during childhood, oftentimes coinciding with the onset of sleepwalking.

He brought his examinations to a quick hiatus in order to completely prepare for the next series of exacting examinations he felt compelled to undertake. Now requiring special laboratory conditions, stringent laboratory conditions, he designed for the accurate qualitative analysis of Od and its various natures. The Baron envisioned a system of special darkrooms by which he could make strict determinations of the luminous Od properties. Absolute blackness would now be required.

He converted several chambers and halls of Castle Reisenberg to this end, providing both for the comfort of his sensitives and for the diverse experimental arrangements which he planned. As designed, all of his experimental apparatus were prepared in an adjoining darkroom. These could be introduced to the darkroom and presented to observers in prearranged sequences. Materials were also carefully laid away in darkrooms for very long time periods. In this manner, any solar stimulated light emissions could be eliminated at once.

In the completely blackened viewing room, a flat rotary table top brought laboratory artifices and material samples to and from the sensitives. Manual introduction of new materials could be effected through thick black velvet curtains. The rotating table had velvet covered windows to insure that no source of distraction could interrupt the sensitives. Floors were blackened and covered with insulative materials. Hallways and rooms were blackened. Windows were blackened and sealed tight all around, covered with thick black layers of cloth.

Outside, the Baron arranged for a rooftop stage, where a great battery of large metal plates could be established, wire braids and solid wires conducting down into the chamber for special experimental observations. When these rooms were finally ready, he discovered an ability which he doubted possible for a time. He himself could now see the luminosities which his sensitives had very easily reported. Now he could corroborate their statements! Od lights appeared without any special stimulations or treatments. Od luminosity differed completely from frictive luminescence. Od was a natural phosphorescence which connected all things together.

It became apparent that even persons of "no sensitivity" could see the Od luminescence. When the proper precautions against light were taken, this rare light was seen pulsating and streaming forth from all substances. Od was a luminous current. In Reichenbach's terminology, it was "a self-luminant". But, watching the wonderful Od luminous display throughout the darkroom, he sensed something which inspired the deepest wonder. It became difficult to differentiate between the experientially induced sensations and the luminous pulsations themselves. Od force was not an inert force, it was a personal force; one whose influences permeated observers. As Od changed, so sensation changed.

Where even the thick masonry and ceilings of several feet thickness each began visibly emitting a rare white light. The Baron correctly hypothesized that, after a suitable saturation period in sunlight, all matter became Od luminous. Faces could be clearly discerned in this strangely wonderful Odic illumination.

The Baron was never able to completely eliminate the permeating Solar Od currents, despite all such precautions. Even during night study sessions, he found that a strange feathery emanation flooded his darkroom. This manifestation was referred to as "ectoplasm" by others. Reichenbach preferred the term Od, recognizing that Od was a world-permeating presence of far greater importance than originally supposed. The relationship of soul and matter seemed not to be a problematic discussion when recognizing that Od force truly permeated all matter. Od represented the world soul, flooded and coursing throughout matter everywhere.

Od saturation in the Baron's darkrooms was so evident that his sensitives were finally able to see everything in the room in a kind of twilight. They even took the Baron by the arm and led him around, with complete visual clarity, among all the assembled scientific apparatus. Identifying each one by this strange visual sense alone, they caught the Baron completely by surprise. He eventually came to discern the peculiar twilight of the dark viewing room. Od vision was most clearly perceived in the darkness. Forms and colors were clearly seen by most sensitives, illuminated by Od light alone.

Saturated in solar Od, though several foot thick stone castle walls intervened between the chambers and sunlight, sensitives reported the presence of a permeating glow throughout the room. He needed to modify the room considerably before reasonably complete "Odic purity" could be observed. Despite the placement of heavy woven Od absorbent tapestries on ceilings and walls, each continued to see the other as a vague and ghostly blue-grey presence. The clearest details of facial features...face, head, shoulders and hands...all exposed parts, could be seen. Solar Od permeated everything with its wonderfully vivifying presence.

These feathery room-permeating emanations provoked numerous discussions on the true nature of vision and mind. The supposed differences between imagination and optical vision began to blur, distinctions becoming vague. Since Od permeated all matter, it also permeated physiology. When considering the active role of Od in both the mind and the eye, it was difficult not to deduce a strong identity. If Od could activate visionary imagination, then it was the actual cause of optical vision. Describing this spontaneous and apparently unending phosphorescence became a fascination from which he never departed to the very end of his life. Earnestly desiring to share this miraculous experience with other colleagues, the Baron warned that none of his precautions were to be violated if success in obtaining Od visual effects was truly desired.

Absolute blackness was the first requirement. Utter and complete blackness. Next was the relaxed preparation in utter darkness for at least one hour. If these steps were omitted in any way, the effects would never be seen. One cannot imagine the Baron's own patience, considering that he performed these experiments for several decades with innumerable repetitions!

SUNLIGHT

In the new darkroom, Reichenbach decided to re-examine the Od content of light with greater attention to spectroscopic detail. A large prism was utilized in order to ascertain which specific portions of solar light actually contained the greatest fund of the mysterious Od current. A special open-walled room just below the Castle roof area was prepared for receiving Od currents from any desired celestial source.

The large glass prism was poised on its metal support, projecting bright rainbows against the weather-beaten brick wall. He arranged for the exposure of conductive wires in each resultant color band. These well-sealed wires brought the Od currents into the darkroom from each color band for examination.

Pure sunlight produced the most powerful Odic manifestations. Refracted sunlight was decidedly weaker. As before, he confirmed that glass-refracted sunlight was cool at the "violet" band, and irritating at the "red" band. But he recognized distinct polarities in the effects which each opposed color band produced. Red irritations were canceled by violet excitations, and vice-versa. He assigned polarities to these spectral bands, giving the violet band a "negative" value and the red band a "positive" value.

Sunlight contained an overwhelming "negative" polarity. Moonlight contained a surplus of the "positive" variety. Od polarities were like nutrients. Depending on their personal energetic deficiencies, people each desired specific colorations.

The dangerous manner in which sleepwalkers often clambered up onto terraces, balconies, or even rooftops could now be understood. Sleepwalkers were mysteriously drawn outside in full moonlight in order to absorb Od-positive rays, to them a nutrient. Moonlight was more agreeable to sleepwalkers. This is why they sought it out even during there light dreamy states. This explained their strange behavior under the moon.

Those who were subject to severe muscular cramping or night fears sought Od-negative rays of sunlight. To them the violet rays of sunlight supplied an energetic nutrient. Sunlight was more agreeable to those afflicted with spasms and cramps. This is why they also abhorred the night, often physically fearing its approach. This explained their "nocturnal phobia", the shaking and fright. Deficient in the ability to saturate violet energies in their own bodies, each sunset left them horribly depleted. The violent shaking and emotions were in no way different than those manifested by undernourished persons. Od-polarizations explained the physiological symptoms of somnambulism, night cramp, night fears, and emotional hysteria in a very concise descriptive manner. The superstitious fear concerning each was effectively dispelled when comprehending Od currents.

He began expanding his examinations of the natural world. He now employed his rooftop stage in a new series of examinations involving celestial light, modifying the experimental apparatus by using several different large metal plates (1 square yard each) of zinc, iron, silver, tin, lead gold leaf. In addition, several strange composites were tested, one such being a sulfur-soaked linen cloth of equal size as the metal plates.

Different metallic conductors were connected to each of these in sequences, the wire terminus examined in the dark room. Conductors were very long, some 40 feet in length. The sensitive was asked to hold this wire during the darkroom examinations. Blinding white solar rays were allowed to fully saturate these plates. Shortly after each connection, sensitives began giving very explicit descriptions of their impressions.

Auric flames, some 12 inches in height, projected from the wire end before their very eyes. The Od cold flooded their bodies and the room. Whenever the air was disturbed by talking, the flame flickered in response. In this latter phenomenon, Reichenbach realized that air was not dispersing the flames. Odic emanations from the mouths of sensitives were blowing out the flames. This vibrant response to speech was a new and thrilling suggestion of telecommunications at an early date.

As solar light shone on the plates, or was alternately blocked by deliberate interpositions, the sensitives reported that projective flames correspondingly rose and fell. An interval of 30 seconds was always required before the effects, caused in the plates, were communicated to the sensitives. The Baron recognized that gradual introductions were necessary in these experiments, since the Od force took time to manifest and experience.

The Baron allowed his daughter to stand in the direct sunlight shaft, grasping the wire in her hand. The sensitives each saw a flame of 9 inches rise from the wire end with a corresponding pleasant Od coolness flooding the room. When she stepped back into the shade, the wire flame diminished, while producing a disagreeable Od heat which they all experienced in their turn.

LIGHT AND METALS

Solar light on copper produced green and blue flames, both gold and silver each produced notable flames of a clear white, lead communicated a grey-blue flame, tin plates produced flames of dull white. Zinc gave reddish white flames. Glass was substituted for metal, producing a white lambent Od flame. Polarized light, through a 35 degree angle window, produced no noticeable differences in Od flame colorations when allowed to fall on the metal plates.

The Baron again arranged his very large glass prism in the open air balcony, projecting its spectrum on the weather worn brick wall. The copper wire touching each color gradually, sensitives reported agreeable sensations from violet to green. Green being the midpoint, disagreeable heating sensations appeared from yellow through red. Spectral green moonlight itself produced strong muscular cramping, as did the green spectrum of sunlight.

Reflected red and infrared spectra always gave nauseating responses, while the violet and ultraviolet spectra gave elevating cool responses. One could therefore fraction Od currents out of light to obtain its vitalizing power. Several inventors later developed other means for selecting and tuning these special energies, founding the science of Radionics.

Night experiments produced equally strange results. The bright moonlight was allowed to saturate his metal plates. Several high sensitives clearly saw a tufted flame, some 10 inches in height and thick, rising from the wire end. Moonlight produced an unexpected attraction in several sensitives, who wished to follow the wire line right out into the night sky! Their hands, arms, and torsos became so rigidified by the wire contact alone that he had to stop the experiment. To give an idea of the Baron's thoroughness in these regards, consider that these experiments were performed through three full moon phases with a specific cluster of sensitives.

Another experimental arrangement was attempted to test to capacitive ability of metals. Their ability to retain moonlight Od utilized German silver, placed in full moonlight. Saturated with this flux for several minutes, sensitives each felt a disagreeable heat. No measurable temperature difference existed. Several other metal plates were allowed to saturate lunar light before being communicated directly to the sensitives. Each produced the disagreeable heating action for a prolonged period of time. In addition, objects so saturated could be identified without question, the characteristic heat of lunar light becoming familiar. There were surprising colorations and auric flames resulting from these lunar exposures. Copper produced red and green flames together, both zinc and silver produced tufted flames of white, while tin extended blue auric flames. The differences between direct natural light and light passed through glass plates was now closely examined.

The Baron allowed wire terminals to touch skylight which had passed through a thick glass block. On touching the other terminal end in the darkroom, the sensitives reported an Od heat. The wire, removed from behind the glass block and placed in direct skylight, produced Od cold. Various aqueous solutions were exposed to both sun and moonlight. Water exposed to direct sunlight had a "different taste", being cool and slightly acidulous. Water exposed to glass filtered sunlight tasted "warm and bitter". Apparently, water was able to retain Od when once exposed to Od sources.

The large prism was now employed in a more detailed investigation of this phenomenon. The various spectra, projected against the wall, each produced remarkable flavorings which were accompanied by "other" sensations. Water tumblers were each placed in its various color bands. Violet color bands producing cool and acidulated qualities, red color bands producing warm and nauseating qualities. Here was yet another qualitative "fete accompli". The Baron challenged fellow chemists to discover the purely "chemical changes" which had been wrought in the water samples. He himself, a "chemist extraordinaire", could in no way find traces of any "chemical" additions by these exposures. The inference that a "pure quality" had entered water, producing these clear and manifest effects, was absolutely abhorrent to his colleagues. The flavors were distinct. The effects sometimes violent. Several sensitives became so nauseous when drinking the "red water" that they began vomiting.

Od "trapped in matter" represented a new phenomenal species. Anyone could taste and sense the stored qualities now. And here is where the Baron began to make strong statements against his academic critics. Here now were means by which forces, which were termed purely qualitative, yielded quantitative effects. Here, qualities were materializing as quantities. They could be retained in water for hours. Water exposed to red light ("amarum") and that exposed to violet light ("acidulum") sparked a new controversy among academicians.

Reichenbach stated the belief that all fluids were subject to those laws by which crystals were regulated and formed. It was possible then to store these Od patterns in fluids of the kind which crystals clearly manifested. This storing of patterns had been cited for the behavior of remedy substances termed "homeopathic". The crossover between two worlds, one of qualities and the other of quantities, proved that one precedes the other. Here it was possible to enhost "mere qualities" into matter. Therefore, qualities themselves were much more than metaphoric realities. It was obvious to the Baron that the qualitative world of Od currents were the fundamental world-permeating power. His colleagues in Berlin were incensed.

Water samples, allowed to stand in pure moonlight for several moments, produced a "mawkish" and nauseating sensation (lunar "amarum"). This sensation ran through the body, producing nauseating tremens and occasional vomiting. These were, in fact, the very symptoms which certain night phobics manifested. Here then was a world of qualities, just as the ancients insisted, where qualities ruled inert matter.

Dr. Buryl Payne recently reproduced these experiments with exacting results, noting the dramatic variations among water samples exposed to specific planetary light emanations. The device he used was a 6 foot metal pipe (5 inches in diameter), mounted on a telescope tripod. Fitted with a light receptive organization, light from various celestial bodies was directed into pure water samples. Solar water produced restful sleep within fifteen minutes' time. Lunar water had a strong unpleasant taste, producing characteristic nausea.

Lunar water tasted as "burnt rubber". Several samples made during a lunar eclipse produced near violent irritability in those who drank the samples. Stopping the light from entering water samples absolutely blocked the unpleasant flavor and sense. Refrigeration eliminated the foul lunar flavorings and the extreme irritability.

Water made by exposure to Venusian light gave a strangely "metallic" flavor, producing an unexpected giddiness. Jupiter exposed water tasted sulphurous, but relieved certain internal upsets with surprising speed. Various other planetary configurations produced specific emotional effects: sadness, weeping, anger, and disorientation. In short, hysterical symptoms.

DARK RAINBOWS

Od was light. A very rare and world-permeating light. It was an unsuspected reality in the heart of nature. What is very conspicuous in all of the Baron's rigorous and lengthy studies is the sheer consistency of reports made by his sensitives. The Baron was a seeker of truth, not fond of self-deception. He already recognized the subtle manner by which the experimenter could contaminate empirical results with "suggestions" and "expectations". He therefore adopted a bland questioning technique by which the new sensitive would be introduced to an experimental arrangement before observations were made. Silence was the rule thereafter.

Except for the sensitive's own verbal descriptions of things experienced, no speaking was ever permitted during the long and arduous observation process. The sensitives were trained in this silent process. None could see the other. Sound mufflers filled the rooms, so that scuffling could not disturb the concentrated observation process.

His sensitives numbered in the many hundreds. They were cases, discovered in his journeys across Europe. Each shared a range of other sensitivities which allowed a deeper kind of Od research. Each of these high sensitives reported the rare ability, personally noted from an early age, of seeing luminous colorations and "rainbow effects" around specific materials. This vision persisted in the night as well as during the sunlit hours. The Baron ascertained from each independently that the auras were visible in both sunlight and in the darkness. Here was a rare opportunity to study the fabled "aura vision" firsthand!

Most sensitives recalled their early concerns when the ability first manifested itself. When neither their parents nor their siblings could "see the pretty rainbows" surrounding everything, these child sensitives felt real pity for the others. With increasing age, however, the ability became a nuisance, especially when social acceptance became the need of greatest personal emphasis. Most dealt with their ability by hiding it, though they were never without the vision.

It was difficult for many of them to associate with persons whose auras were unsightly! While the Baron was truly fascinated, he focussed primarily on the more natural manifestations of the auric vision. The colorations which they saw in and around each viewed object always maintained a fixed identity. Auras differed among objects and materials, providing a means for distinguishing among the objects viewed. Sensitives reported that material auras each were possessed of distinctly "soft, striated, and harsh" qualities. Sometimes these auras "pulsated, throbbed, and streamed" into space. Reichenbach was completely enthralled by this new discovery. Sensitives could enable the accurate exploration of the Od world.

The Baron studied the human atmosphere or aura, speaking of its "radiant light which, undetected, sweeps into space". He carefully noted distinct differences between the Odic luminosity of male and female auras. Auric differences among persons of different age and temperament were distinguished. The Baron stated that, by experimental examinations, the aura of each individual differs "as perfumes differ...as various tones differ...as various colors differ". States of health and illness could be correctly diagnosed by the auric observation high-sensitives on infirm persons. Sensitives could actually "see into" the auric bodies and anatomical chambers of others, discerning states of vitality or illness, even detecting "lesions" or other such "dark markings". There were frequent corroborations with physicians, who isolated the very disorders described by the sensitives. Their vision was indeed accurate. This was no superstitious activity. To the Baron, this was a case of superior and mysterious vision. Surpassing Od vision.

Some sensitives exhibited extraordinary sense receptions: hearing and even seeing through the hands and stomach area (solar plexus). Notable degenerate conditions of auric color and form indicated disease states. During the early twentieth century, Dr. Walter Kilner developed diagnostic techniques which derived from Reichenbach's studies of the human aura. Dr. Kilner's method employed special glass filters of dicyanin, a liquid through which the aura could be clearly visualized. It was possible for Dr. Kilner to make detailed examinations of human auras against sunshine in special examination chambers.

He now planned now to expose each of his special sensitives to successive presentations of materials. A whole range of different materials were brought together in order to record descriptions of Od emanations by sensitives. Chemical solutions, chemical powders, metal plates, cloth composites, organic matter, stones, plant matter, the human body itself; all were drawn into the viewing chamber to be studied and re-studied several thousands of times.

Within this wonderfully relaxing environment his results would be more highly considered by other colleagues. In the Baron's darkrooms, natural phenomena could freely express themselves without hindrance. His experiments began with study of the "dark rainbows" and their relationship with mineral matter. An incredible array of substances were introduced into the preparation chamber.

Samples were placed on the small circular revolving table in the adjoining equally blackened preparation room. Rotation of this table would silently send it into the viewing room. The signal bell being struck, each sensitive would examine the viewing space for luminous radiations. As each empirical description was made in detail, the former materials were removed. New materials were continually and gradually introduced on this revolving table. Each toll of the bell signalled a new "dark visual" examination.

The process was arranged in a selected sequence. The Baron had a vast chemical and mineral collection at his disposal. Through his numerous mines and industrial refineries he was privy to otherwise inaccessible minerals and chemical samples. Each of his rare chemical, metal, and mineral specimens were carefully brought through the window, Od emanations being described. These sessions each took many hours. The number of substances actually employed during each darkroom examination period exceeded six hundred. Along with these, came numerous other element "composites" in which material combinations were studied. Viewing each of these samples, his sensitives reported astounding variations in Od color and intensities. These colors were wonderfully brilliant when once the eyes had grown accustomed to the absolute blackness, in no way resembling the familiar phosphorescence of solar stimulated rocks and chemicals. These new manifestations were flamelike and complex, possessed of defined structure and polar differentiation.

The Baron himself grew accustomed to recognizing the same luminations which sensitives reported. After several hours of patient observation in utter blackness, the wonderful ability became effortless. He then realized that sensitives were neurologically gifted individuals who were capable of sensing the exceeding faint influences of Od. Persons who never realized their own sensitivity to such energies were brought into the black rooms. As soon as they too grew accustomed to the blackness, they each actually saw the Od lights emanating from each sample. The Baron stated that most people can see Odic phosphorescence, but may never have had the experience because of the total blackness and lengthy preparation time required before seeing the displays.

ODIC CHEMISTRY

Sensitives examined elemental groups, reporting their color impressions. Those elements which had multiple colorations displayed multilayered flamelike emanations. The Baron listed what the group had independently and painstakingly affirmed in his "Od Elemental Tables": Cadmium (white, blue), Cobalt (blue), Silver (white), Gold (white), Palladium (blue), Rhodium (blue), Chromium (green, yellow), Titanium (red, violet), Arsenic (blue-red), Osmium (red, grey), Potassium (red, yellow), Nickel (red, yellow, green), Sulfur (blue), Selenium (blue, red), Tin (blue, white), Copper (red, green), and so forth.

In addition, it was reported that strong Od cold was produced by sulphur, bromine, graphite, arsenic, tellurium, iodine, selenium, phosphorus, oxygen, arsenic, and manganese. Strong Od heat was produced by gold, platinum, potassium, mercury, tin, cadmium, iridium, iron, oxygen, rhodium, lithium, zinc, osmium, lead, nickel, sodium, antimony, cobalt, silver, titanium, copper, and palladium.

The flames of certain elements expelled their Od flames in curious tensions. Points on the samples produced more intense colorations. Variously colored, and decidedly high pressured in appearance, these auras each resembled electrical "brush" discharges. With these fundamental pieces of data, he was able to arrange an "Od-periodic table". The Baron found that strongly electropositive elements (alkali metals) produced Od heat, and strongly electronegative elements (halogens) produced Od cold. These known Od sources now provided the Baron with an experiential reference through which new comparisons could be made. How Od was modified when passing through different materials or configurations would teach more about the Od nature.

Compound metal blocks (welded copper and zinc) gave no compound effects in sensitives, whose senses detected the independent emanations of each individual element. Soldered copper-zinc blocks produced only the "copper warmth" and the "zinc cold".

Touched for a few seconds with Od-positive and Od-negative elements, it was found that glass rods and metal rods retained the Od "charge" condition for much longer times than any other tested materials. Thus, Od could be transferred as a slow-moving current through any conductor.

There were few materials which did not conduct Od: paper (especially in layers), leather, and cloths of excessive weave and thickness. Od could jump across gaps and resistant barriers, forming its own articulate circuits directly through matter. Od did not behave as electricity often does, surrounding and avoiding the interiors of materials and geometric forms. Od penetrated and articulated throughout the body of most materials.

The emanations themselves were projected from each element in a variety of strengths. These usually took the appearance of a soft flame which took an ascendant path toward the ceiling. Strong blowing against these flames successfully divided these flames. Their progress toward the zenith was not hindered afterwards, the flames rejoining and ascending in full force once again.

There were times when these flames were influenced, wavering rapidly by unknown external causes. These breezy effects were never quite understood by the Baron, there being no knowable cause for this at that time. It was obvious that some insensible regional disturbance caused these wavering drifts. He found it possible to cause auric divergence by rapidly rotating the samples. Auric flames were blown outward by this rotary action.

The Baron examined auric emanations proceeded from chemical solutions, frictive actions, and chemical reactions in various containers. Chemical solutions were mixed in the preparation room and instantly introduced before the sensitives. Chemical reactions produced bright colorations whose luminosity gradually faded with time. Sulphuric acid, poured into water and stirred produced a red tufted flame which rose straight up the glass stirring rod. When an iron wire was placed into a concealed sulphuric acid solution, sensitives visually identified auric flames from the wire terminal brought out before them.

Sensitives accurately reported when the wire in the concealed chamber was lowered into and out of the solution. Each time a large flame rose up the wire and projected visibly into the air for several inches. With these luminations came also projected sensations of heat or cold. Iron wire flames in sulphuric acid, were white and red-blue. Brass wires in the same solution gave white and green light flames.

Various combinations of chemicals were stirred in the preparation room. Wires, plunged into these freshly stirred solution, brought no measurable thermic effects through many yards of wire length, while Od cold and Od heat was powerfully sensed in the viewing room. The Baron wrapped the wire around a long glass stirring rod, plunging it into various solutions. Od travelled through the glass rod and passed through the long wire to the sensitives in the room beyond.

Iron filings in water, stirred with the glass rod, gave Od warmth. Iron filings in vinegar, when so stirred, produced Od cold. Touching their end of the long connective wire, certain sensitives could feel the bubbling of chemically active solutions as real "shocks", which travelled through their bodies in a disagreeable manner. Ordinary candle flames were touched to the long conductive test wire, producing an amazing Od cold in sensitives. Many sensitives reported that, when attending church mass, they were often made uncomfortably cold by "frigid draughts" exuded from candle stands!

A hot iron was passed over a copper plate until warm. The wire being connected to this plate and the terminus held by the sensitive produced two unexpected reactions. First, the unmistakable heat seemed too intense to bear, though the plate itself was only warmed. Second, the sensitive remarked that her hand had become noticeably "heavier".

Porcelain and wooden rods, when heated at one end, felt "cold" to sensitives. This remarkable phenomenon of Od cold in otherwise strong heat insulators defies academically expressed physical principles. Fire itself, touched with very long metal wires and held at their terminus, felt "cold" to sensitives.

Copper plates were rubbed vigorously with wood, producing Od heat in thirty seconds through 21 feet of wire. Wool was used to rub the same copper plate, producing a strengthened Od heat effect. A silk handkerchief produced the strongest reaction of the previous two trials. Rubbed tin plates produced weak Od heating effects. A saw blade ripped through some wood in the dark. The sensitive saw nothing of the possible lights produced by this friction, but rather identified definite reddish flames coming from every tooth of the blade shortly after it was drawn through wood.

Iron rods were rubbed together until the sensitives saw bright flames with a warm wind projecting outward from the extreme rod tips. Charcoal pieces were rubbed together, producing dark red flames throughout their substance. Glass rods were rubbed in a covered cloth, the free ends of which glowed with the unearthly light. These flames, along with the heating emanations, disappeared a while after the rubbing ceased. Iron and steel demonstrated an amazing ability to retain Od effects for upwards of two hours.

He wished to test the effect both of heat and electrification on his sensitives. The Baron decided to try rubbing cloths with an electrophorous, a sulphur cake. This primitive electrostatic charging method was a fricative source. He found that sensitives consistently reported Od cold with positive electrifications, Od heat with negative charges. Od sensations out-proportioned both frictive and electric forces in these cases. In addition, the Baron noticed that the luminous effects of electrification proceed independently of those produced by Od effects, the two never interfering.

Silk offered special and surprising conductive qualities not manifested with other woven materials. Reichenbach suddenly comprehended the success of early electrostatics researchers while using silk threads as conductive lines. In addition, experimental researches of the "forbidden" Franz Anton Mesmer were now equally fathomed. The Baron examined the earliest designs of Mesmer, his special ground battery, the "baquet". Analyzing its construction from early descriptions available to him, he comprehended why the device so powerfully acted on subjects. The Baron knew why it operated, and why it was able to send "thrill shocks" into those who touched its single iron terminal.

The Mesmer battery was a powerful chemical Od generator, having several layers of Od reactants. He clearly comprehended why stirring the battery occasionally would "restore its activity".

MAGNETIC OD

He presented a small bar magnet along with his mineral samples once. When this magnet came through the dark window into the viewing room, the sensitive peeled with an ecstatic delight. Declaring the colors and intensity of Od light from the magnet to exceed all other displays, the Baron was truly impressed. Reichenbach next examined magnets of great strength. His hopes of comprehending something of the terrestrial condition would not be disappointed. Large magnets of various sizes, shapes, and symmetries were now examined. With these came perhaps the most startling colorations and focussed Od effects.

All of his sensitives were now exposed to a great variety of magnets. Each exclaimed, in loud and prolific exclamations, the sheer beauty and elegance of these displays. Weak sensitives saw a bright blue flame which issued from the magnetic north pole, while a red flame issued from the polar south. The Baron was excited by these declarations because he too could see them. He wished to produce the effect with greater power. Employing a 100 pound horseshoe magnet of numerous layers, poles pointing upward, each sensitive clearly saw a powerful luminosity which filled the viewing room.

Flickering magnetic flames issued upward, apparently under pressure, from each pole. Poles were fairly covered with little flame tongues which incessantly moved over the pole surface. The colorations were truly spectacular. These grand luminous magnetic displays projected other sensual attributes besides color. North polar blue flames were cool and soothing, while south polar red flames were hot and irritating. The differences were striking, corresponding pretty nearly with the solar spectrum colors. The Baron perceived a possible world unity in this Od force phenomenon.

The high sensitives each perceived much more, however, than the simple blue north and red south poles. They saw glittering rainbows spontaneously surrounding the magnets in all directions. Dark rainbows. These rainbow colorations held their structure in space. Despite attempts at disturbing this apparent radiant structure, none could force the rainbow colorations to shift in any way from their structured form.

This light proceeded in its multicolored splendors from all sides of the large magnet. The rainbow effect from the magnet reached two yards! The steel of the magnet body between poles was awash in a flowing white light. This whiteness was filled with numerous sparklets and flickering flames all across the gap and iron surfaces. Flames, issuing in tufts from the poles, did not attract or influence one another. Each behaved independently, neither combining or repulsing. Magnets swarmed with tiny flickers of white light here and there, a mystifying "combustion".

A grey-white "Od smoke" rose to the ceiling. Blowing on these flames made the displays flicker. Unlike the continued flickering of ordinary flames after such draughts, the magnetic flame quickly reassumed its shape and intensity. The Baron had already realized that the act of blowing on these Od flames was not the result of air draughts at all. Magnetic Od flames could be extinguished momentarily, while not affecting the magnetic strength at all.

Reichenbach took several large bar magnets and stood them on end. The south pole in the vertical position, sensitives visibly identified a radiating wheel of colors with greenish yellow (north), violet blue (east), red (south), and yellow (west). The bar magnet having north pole vertical gave greenish yellow (north), violet blue (east), red (south), and orange (west). The fixed structure and form of the whole display was the startling feature. Considering that the discharges were a continuous rushing Od flame, it was difficult to comprehend the actual source of this structure. How was its complex coloration maintained in such a fixed organization?

Copper coils were wrapped around magnetic poles and long conductors were drawn into the dark viewing room, whereupon sensitives saw large magnetic flames issuing from the free wire ends. Terrestrial magnetism did not influence or modify the polarity of Od phenomena. Large soft iron bars were used to probe the terrestrial magnetic field in the darkroom, being poised in various compass directions. The earth magnetic colors, though dull, yet followed the same "rule" of color and form. No quantitative analysis ever discerned such distinctions as magnetic "east" and magnetic "west"!

Various soft iron shapes were examined: plates, discs, cones, cylinders, and spheres. Exposed to bar magnets, these iron forms displayed Od color effects due to magnetic induction. Streams of Od currents were visibly identified traversing the spaces between magnet and iron, a fluidic transfer of Od radiance. The blue north pole induced a corresponding red south pole on the exposed face of iron forms, the blue northern flames appearing on the opposite iron faces. Multicolored wreaths covered the surfaces in a continual flickering flow of light. Induced flames, when blown upon by the sensitives, grew brighter, divided, and then resumed their original appearance.

He tried to induce electricity and magnetism with Od. Taken from elements and minerals through wire connections, he found it impossible to produce either electrostatic charges or magnetic polarities. This brought him to the realization that Od existed independently of its sources. Materials where Od was found were simply Od concentrators or Od "foci". Od did not convert into other energies, although it was present where other energies manifested themselves. Od luminous phenomena appeared when electricity and magnetism did not: in sunlight, moonlight, elements, and minerals.

FOCUSSING OD

Magnetic materials clearly focussed Od, but Od did not produce magnetism. Glass lenses, placed near the poles, actually focussed the colored flames! The blue north pole flame was gathered into a tight bundle and focussed in the space beyond the lens as a bright white glow in mid-space. A distinct cone of Od light was seen when a white card was placed along the lens focal axis. Prismatic rings could be seen at various distances along the axis, produced by a veritable Od cone of light.

At the prime focus, the Baron observed a radiating series of sharp lines. This iris was the result of an inductive effect. Focussed Od seemed to be projected outward from the large lens with some acceleration. Focussed Od induced new polarities on striking the card target. Strong magnet poles projected Od light out to walls and floors when so directed. Sensitives saw large spots of blue or red polar Od light which appeared several yards away from the larger magnets in midair! Here was a new world of optics which few of his colleagues would ever recognize.

The Baron attached a single silver-zinc battery terminal to an iron disc which was suspended by a silk thread. With only the silver terminal connected, the red coloration at once appeared. When the zinc terminal was applied, the blue coloration appeared. With both connected, the blue-red coloration swelled from the disc. By this experiment, he showed that electrical polarities could focus Od.

He tried reflecting the magnetic Od light in a large mercury mirror. At an angle, sensitives could see the reflected Od light. Mirror-reflected Od often allowed only the irritating Od heat rays to be transmitted. Such Od temperature effects were felt so intensely, that perspiration often grew upon sensitives. The measured temperature had not changed.

Though these Od rays were refracted and projected through lenses out to several feet distance, magnetometer studies made by Haldat (1846) proved that the magnetic rays were neither refracted nor reflected. Moreover, when magnetic rays were interrupted in space by a suitable iron armature, the Od rays continued right through the armature to the space beyond and could be seen.

Electromagnets produced identical Od luminous effects as when sensitives examined permanent magnets. Electromagnets permitted the controlled observation of "Od lag" effects. When the electromagnets were electrified, Od required some thirty seconds for its appearance and disappearance, thirty seconds after current was removed. While electromagnetic effects appeared and disappeared instantaneously by the closing of a switch, Od charge and discharge lagged considerably behind the initial impulse. In addition, Od maintained its polarity when electromagnets were "impulsed" with DC current, continuing to flow between the impulses.

AURORA BOREALIS

In all of this, the Baron was progressively moving toward an astounding demonstration which, he believed, would give an unequivocal explanation for the Aurora Borealis. An electromagnet, placed within a large hollow iron sphere, was examined in the darkroom under varying degrees of electrification. The Baron referred to the iron globe as his "terrella", or, "little earth". The electromagnet poised within this globe, he raised the rheostat in degrees. Sensitives clearly saw a very intensified color display which proceeded from both poles toward the center. These intensely colored flames struck out across the outer globe surface in sharp, very bright flares. Observation taught that Od lights of such great extent did not adhere, but freely flowed over the surface of conductive materials.

Each such flare occurred as a sharp discharge from the pole surface, proceeding in radial directions. The colors varied across the polar surface along succinct radial directions outward: Od light meridians. Moreover, isolated filaments gathered the Od discharges into distinct bundles. These wandered over the outer globe surface in meandering flares, flickering to and fro like discharges. Together, these meandering radial flares produced a flashing multicolored display. He was convinced that magnetic Od produced the Aurora Borealis.

Just as magnetic Od was blown about by unknown winds, the Aurora was blown about by the high winds of the uppermost atmosphere. This explained the erratic undulations, rolls, and serpentine ripples of the Aurora. When strong magnets were placed in bell jars and evacuated in degrees for observation, the Od luminosity expanded in wonderful colors. Each stage of evacuation gave a greatly enlarged luminescence. The rainbow colored Od light also grew brighter with increasing vacuum. No such phenomenon was ever reported in the academic circles.

Higher vacuum states produced brighter and more intensely colored Od flames. He now hypothesized that at near space altitudes, the corresponding expansion of Odic lights would be observed. His theory also explained the strange hue-tinted white glow, often seen covering the night sky. Focussed in rare night cirrus clouds, Od could adequately explain their wondrous appearance.

Bearing the characteristic of flames, capable of being mechanically moved by special winds, he believed that he had discovered the true and fundamental cause of Auroral effects. The Baron's own novel theory of the Aurora Borealis, an Od light display of great magnitude, has never really been appreciated for its rare merits. Explaining not only the sudden meandering flares of polar light, his theory alone explains the wide variations of color and remarkable transformations of shape seen when polar lights are active. It is interesting to note that none of the color or shape phenomena are ever successfully explained by the "electrical aurora" theory of Bjerknes.

Erratic movements of magnetic needles, which always preceded auroral phenomena, evidenced the natural fluctuation of radial Od flares from the poles. The weakening of magnetic needles during violent auroral episodes could be explained by the erratic motion of Od filaments, clustering Od force in a singular channel.

OD WATER

The nature of Mesmer's famous "magnetic water" was sought. A tumbler full of water was stirred with a clean permanent magnet. The result was a real and lasting sensate effect, especially when sensitives tasted the samples. Water which was stirred with either pole gave characteristically different sensations. North poles produced a water stimulant, while south poles produced an irritant. The sensations being real, there had to be an equally real reason for the changes brought about by the stirring action.

Several sensitives actually fell asleep when approached by magnets. The responses were truly bizarre. Though completely unconscious, the hands of some sensitives actually adhered to the magnets. Others, though remaining awake, would fall into painful tonic spasms when merely approached with a magnet. Sensitives could tell whether tumblers of water had been stirred with magnets at all. They also learned to tell whether north or south poles had been employed in the stirring.

It was found that the water samples were not chemically changed. The change was a deeper one than supposed. Here again, Od energy had successfully brought new qualities to an inert substance. The effect was tested against neutral samples. The result of the Baron's "blind" and "double blind" examinations proved that there was indeed a sensible change of taste after magnetic stirring.

Baron von Reichenbach measured the "permanence" of Od charged water. Od charged water remained active even when poured among numerous glasses. Luminous Od remained in water samples for long time periods. North pole-stirred water remained a tonic stimulant, while south pole-stirred water remained an irritant of nauseating proportion. In addition, the Baron found that hand-held water retained Od. There was a strict polarity transference from hand to sample. Water, activated by the right hand produced energetic responses. Water held in the left hand was nauseating.

ELECTRIC OD

Electrophoric cakes of sulphur were rubbed, emitting white flames of light as well as Odic smoke in the utter blackness of the Castle darkrooms. These white flames disappeared after a few minutes. The white luminous smoke reached the ceiling, where it curled in luminous billows upon itself. Od smoke was a phenomenon as mystifying as all the other Od lights. Od smoke was like the "ectoplasm" and "ghostlights" which certain researchers actively hunted.

Because of these numerous observations, and because of the frictive manner in which Od phenomena were made manifest, Baron von Reichenbach surmised that Od was a material "dust" of superfine degree. He also suggested that Od was the very material which luminesced in the electrified vacuum tubes of Plucker and Geissler. A young Sir William Crookes, no doubt, studied and endorsed these Od treatises with great delight. Tesla would later echo these very same statements in greatly advanced form.

Od would only make its appearance in electrical devices a short time after they had reached their maximum strength. The lag was considerable, usually taking 30 or more seconds to manifest. Completely grounded wires, connected to large electrostatic machines, continued radiating Od light for 60 or more seconds after the machines ceased revolving. Negative electrostatic charges gave Od heat. Positive electrostatic charges gave Od cold. Several hours were required to remove the Od content of electrically activated substances, even after continual contact with neutral bodies. Induced by electrostatic charges, Od seemed to have a deeply penetrating nature. Induction of Od through electrical means seemed to be more the result of an internal frictive action; a thorough friction caused by electrostatic stimulation of metals and insulators alike.

The manifestation of Od currents, stimulated by electrostatic energy, required 30 seconds or more before becoming visible. It was obvious that this Od light was not simple electrostatic luminescence. Wires, Odically charged by electrical means, continued to emanate a strong white light for 60 seconds or more after being completely removed from the circuit! Charged Leyden jars glowed with Od currents for 120 seconds or more after being electrically neutralized. Dark viewing room wires were connected to Leyden jars, glowing with a continuous and intense white light when discharged through sealed sparkgaps. This intense white Od light continued to radiate for 300 seconds with EACH spark discharge.

These electrostatic inductions of Od were truly anomalous from an electrical viewpoint, since the wires used all ran along the castle floor! Reminiscent of work later done by Tesla, this Od luminous momentum revealed how different Od and electricity were. Voltaic piles projected strong Od lumination of peculiar motion throughout the entire battery. In addition to this luminous display, the Od which radiated from voltaic columns maintained a defined and continual columnar rotation! Od luminations, though caused by electrical shocks, behaved as completely independent of their causative electrical charges. Where electrical activity only occurred when circuits are closed, Od luminations continued for hundreds of seconds.

High voltage electrical induction coils also produced Od luminosity after 45 seconds, continuing to do so long after the electrical currents had measurably ceased. The Baron turned his thoughts on the possibility that Od might induce electromagnetic effects in suitable designs, and set about to test these possibilities. It seemed at the time that electricity and magnetism were able to induce Od at a distance, but Od could in no manner induce electricity or magnetism.

Od forces did not give reciprocal electromagnetic effects. The "traction" of Od by electric and magnetic means was clearly observed, but the induction of electromagnetic charges by Od was never determined by the Baron. In addition, both electrical and magnetic phenomena available to the Baron at the time produced Od effects in weakest possible manner, being more the result of frictive effects than of true Od induction. He encountered numerous setbacks in his hopes of demonstrating the mutual "transformation" of both Od and electromagnetic energies.

"Powerful Od sources fail to induce electrical or magnetic effects, when a magnetic needle possessing one-hundredth the Odyllic power will do so instantly". There was some obvious reluctance which Od exhibited when being forced to become electrostatic or magnetic force.

While this great "failure" seemed to close a curtain, the Baron saw a far greater curtain opening. Goethe's prescription was to look deeply into the face of nature for answers. He soon discovered, quite in the course of his continuing research, that there were far more potent "portable" Od sources in nature than even permanent magnets. For Od itself was a new force, a new world energy. Od could be used as it was to perform hitherto unforeseeable activities in the benefit of humanity.

In fact, nonactive electrical apparatus and components exhibited remarkable ability in storing, conducting, and modifying Od. This fact was developed into the science of radionics independently by Dr. George S. White and Dr. Albert Abrams, both medical doctors.

CRYSTAL OD

Permanent magnetism was as fundamental a force concentrator as could be found, until crystal Od was discovered. A much more fundamental Od activity was observed in chemicals and minerals. The Baron recognized that, far beyond the magnetic activities, minerals represented nature's most fundamental energies. Magnets were artificial materials, being created in electrical fields. Minerals were found in their native state. Strong Od sources were found among the minerals and natural crystals.

In May of 1844, the Baron brought forth a large mountain crystal from his collection for viewing. The sensitive reported that the entire crystal appeared to be streaming with a fine white Od light of great power. When asked to describe the light distribution, the sensitive described colorations not unlike those produced by large magnets. The sharp crystal point projected a deep blue Od jet, some eight inches long. This bright blue projection was in constant motion. Emitting numerous sparks, the flame-jet in the space beyond the sharp point was tulip shaped. Turning the crystal around, the broken crystal base revealed a dense red and yellow smoke.

The phenomena of blue and red-yellow polarities in crystals were accompanied by visual and visceral sensations. Blue jets were cool and red-yellow jets were hot. Crystal Od did not effect changes in electroscopes. Crystals not being magnetic, this new Od phenomenon fairly astounded the Baron. Only crystals whose principle axis of symmetry was singular could produce this rare force with an unexpected power. Many large and splendid crystal specimens were obtained from the Imperial collection in Vienna for his use. With these large Od sources, sensitives experienced not only blue and red Od flames, but real breezes and (more spectacularly) convulsive muscular spasms in defined degrees. These powerful manifestations so distinguished the crystallic force from all other Od manifestations that the Baron considered it to be the most fundamental Od focus in nature.

Some specimens compelled the hands of sensitives to grasp the crystal terminals in a strange display of physiological traction. Reichenbach referred to certain crystals as "detectors of Od", inferring by this that world-Od actually focuses in crystals.

Several crystals produced uncontrolled "grasping responses" and painful "cramping" effects in the fingers of sensitives. These included diamond, antimony, witherite, cassiterite, corundum, hornblende, staurolite, copper sulphate, tungsten, mica, potassium ferrocyanide, witherite.

Some crystals which caused the convulsive and painful "hand closing" effects did so without attracting the hands of sensitives. These included rock salt, quartz, sphaelerite, cobalt glance, iron ore. Crystals which caused strong hand clenching and painfully violent spasms included meteoritic iron, French quartz crystal, calcite, arragonite, tourmaline, beryl, selenite, fluorspar, and barite. Crystals lacking optical axes (natrolite, zeolite, arragonite, stalk-like crystalline heaps, globular crystallizations, all nucleated crystal masses) did not produce the crystallic Od force, emanating only the amorphous Od current observed in other minerals. A quartz crystal, some 2 inches thick and 8 inches long, was drawn down the arm of a sensitive. The recipient had the sensation of a cool and luxuriantly comfortable breeze. A "reverse pass" beginning from hand to elbow produced the disagreeable hot Od sensation. A crystal three times the size of this produced violent spasmodic effects on the physiology of exposed sensitives. These effects often approached the artificial evocation of seizures.

Crystallic points produced "thrilling" stimulations and wakefulness, while crystallic bases produced deep and sudden sleep. Crystal Od was powerful. Its permeating force supervened the neurology of sensitives, causing physiological tractions and convulsions. The bodies of sensitives stored up crystallic force. Numerous passes across the body of sensitives and non-sensitives alike produced violent cramping, the force being the collective sum of each pass. This was not observed with magnets. Numerous passes produced the same response as single passes. Darkroom examinations revealed that crystals visibly expanded the human aura when grasped at the base.

It was thought that traces of iron in the crystals produced the crystallic Od force. Wishing to determine whether crystallic Od was not some variant of magnetic Od, the Baron took the large selenite crystal and suspended it within the gap of a large horseshoe magnet. Even when made to oscillate, no divergence or variations could be observed. The magnetic activity of the crystal was thus eliminated as a possible "ordinary" cause.

He purchased a very large quartz crystal prism from Gotthard, "measuring eight inches in diameter...a six-sided colossus having pyramidal ends...which I had great difficulty in using". Sensitives could not approach this splendid mountain crystal without the immediate sensation of a cool wind projecting from the sharp pointed end. One sensitive described the wind "as if cool air were gently blown upon him through a straw". In another series of experiments, the large crystal was used to induce instantaneous sleep in sensitives at a distance of 42 feet! Od force apparently increased with the size of the crystal.

He tried raising iron filings with it, though none could be raised even with the crystal point. He tried using it to magnetize iron needles without effect. Transference of the Od charge did not degenerate into magnetic or electrostatic charge. He suspended needles by silk strands, but no induced motions could be observed in these. The giant crystal apparently had no power to induce physical movements.

A large selenite crystal was suspended by silk, placed under a bell jar, and approached by a large magnet. No motions were observed in the crystal Od. The same selenite crystal was suspended near an electrified wire without resultant motions. He took much smaller crystals, suspending each by fine silk threads and observing their orientations in the terrestrial magnetic field. No general orientation was ever discovered among the crystals.

Furthermore, crystal concentrated Od activated electrical apparatus and components in unexpected ways. A large copper coil was wound, into which the Baron thrust the selenite crystal, but no induced currents were ever detected in the most sensitive astatic galvanometer. Despite these inabilities to measure interactions among the known forces and Od force, other laboratory electrical components exhibited remarkable ability in storing, conducting, and modifying crystallic Od.

Placed on metal plates or in large metal coils, crystals produced a penetrating and highly focussed Od polarity. Representing more potent natural Od states than even lodestone, crystallic force powerful compelled the Baron into new investigations. With exception of a few organic materials, crystallic Od charged all contacted objects with ease and power.

Metals conducted crystallic Od with a sensible "shock" which sensitives could feel. Capable of penetrating the body on contact, crystallic Od evidenced its natural predominance over all other Od sources. With crystallic Od, the propagation time in wire conductors was nearly instantaneous! Crystal-transferred Od was intense, penetrating, and complete. Od-charged metal objects retained their flow for longer time periods, springing with prolific Od flames.

Crystallic Od transference intensified the natural Od radiance of elements and minerals. Crystallic Od was the quintessential natural Od focus, permeating all matter. The degree of force from crystals surpassed all other Od agencies. A brief contact with a crystal point was sufficient to bring about an enormous Od current. Charges conveyed to copper, zinc, silver, iron, linen, silk, water, and every other material which the Baron had earlier tested gave greatly intensified Od radiance.

It seemed difficult to isolate crystallic Od. Tin plates seemed able to block some of this force, but never completely. Organic materials (wood, glass, leather, paper) classified as Od insulators, offered resistance to crystallic Od. Multilayered paper was found to be an absolute Od insulator and resistor. The multiple layers of paper in a thick book resisted Od charging regardless of exposure time near the crystal point.

Crystallic Od had difficulty passing through the one inch thick book, and did not succeed in passing after being moved over the large crystal point several times. Od propagation was blocked in organic matter. Deal board (a composite) required a long time and numerous movements before a weak crystallic force could be felt in the hand. Curiously, the giant mountain crystal was incapable of penetrating paper, while in free space it induced catatonic sleep at 42 feet. Wood, touched to the crystal point for a short time, induced both a sudden "shock" and spasm.

A 40 foot woolen cord was traversed by crystallic Od in a very short time, again the result of its longitudinal strands. Silk and glass seemed more perfect crystal Od conductors when compared to paper, wool, and wood. Long exposure times did not increase the Od charge of a metal object after its saturation point was reached. Saturated objects radiated their Od surplus into space. The critical time for saturation was tested among several metals.

Crystals could charge water samples as did magnets and sunlight. Crystal points projected coolness into water, the acidulous taste resembling that produced by magnetic north poles and violet spectrum sunlight. Crystal bases gave the same nauseating flavor produced through both south magnetic poles and red spectrum sunlight.

After sufficient examination, it was found that crystallic force resided primarily in the optical axis of crystals. Those crystals which lacked strong optical axes also lacked the strong crystallic focus. Blue Od light, observed in darkrooms from crystal points, was filled with sparks and dartlets. The blue Od light passed upwards into a white light. Interior crystal light appeared wondrous to both sensitives and well prepared darkroom observers, evidencing movements, sparks, starlike formations, all of which exceeded the magnetic Od in both color and intensity.

The discovery of crystal Od, and its ancillary phenomena, so impressed him that he spent a great long while both studying and marvelling over its very existence. Crystallic Od was the luminous source which "beams like sunlight". He saw the crystallic force as a detector and concentrator of world-force. It did not require the expenditure of energy. Reichenbach emphasized the term "concentrator" when speaking of Od sources, recognizing that Od was not generated by these objects at all. Od was conducted and concentrated in "specific foci" as it flowed through the world. He declared that the Odic streams "flowed on eternally".

With this crystal giant, the Baron launched on an amazing series of experiments designed to determine the usefulness of crystallic force in a new technology. Modern electrotechnology itself is predicated on the use of electron currents. It is completely dependent on interactions which these currents exhibit when made to move through special structures. It became obvious to researchers such as Reichenbach that there were indeed "other currents", Od currents, whose use would release new potentials to humanity. The wonderful aspect evidenced in Od currents is their obvious ease in effecting subjective sensations and impressions.

Dr. Ashburner, Reichenbach's English translator, produced several remarkable crystallic Od detectors, having eight or more large mountain crystals of quartz wrapped in great silk-insulated copper coils. These, closed with a platinum "keeper" produce a prodigious Od charge of shocking power. There were researchers who later managed the extraction of Od-like energies from certain rare minerals and specially synthesized crystals (Moray).

Baron von Reichenbach speculated that the crystallic Od is to crystals what vital force is to organisms. In other words, he speculated on the intelligent living aspects of the Od force. Here then was the beginning of Radionic Science, which Twentieth Century investigators would later privately implement in exceptional analyzers and "tuners".

CELESTIAL OD

The Baron examined the Odic influences of both solar and lunar light and their ability to modify crystal Od sources. This was the first in a series of geophysical Od studies. The Baron monitored terrestrial Od flow by observing crystal Od variations during specific seasons. It was found that an uninterrupted supply of celestial Od currents charged both the atmosphere and ground. Such celestial Od streams traversed the ground surface, seeking specific ground sinks. This flowing concentration of Od in groundpoints was associated with numerous surface "ghost lights", noted in specific geological points by local inhabitants. In fact, the Baron had occasion to discover the true source of legendary "grave lights".

Taking several of his sensitives out in the night air, he led them into a new cemetery one at a time. Each sensitive clearly beheld the Od light eerily hovering over the ground where new burials had taken place. He explained these remarkable phenomena as the combined effects of earth magnetic and chemical reactions taking place in the tilled soil. He also found other sites where this Od luminosity had lent a "haunted" atmosphere to otherwise lovely locales. The Baron cited the prevalence of ground breezes which blew the Od lights around, especially in cases where these "grave lights" or "ghost lights" wavered about.

Reichenbach sought to probe the mysteries of Od in space now. Od traversed great lateral distances of 160 or more feet across laboratory spaces, effecting powerful sensations across these distances without appreciable loss. Od did not apparently weaken when once in the radiant form. Why should it weaken when traversing space? Od came to earth directly from purely celestial influences. He had proved this with moonlight, and then with sunlight. But the discovery that starlight could also charge objects with Od suddenly became a theme of major importance. Here was an energetic charging effect which occurred across vast stellar reaches. It therefore represented a unique energy source which offered humanity a possible new means for broadcasting a usable power. More would have to be learned about its conversions and modes of propagation before any technological advances could take place.

Nikola Tesla advanced the knowledge of solar Od, permitting the true transformation of Od into electrostatic charges. In a notable patent (685.958) he describes the powerfully transformative results obtained when solar light is conducted into a specially prepared metallic plate, vertically poised within a high vacuum tube. The device, grounded through a heavy duty mica capacitor, produced prodigious amounts of electrostatic energy when illuminated by strong sunlight.

The Baron observed diurnal pulsations of Od potential from various physiological centers, noting and charting their natural regularity. He noted the strong association of physiological Od with nerve ganglia. Od was especially concentrated in the solar plexus. Od pulsations in the body followed Od pulsations throughout nature, a remarkable and unsuspected circadian rhythm.

The most sensitive body spots showed the most concentrated proportions of Od: the lips, face, finger tips, erogenous zones, feet, and toes. Each of these energy spots surged with the solar Od rhythm. The curious correspondence of biological Od force with solar rhythms was especially fascinating. Od strength increased at sunrise and decreased at sunset. The Baron noted that every part of the natural environment directly responded to the solar Od supply. Soils, minerals, lakes, trees, animals, and other humans revealed a simultaneous response to solar Od energies. Od was now understood to be a shared force, the world unifying agent. This fact was truly appreciated by ancient scientists.

In solar energy, ancient natural philosophers recognized wonderful metaphysical activities. The solar Od rhythm revealed an essential world dynamism, a "reciprocation" which had long been forgotten. Despite the fluctuating charging and discharging of solar Od absorbers there was a constant terrestrial Od foundation which did not wax and wane with the sun. Residing continuously and without diminished force in crystals, the Od supply remained constant throughout the night. This essential and mysterious function of crystalline basement rock provided a special and rare Od supply. Was it any wonder that myth and legend spoke of subterranean jewels, the magickal and luminous sources of living energy. Here, Reichenbach was realizing the fundamental energetic structures of the world, learning of an essential "metabolic process" by which solar Od was absorbed during the daytime and crystal-discharged during the night. The Baron was now convinced more than ever before that Od was the fundamental world-force, preceding even magnetism and electricity in natural origin.

The deep blue night sky revealed a wide variety of Od currents. Several of his sensitives had shown great attraction to specific sectors of the night sky. Whorl-like sectors of space displayed Od currents of different polarities. Od currents were mapped, coursing among the planets and stars. There were groundpoints, marked by sensitives, where celestial Od currents ran into and through the earth. In this he glimpsed something of the forgotten technologies with which the ancients had ability. Why they labored so much in marking special groundpoints with tall stones now became obvious.

Several of his sensitives always mentioned that portions of the night sky seemed especially attractive, while others were disagreeable. These conditions remained fixed through time and season, except when modified by progressive lunar and planetary movements. When closely examined it was also discovered that select portions of the western sky gave a vivid "cold" just after sunset. At nine in the evening this western coldness shifted toward the northwest, during which also south and southwestern skies were most warm. At midnight, the north sky became cold, the south warm. At four in the morning, north and northeast became cold, south and southeast warm. Finally, just before sunrise, eastern skies appeared to be most cold. The anomalous persistence of solar cold in the west, its sudden disappearance and reappearance in the eastern sky at three in the morning is most mystifying. The throbbing Od pulsations in the sky mystified him.

Blindfolded and asked to seek that constant cold of the night sky, sensitives invariably located the magnetic meridian toward the north. The Milky Way produced the most delightful coolness, the Pleiades excelling in these soothing and cold sensations. Planets each gave a strange and disagreeable warmth, despite the general coldness noted in the entire starry vault. Reichenbach and his sensitives charted the dominant Od paths between the stars, and from stars to earth. Everything seemed effulgent in Od, the flowing vivifying light which bound all natural things together.

He allowed stellar light to fall upon a copper plate. To this was attached a long conducting wire. Held by the sensitive, a slender white light of 24 inches rose from the wire end, becoming very cold and invigorating. The slender light rose and fell when the Baron repeatedly moved the copper receiver in and out of the starlight. He chose a zinc plate to receive the stellar whiteness, obtaining the exact results but with diminished intensity. In addition, the stars collectively acted upon the sensitives as a rather weak magnet, producing effects on the head and spine.

The Baron accidentally discovered that planetary light, even from a single planet, completely absorbed the collective invigorating action of the stars. Jupiter became an unbearable sight for certain sensitives. Planetary light, on copper collecting plates, neutralized Od sensations completely. When this occurred, sensitives could not hold the wire terminals. It was obvious that planets had opposed Od polarities when compared with the polarity of stars. The neutralizing effect was not pleasant to personally contact, rather like a deeply disagreeable electrical shock. Somehow, planets were drawing off the cool, invigorating light of the stars.

In his most remarkable statement in these regards, the Baron alluded to the true energy behind astrological configurations: "we stand in connection with the universe by a new and hitherto unsuspected reciprocation ... consequently, the stars are not without influence upon our sublunary (and perhaps) practical world, and the proceedings of many heads".

VEGETATIVE OD

Sensitives were able to detect metals despite their covering in thick insulators. Because of this, they were very capable as human ore detectors. He employed them on several occasions to explore on behalf of the local mines. Coal, zinc, lead, copper pyrites, water, all were sensed by their finely tuned nervous systems. Walking outside at night, sensitives felt the Od emanations of trees, through intervening spaces up to 400 feet. The world of "vegetable" now became a special fascination to the Baron, who by now had developed sufficient data on the world of "mineral".

His examination of "living organic structures" commenced with the acquisition of several potted plants. Copper wire coils were wrapped around potted plants, the long free terminus grasped by the sensitive in the darkroom. This method was later adopted by Georges Lahkovsky in studying celestial radiations and the growth of plants. The copper receptor coil was placed on a Calla Aethiopica, producing an unexpectedly rapid and vivid reaction. A penetrating and excessive heat suddenly permeated throughout the sensitive's body. An Aloe Vera plant was then sampled. Its effects were similar, though weaker in contrast.

Walking through opened fields in sunlight, a high sensitive began examining the numerous flowering plants near Castle Reisenberg. Most of the blooms gave warmth in the stem and cold in the flowers. After sufficient examination of many flowering plants it was found that the growth rate of a plant was an accurate measure of its penetrating efficacy as a medicinal agent. Certain ray-flowers gave warmth except in their cool central discs, a reversal of Od effects.

Here, the Baron recognized a long forgotten sensitivity which physicians had anciently known. The true value of medicinals now was comprehended as the radiant emanation of such materials, not the bulk substance of the same. Od radiance, the true rationale of "medicines", divided the ancient reliance on herbals above minerals in this regard. Herbal medicinals, far above the use of mineral medicines, gave more penetrating and rapid Od effects. Roses, pear-blossoms, and apple blossoms each produced tranquil sleep in sensitives. The Baron determined their common chemical (phlorhisine), recognizing in it a most concentrated Od polarity.

Indeed, certain plants gave completely contrary Od temperature effects because of the alkaloidal chemicals manufactured by them. In these cases, the purely Od chemical predominated the Od biological states. Tall plants gave Od polar reversals which remained fixed in their various segments all along their length. Flower, stem, leaf, fruit, root, or tuber...each gave differing Od polarities and intensities of the same according to the species.

A general rule for vegetative Od polarity emerged; showing that vegetative parts of lethargic growth were always Od negative (cool) effects, while vegetative parts of rapid growth gave Od positive (hot) effects. Thus were vascular bundles found to always be most Od positive, while upper leaf faces were most Od negative.

Trees were also Od cold in their upper parts and Od warm near the roots. Such facts later reappeared in the designs of Nathan Stubblefield, who buried special compound bimetallic coils at the roots of trees to obtain commercial currents of energy.

The Baron sought to examine insects by his method. A little rose beetle, a moth, and other garden variety insects were placed on a copper plate to which a long wire was connected in the familiar manner. The sensitives invariably felt this presence as Od heat after a few seconds. When a small animal (a field mouse) was placed on the plate, the sensitives felt the od heat most vividly. This reaction was also identically felt to greater degree when the Baron placed a kitten on the copper sensing plate. Both the copper plate and copper coil sampling technique, employed in these latter experiments, became the regular component of Twentieth Century's Radionics.

Goethe had once described the world as a "process" whose transformations relied upon a mysterious world-force. This vivifying power drove the development of all created things along their various "metamorphoses". Each individual metamorphosis had as its aim a curious and mystical conformity with "ultimate forms" which existed in metaphysical space. The "world-process" drove changes in space, mineral, vegetable, and animal worlds until each conformed to the likeness by which each was formed.

Young Karl Reichenbach studied Goethe with a deeply passionate regard. His new studies reminded him of Goethe's statements concerning world-process. Now realizing that Od is a general world-condition, he suddenly glimpsed the far reaching aspects of the "luminous world". In this conviction, the natural world was seen to be a natural and organismic system where Od was the "blood" and "vitalizing fluid". Here, at long last, the Baron had found the physical proof of Goethe's "world process"! Od was the closest which scientific research had then come toward knowing the "life force".

ODOGRAPHS

His articles on Odic or Odyllic light were prized by academic notables such as Liebig and Wohler, who published them in their famous "Annalen der Chemie". The many thousands of observations made by Reichenbach through the course of more than a decade were to comprise a massive volume of works. The greatest of these is entitled "The Dynamics of Vital Force". In this tome, Reichenbach enumerates the names and reports of several hundred sensitives. He lists their personal sensitivities, social rank, numerous incidents from his own diaries, and a complete summary of his major experimental results.

European scientists of high rank found the work done by Reichenbach to be of sound repute. Liebig, Wohler, Berzelius, Dalton, Poggendorff and others like these notables agreed that Reichenbach had discovered a new energy form. But they clamored for objective proof. Reichenbach had already found that lens focussed Od light could produce images on daguerreotype plates. He therefore began producing photographic proof of perceptions made by his sensitives for his colleagues. The first of these "Odographs" were published in 1861, to the astonishment of critical academes.

These photographs were made by the Od light projected from fingertips, crystal points, magnets, metals, chemicals, frictive effects, and heat effects. Several photographs demonstrated the transference of Od light from one material to another. Lenses were used to intensify these effects in certain plates, a fact which Dr. Abrams and Dr. Kilner later elicited in studying "human" (auric) energy.

In his photographic experiments, Reichenbach sought the aid of the Royal Photographer Gunther (Berlin). With Gunther, he discovered that Od light was in fact a completely different energetic species than ordinary light. The two established that strong sunlight could not produce photographic images through 6.5 inches of glass, while the eye could clearly perceive images through the same. Reichenbach realized that, while the chemically active rays of sunlight could not penetrate the glass block, the "illuminating rays" did. He cited instances where deep sea divers could not see sunlight, yet could see objects. At a critical sea depth, neither sunlight nor objects could be seen, a radiant blackness flooding the eyes. The essential difference, between perceptible Od light and measurable optical light became apparent.

In 1930, Dr. Ruth Drown produced a remarkable device which was capable of producing anatomically interior photographs through radionic tuning systems and photographic plates. She produced a catalogue containing many thousands of these "radiovision" prints. With photographic evidence of radionic energies, she toured European medical circles. Physicians were eager to both learn and implement the revolutionary technology. DeLaWarr (1948, Oxford) continued these researches, producing similar photographic results.

EXPERIENCE

Quantitative science is entirely based on data derived through the use of "objective measuring devices". Academic science has produced a mechanistic worldview because of its reliance on measuring devices. This worldview reduces all natural dynamics to the collective activity of four fundamental forces. The quantified world is therefore framed by academicians as a "field of forces" whose collective blend produces "force patterns". When these "force patterns" are encountered by sensing organisms, they are "locally interpreted" in stimulated nervous systems as "the world".

This quantified worldview cannot describe sensations, qualities, or metacognition; it cannot directly describe our senses, sensation, or awareness. These experiential realities are fundamentally distinct energies which have extent and continuity in space. We see the utter collapse of quantitative analysis when it attempts to analyze experiential realities. Capable only of describing energy field epiphenomena, it selectively filters out the very center of what it examines. There are some academicians who cannot comprehend why this occurs.

It is no coincidence that the quantitative worldview works very well in describing superficial force dynamics of inertial space, employing force instruments to obtain its data. But experience, in its fundamental core, is no collection of quantifiable forces. The force analytic method cannot mechanistically explain consciousness, failing miserably when extending its analytic prowess into experiential phenomena. All it manages to do in this regard is describe the epiphenomena which accompany consciousness: magnetic, electric, and chemical "fields". This diversion results because measuring instruments cannot conduct or respond with the sensate energies themselves. Quantitative science has been continually constricting its worldview by adopting the inferior examination methods which instruments provide, misguiding its own consciousness from the most fundamental and accessible worldview. Consciousness is a non-inertial realm into which measuring devices do not enter. There is no way to reduce consciousness to a force model.

Each of the historic ruling paradigms has so screened, filtered, and divided Nature that each derived worldview no longer holds a meaningful place in human consciousness.

Because of the blind insistence that all things are force-reducible, modern quantitative science ultimately moves toward a completely alienated treatment of Nature, where sense-substituting instruments filter specific natural signals and thought-substituting statistics interpret data. Quantitative Science is very good at achieving one objective: the kaleidoscopic fragmentation of experience. Instrumental examinations begin with the general acceptance that experience is "faulty" and consciousness is "biased", both being judged and condemned as "invalid". Of late, the conscious interpretation of instrument-derived data is undertaken by statistical analysis, conscious interpretations also considered "invalid".

This self-destructive and self-annihilating devaluation of consciousness is abhorrent and unacceptable. Because so much emphasis has been placed on instruments, quantitative science has formulated a special world-interpretation which adds nothing to the experiential dialogue. Quantitative science projects a worldview so alien from consciousness that its statements become abhorrent and humanly inconsequential.

Force measuring instruments have bound quantitative science into a closure from which it actually desires no escape. Like Narcissus, fixated on its own reflection, quantitative science also imagines that what it sees "is the only face in the world". By projecting its inertial measurements upon the world experience, quantitative science effectively filters out Nature's fundamental language: its projective meanings and consciousness.

If human experience is invalid, then who decided whether or not instrumental examinations of Nature were more valid? More valid compared to what or whom? Whose "faulty" consciousness decided this ridiculous contradiction? These continually emerging inconsistencies are the embarrassing "warning signs" to quantitative method.

The scientists of former times were not the mere force-measuring technicians of today. They were truly Doctors of Philosophy. These natural philosophers employed the powerful tool of consciousness in a metacognitive consideration, reflection, and interpretation of Natural behaviors. Data, and the accuracy of its acquisition was never considered more valid than the philosophical considerations concerning Natural meanings and their interpretations. Consciousness was the medium in which they wondered at, gloried over, and pondered the natural behaviors. Their advanced science was one in which natural phenomena were assessed against the "dream sea", the collective fund of archetypes. Victorian Qualitative Science required a special sensitivity and process.

Data, however accurate, was often disqualified as inadmissible and inconsequential, when far greater principles of metacognitive import were weighed. The genteel elegance of their philosophical art has been replaced by a base and mindless technical threshing tool which is incapable of discerning between mind and object. If the scientific philosophy of the Victorians has been considered outmoded and classified as "pseudo-scientific", whose philosophical contentions decided that instruments would do a better job? The continual removal of human experience by the invalidation principle is now producing an amusing consequence, by which quantitative science is eliminating the very consciousness which framed its rules!

Former science, qualitative science, viewed the world as a vast potential of experiential possibilities. Qualitative Science understood that nature was fundamentally an experiential reality. Viewed as an exceedingly complex collection of qualities and aspects, ancient naturalists employed metacognitive process to interpret the various meanings of these qualities and aspects.

Qualitative Science studied the language of Nature, interpreting its meanings and enjoying the consciousness so evidently flowing throughout the world. Qualitative Science recognized that Nature was a sea of experiential realities. Measurements had no place in the sciences because measurements were not experiences. Experiential states were the means by which Nature was examined and intimately known by Qualitative Science.

Qualities and aspects are the innumerable "experiential potentials". Consciousness, the mysterious energy, is the means by which we may know Nature intimately. What we need to do is rediscover the lost means for examining Nature by direct experience at a magnified potential. Experience is the heart and core of being. In fact, we all wish to "merge with Nature" completely. This is the true motivation beneath the scientific study of Nature.

Quantitative science does not give this. Art more completely achieves this purpose. But Qualitative Science most completely achieved this in the development of experience magnifying instruments, the Science of Radionics.

Sensation remains the only possible window through which consciousness accesses Nature directly. It is, in fact, the only window through which meaning is acquired. Sensation is the only accessible means through which a larger world-language of meaning and symbol is engaged by the ones who experience. Through the gates of perception, the experimenter realizes the deepest and most world-suffusing space, where meanings and symbols rule inertial realities.

It was in this world of pure experience, sensation, and consciousness that Baron Karl von Reichenbach found answers which have yet to be considered by his academic progeny. Perhaps the time has again come when the results of his massive research will again be studied with a view toward a new Qualitative science. It was in the study of Od phenomena and sensations that new consciousness-extending apparatus were developed in the Castle Reisenberg, laboratory fortress of Baron von Reichenbach.

WORLD LIGHT

In the forty years during which these meticulous and thorough investigations were conducted, Baron von Reichenbach managed to collate an incredibly massive data bank. His prolific and creative writing style were closely followed by devotees. His copious publications flooded European academic circles, becoming the prized possessions of notable minds; Crookes, and Tesla being but two such personalities. His separate volumes, letters, lectures, and unpublished notebooks could corporately fill a small library.

Few modern researchers ever manage the acquisition of lost Victorian knowledge. Few manage the sublime realization of lost technology which our Victorian predecessors developed. Fewer yet develop the qualitative experiential modes by which the foundational worldview is realized. The world is founded in consciousness, exchanging conscious energies among its structured parts.

Those who received the Victorian treasurehouse of knowledge made their own thrilling discoveries upon that foundation. Researchers of the early Twentieth Century would later duplicate the experiments of Baron von Reichenbach (G.S. White, Tesla, Le Bon, Abrams, Drown). After the aged Baron passed away, Fechner himself published a treatise, a silent tribute to Reichenbach:

"It is in a dark and cold world we sit, if we will not open the inward eyes of the spirit to the inward flame of Nature".

The words of the truly great Baron Karl von Reichenbach yet resound in the minds of those who know a secret of lost world-vision, a secret of lost science...

> "Everything then, emits LIGHT...everything...everything
> We live in a world full of SHINING matter!"

CHAPTER 2

HEARING THROUGH WIRES: ANTONIO MEUCCI

ARRIVALS

Antonio Meucci is the forgotten and humble genius whose inventions precede every revolution in communication arts which were achieved during this century. The time frame during which his notable discoveries were made is a most remarkable revelation. How Meucci developed his accidental discoveries into full scale working systems is a true wonder in view of this time reference.

The culturing of technology from the simple sparks of vision is a feat of its own distinct kind. As the earliest chronicled inventor of telephonic arts he is justly applauded as the true father of telephony by afficionadi who know his wonderfully touching biography. But he invented far more than the telephone with which we are familiar. Meucci discovered two separate telephonic systems. His first and most astounding discovery is known as physiophony, telephoning through the body...hearing through wires. His second development was acoustic telephony, preceding every other legendary inventor in this art by several decades.

Meucci powered telephones with electricity taken from the ground through special earth batteries, and from the sky by using large surface area diodes to draw static from the air. Eliminating the need for employing batteries in his telephonic systems, Meucci first conceived of a transoceanic vocal communication system. His notion was grand and achievable. Marconi later employed methods pioneered by the forgotten Meucci.

He developed ferrites, with which he constructed true audio transformers and loudspeaking transceivers. He invented marine ranging and undersea communication systems. His numerous achievements in chemical processing and industrial chemistry are too numerous to mention in such a brief treatise. All of these wonders were conceived and demonstrated well before 1857.

Sr. Meucci was a prolific inventor, engineer, and practical chemist. Living in Florence, he worked as a stage designer and technician in various theaters. Antonio Meucci and his wife left Florence to flee the violence of the civil insurrections which raged throughout Italy. Many immigrants who wished for a peaceful life thought they might find some measure of solace in the New Land which lay to the west.

Unhappily restricted by law from entering The United States, persons such as Meucci and his family chose the route into which most other Mediterraneans were forced at the time. Being turned southward, they were literally compelled to dock in Caribbean or South American ports. There sizable populations of European immigrants remain to this day, legally restricted from North American shores. Most found that their presence there was received with an acceptance and warmth equal to a homecoming. It should have been in these lands that their legacies were written.

New arrivals in Cuba, the Meucci family made Havana their home. They found the warm and friendly nation a place for new and wonderful opportunities. Sr. Meucci pursued numerous experimental lines of research while living in Havana, developing a new method for electroplating metals. This new art was applied to all sorts of Cuban military equipment, Meucci gaining fame and recognition in Havana as a scientific researcher and developer of new technologies.

Several special electrical control systems were designed by him specifically for stage production in the Teatro Tacon, the Havana Opera. Electrical rheostats served the safe and controlled operation of enclosed carbon arclamps. Mechanical contrivances hoisted, lowered, parted, and closed heavy curtains. The automatic systems were a wonder to behold.

A young and dreamy romantic, Meucci found the beauty of theater work quite entrancing and inspirational. There, dreams became realities, if only for the short time during which hardened pragmatism was suspended. Fantasy and wonder were magickal liquids which perfumed the soul and opened the mind's eyes. As in childhood, one could receive the elevating epiphanies of revelation necessary for discovering unexpected phenomena, and for developing unequalled technologies.

The decision to move to Havana was indeed a good one. Genuine acceptance, and loving recognition added joy the lives of the bittersweet exiles. Meucci's wife was often amused by his more outlandish inventive notions. But, as their stay in Havana continued, she scolded that he had better develop something solidly practical on which to "make a living".

A long time fascination with physiological conditions and their electrical responses, Meucci was prompted to begin study of electromedicine. With just such a practical view in mind, he established and maintained an experimental electromedical laboratory in backrooms of the Opera House. Investigating the art of "electro-medicine", as popularly practiced throughout both Europe and the Americas, Meucci investigated the curative abilities of electrical impulse. Applying moderate electrical impulses from small induction coils to patients in hope of alleviate illness, Meucci learned that precise control of both the "strength and length" of electrical impulse held the true secret of the art.

As viewed by Meucci, pain and certain physical conditions were treatable by these electrical methods provided that very short impulses of insignificant voltage were employed. Impulses of specific length and power were necessary to rid suffering patients of their pain. In addition, Meucci imagined that tissue and bone regeneration could be stimulated by such means.

What really intrigued Sr. Meucci was the length of impulse time involved in body-applied electricity. To this end, he developed special slide switches which were capable of specifying the impulse length. It was possible to slide a zig-zag contact surface over a fixed electrical source. By varying the spacings between such slide contacts, Meucci could mechanically generate very short electrical impulses.

Rheostats could also be employed to control the current intensity. By the employment of these two control features, he was able to apply the proper impulse "strength and length". Meucci wished to chart a specific impulse series which would neutralize each specific kind of pain or illness. Developing catalogues of electrical impulse cures was his real aim. Such a technology, if developed thoroughly, could arm medical practitioners with new curative powers.

Sr. Meucci applied continual experimental effort toward these medical goals. He often applied these same impulses to theater employees and stage artists alike. These people came to regard such electric cures as definitive. Meucci's method was known to reverse conditions completely. He paid special attention to the placement and size of electrodes on the body. Tiny point-contacts were often held to the body at specific neural points, effecting their analgesic effects. He was especially careful with "shock strength", applying only millivolt surges to his patients. Pain could be gradually made to retreat by the proper impulse administration.

Meucci had already developed fine rheostatic tuners for limiting the output power of his electrical device. He always applied the current to his own body in order to give completely "measured" electro-treatments. In this manner he was able to judge the parameters more personally and responsibly. It was his habit to administer treatments of this kind to his ailing wife, Esther. Crippling arthritis was becoming her personal prison, and Sr. Meucci wished to cure her completely of the malady. Watching and praying through until the dawn, Antonio struggled to perfect a means by which cures could be effected with selective impulse articulation.

As with each of Meucci's developments, the fulfillment of his advanced medical ideas are found throughout the early twentieth century. Each researcher in this field of medical study employed very short impulses of controlled voltage to alleviate a wide variety of maladies. Independently rediscovering the Meucci electro-medical method throughout the early twentieth century were such persons as Nikola Tesla, Dr. A. Abrams, G. Lahkovsky, Dr. T. Colson. Each developed catalogues by which specific impulses were methodically directed to cure their associate illness. Each researcher developed a method for applying impulses of specifically controlled length and intensity to suffering patients, effecting historical cures.

More recently, several medical researchers have employed impulse generators to effect dramatic bone and tissue regenerations. They affirm that human physiology responds with rapidity when proper electroimpulses are applied to conditions of illness. These were closely regarded by government officials, eager to regulate the new science.

Most medical bureaucrats, fearing the elimination of their own pharmaceutical monopolies, sought opportunity to eradicate these revolutionary electromedical arts. Upton Sinclair obtained personal experience with these curative systems and the physicians who devised these methodologies. He championed their cause in numerous national publications with an aim toward exposing those who would suppress their work.

Sinclair pointed out the social revolution which would necessarily follow such discoveries. He was quick to mention that proliferations of new technologies would not come without a dramatic battle. Fought in the innermost boardrooms of intrigue, Sinclair underestimated the ability of regulators to eradicate technologies of social benefit.

This notable literary personage wrote extensively on the work of Dr. Abrams, who was later vilified by both the FDA and the AMA. An outlandish national purge quickly mounted into a fullscale assault on these methods. But this is a story best told in several other biographies. Meucci's electromedical methods would soon be transformed into a revolutionary means for communicating with others at long distances.

SHOCK

The most central episode of Meucci's life now unfolded. It was to be a serendipity of the most remarkable kind. Throughout his later years, Meucci recounted the following story which occurred in 1849, when he was forty-one years of age. A certain gentleman was suffering from an unbearable migraine headache. Since it was known to many that Meucci's electromedical methods possessed definite curative ability, Sr. Meucci's medical attention was sought.

Meucci placed the weak, suffering man on a chair in a nearby room. His weakened condition inspired an easy pity. Antonio had already felt the thorns of his beloved wife's pain. Her eyes, like the man before him now, begged for the cure which lay hidden in mystery. Carefully, caringly, Antonio now sought to ease this man's suffering.

In this severe instance, Meucci placed a small copper electrode in the patient's mouth and asked him to hold the other (a copper rod) in his hand. The electro-impulse device was in an adjoining room. Meucci went into this room, placed an identical copper electrode in his own mouth, and held the other copper electrode to find the weakest possible impulse strength. Meucci told his patient to relax and to expect pain relief momentarily, making small incremental adjustments on the induction coil.

Migraines of severe intensity characteristically produce equally severe reaction to the slightest irritation. The man being now highly sensitive to pain, Meucci's insignificant (though stimulating) current impulses were felt. The patient, anticipating some horrible shock, cried out in the other room with surprise at the very first slight tickle.

Momentarily, Meucci forgot the hurtful sympathy which he naturally felt in assisting this poor soul who sat across the hall. His focussed attention was suddenly diverted as an astounding empathy manifested itself: he had actually "felt" the sound of the man's cry in his own mouth! After absorbing the surprise, he burst into the adjoining room to see why the man had so yelled. Glad the poor fellow had not run out on him, Meucci replaced the oral electrode of his suffering patient and went into the other room to perform the same adjustments...through closed doors this time. He asked the gentleman to talk louder, while he himself again held the electrode in his mouth.

Once more, to his own great shock, Meucci actually heard the distant voice "in his own mouth". This vocalization was clear, distinct, and completely different from the muffled voice heard through the doors. This was a true discovery. Here, Antonio Meucci discovered what would later be known as the "electrophonic" effect.

The phenomenon, later known as physiophony, employs nerve responses to applied currents of very specific nature. As the neural mechanism in the body employs impulses of infinitesimal strengths, so Meucci had accidentally introduced similar "conformant" currents. These conformant currents contained auditory signals: sounds. The strange method of "hearing through the body" bypassed the ears completely and resounded throughout the delicate tissues of the contact point. In this case, it was the delicate tissues of the mouth.

Each expressed their thanks to the other, and the relieved patient went home. The impulse cure had managed to "break up" the migraine condition. Meucci's reward was not monetary. It was found in a miraculous accident; the transmission of the human voice along a charged wire. In these several little experiments, Meucci had determined and defined the future history of all telephonic arts.

VOICES

Excited and elated Antonio asked certain friends to indulge his patience with similar experiments. He gave individual oral electrodes to each and asked that his friends each speak or yell. Meucci, seated behind a sealed door, touched his electrode to the corner of his mouth. As each person spoke or yelled, Meucci clearly heard speech again. Internal sound reception in the very tissues of the mouth. An astounding discovery.

Without question, Meucci's most notable discovery in telephonics is physiophony. Meucci did not foresee this strange and wonderful discovery. Think of it. Hearing without the ears. Hearing through the nerves directly! The implications are just as enormous as the possible applications. Would it be possible for deaf persons to hear sound once again? Meucci knew it was possible.

His first series of new experiments would seek improvement of the electrophonic effect. To this end Meucci designed a preliminary set of paired electrodes. The appearance of these devices was strange to both the people of his time and those of own. Each device was made of small cork cylinders fitted with smooth copper discs. Designed as personalized transmitters, each person was to place their own transmitter directly in the mouth! The other electrode was to be hand-held.

Meucci verified the physiophonic phenomenon repeatedly. Upon experiencing the now-famed effect, visitors were awed. Furthermore, it was possible to greatly extend the line length to many hundreds of feet and yet "hear" sounds. The sounds were clearly heard "in the nerves" with a very small applied voltage. Sounds were being deliberately transmitted along charged wires for the first recorded time in modern history.

The auditory organs were not in any way involved. Meucci discovered that oral vibrations were varying the resistance of the circuit: oral muscles were vibrating the current supply. Spoken sounds were reproduced as a vibrating electric current in the charged line which can be sensed and "heard" in the nerveworks and muscular tissues.

With very great care for obvious injuries, it is possible to reproduce these remarkable results to satisfaction. The voltages must be infinitesimal. When properly conducted through the tissues, sounds are heard near the contact point the body. No doubt, the impulsed signal reproduces identical audio contractions in sensitive tissues. This is one source of the sounds internally "heard". Nerves actually form the greater channel when impulses are arranged properly, directly transmitting their auditory contents without the inner ear.

Physiophony is Meucci's greatest discovery, one which he should have pursued before also developing mere acoustic telephony. Twenty-five years later in America, an elated Elisha Gray would rediscover the physiophonic phenomenon. He would develop physiophony into a major scientific theme. Long after this time, these identical experimental demonstrations conspicuously appear in Bell's letters; copying the identical experiments taken first from Meucci, then from Gray, and Reis.

During the early twentieth century, music halls for deaf persons were once found in certain metropolitan centers. These recital halls enabled nerve-deaf persons to hear music through handheld electrodes. Modifying the appliances in order to allow considerable freedom of movement, several such places allowed deaf people to dance. Holding the small copper rods, wired to a network on the ceiling, musical sounds and rhythms could be felt and heard directly. Physiophony, more recently termed "neurophony" holds the secret of a new technology. Physiophony, rediscovered of late, facilitates hearing in those afflicted with nerve-deafness.

Meucci discovered two distinct forms of vocal communication: physiophony and acoustic telephony. Meucci's next experiments dealt with the development of a means for separating the physiophonic action from the human body entirely. He developed working systems to serve each of these modes, with primary emphasis on acoustic telephony. Replacing tissues of the mouth with a separate vibrating medium required extending the cork-fixed electrodes.

Meucci coiled thin and flexible copper wire so that it could freely vibrate in a heavy paper cone. Once more, Meucci varied the experiment. This time his own oral electrode would be enclosed in a heavy paper cone. Again each subject was asked to talk into the first cone-encased electrode as Meucci listened at the other terminal. Each time, speech was heard as vibrating air. This was his first acoustic transmitter-receiver.

Meucci wrote up all these findings in 1849...when Alexander Graham Bell was just 2 years old. Living in Havana at the time, Meucci conceived of the first telephonic system. He imagined that American industry would allow infinite production of his new technology. A telephonic system would revolutionize any nation which engineered its proliferation.

CANDLES

Freedom doors were not swung open in wide and unconditional welcome for Europeans during the latter 1800's. Strict immigration laws forbade Europeans from even entering New York Harbor. It was more difficult, if not impossible, to find employment. New arrivals in America faced difficult, almost inhuman conditions. No support systems existed in the land of free-enterprise. No catch-nets for failed attempts in the land of the free.

True and unresisted freedom was reserved only for the upper class, who had already begun regulating and eliminating their possible competitors. Every means by which that prized upper position might be usurped was destroyed. Forgotten discoveries and inventions flowed like blood under the heavy arm of the robber baron.

The "New World" was not anxious to welcome these people. Discrimination against European immigrants went unbridled, unrepresented, and unchallenged. When American doors finally did open, there were no sureties for those who came to work and live in the New World. There was no promise, no meal, no housing, no job, no emergency support. To be in America meant to be on your own in America.

Prejudice against the "foreigners" was vicious during this time period. Immigrants who imagined a better life to the northlands would be sadly disappointed at first. Many of these newcomers preferred the temporary pain of atrocious city ghettoes simply because their eyes were on the future.

Europeans arriving in America came with trades and skills. Master craftsmen and technicians in their Old World guilds, these "unwelcomed" eventually won the hardened industrial establishment with their good works, many of them later forming the real core of American Industry. It is not accidental that Thomas Edison hired European craftsmen exclusively. In less than two generations the children of these brave individuals became leaders of their professions, giving the leukemic nation its periodically required red blood.

Established families despised the newcomers, who were regarded first with dread, then with resentment, and finally with a firm resolve. After ruthless campaigns by bureaucrats and moguls to eliminate the foreign presence in North America, wealthy puritanical antagonists sought the supposed surety of legislation to achieve elitist isolation. Neither cultivated nor creative, this ability to manipulate the tools of liberty for the sake of domination became a theme which continually stains their history. The unbridled and impassioned expansionism of these "foreign people" was so threatening to the impotent bureaucrats that legislation was installed for the expressed purpose of limiting their unstoppable movement. Sure that these were in fact the feared usurpers of a young and recently consolidated Republic, financiers impelled legislators to create a "middle class" economic stratum which has remained in force to this very day.

Bound to a life of tireless work and taxations, the children of immigrants no longer question the barriers to limitless personal achievement. While a very few wonder why their frustrations rarely allow escape into the true individual freedom of which America boasts, most simply satisfy themselves with banal consumer temptations.

Nevertheless, the "American" explosion in music, art, crafts, and technological arts followed the immigrants wherever they were forced to flee. When Antonio and Esther Meucci arrived in New York City, he was now forty-two. They made their home near Clifton, Staten Island.

Clifton was once a picturesque little town, nestled on a rocky ridge and surrounded by babbling brooks and lush forests. The year was 1850. The Meucci's acquired a large and spacious house, filled with windows. Golden bright sunlight flooded the home in which Antonio devised the technology of the future. The rooms contained numerous pieces of striking art nouveau furniture which Meucci himself handcrafted. A beautiful four octave piano and several of these furniture pieces yet remain, the house itself having been declared a national monument.

His poor wife, now crippled completely, was confined to their second floor bedroom. It was there in Old Clifton that Sr. Meucci developed his "teletrofono". The device was successively redesigned and improved until several distinct and original models emerged. Mundane needs being the primary necessity, Meucci developed a chemical formula for making special chemically formulated candles and opened a small factory for their production. His smokeless candles earned a moderate income by which the small family could maintain their place in the New World. Throughout the long years to come, he also supported countless others who were in need.

He patented this smokeless candle formula, along with several other chemical processes related to his small industry. Soon, Antonio found that his candles were sought by neighbors, parish churches, and small general stores. He therefore took his devotions, and went into production of the same. Marketing the product locally, he was now again able to supply his experimental facility. This was his encouragement. The inventions began flowing again like rich red wine.

Meucci installed a small teletrofonic system in his Clifton house, as he had done in Havana. Esther Meucci was now completely crippled with arthritis. Connecting his wife's room to his small candle factory, Antonio could now speak throughout the day with his wife. The system lines were loosely wrapped up and around staircase banisters, through halls, across walls, and finally spanned the long distance to the factory building, naturally running slack in several locations.

Meucci made sure that the lines did not run tight in order to prevent wire stretching and cracking during winter seasons. In every model aspect, Meucci's system was the prototype. Everyone of his surrounding neighbors had become personally familiar with his system, having been allowed to try "speaking over the wire."

Meucci and his wife took boarders from time to time in order to afford minimum luxuries...the luxuries of ordinary people. When Garibaldi was exiled from Italy as an insurrectionist, he sought out Meucci. A small factory was established near his home for the manufacture of his chemically treated candles.

With this, his sole and sturdy financial source, Meucci continued his other beloved experiments. He had already established and regularly used several teletrofonic systems throughout his home and factory by 1852. Both he and Garibaldi walked, hunted, and fished in the lush greenery and flowing flowered hills of old Dutch Staten Island.

Each new teletrofonic design eventually was added to a growing collection box in the timber lined cellar. Improved models were made and brought into the general use of his system. With these modified devices it was effortless to communicate with his ailing wife, employees, and friends. Distances posed no problem for Meucci. His system could bring sound to any location. Numerous credible witnesses actually used his remarkably extensive telephonic system across the neighborhood. One such highly credible witness was Giuseppe Garibaldi himself.

Garibaldi was welcomed to live with the Meucci family in their modest Staten Island home for as long as he wished. Garibaldi, Meucci, and his wife vanquished sorrow and poverty with faith, hope, and love expressed in a myriad of ways. Each supported the other in the struggle against indignity, accusation, outrage, and all the particular little alienations imposed upon them. The Meucci household not unaccustomed to the deprivations through which character is developed.

Both Srs. Meucci and Garibaldi continued manufacturing candles and other such products of commercial value, supporting themselves and the needs of others in the new land. Frequent financial crisis never deterred his dream quest. Never did such reversals place a halt on Meucci's laboratory experimentation or any of his devoted attentions.

As it happens in the course of time, new changes bring fresh opportunities and joys to lift tired hearts. The sun rose in the little windows after a long winter's dream. An old friend from Havana came to visit Meucci and his wife. Carlos Pader wished to know whether Meucci had continued experimenting with his now famous "teletrofono".

Pader was shown the results, but Antonio confessed the need for new materials. Both Sr. Pader and another friend, Gaetano Negretti, informed their friend Antonio that there was an excellent manufacturer of telegraphic instruments on Centre Street in Manhatten. And so, Sr. Meucci was introduced to a certain Mr. Chester, a maker of telegraphic instruments.

Mr. Chester was an enthusiastic and friendly tradesmen. He enjoyed speaking with Antonio. The two shared their technical skills in broken dialects. Meucci was always welcomed there on Centre Street. Meucci visited this establishment on several occasions to purchase parts and observe the latest telegraphic arts. It was here that Meucci "gained new knowledge". He set to work, purchasing materials for new experiments. New and improved teletrofonic models began appearing in the neighborhood.

Meucci was methodical, thorough, and attentive to the unfolding details of his experiments. Meucci kept meticulous notes; a feature which later worked to vindicate his honor. He worked incessantly on a single device before making any new design modifications. Meucci's creative talent and familiarity with materials allowed him to recognize and anticipate the inventive "next move". In observational acuity, inventive skill, and development of practical products he was unmatched.

Thomas Edison, after him, most nearly imitated Meucci's methods. Meucci searched by trial and error at times when reason alone brought no fruit. It was, after all, an accident which revealed the teletrofonic principles to him. Providence itself in action.

TELETROFONO

Meucci methodically explored different means for vibrating electric current with speech. From 1850 to 1862 he developed over 30 different models, with twelve distinct variations. His first models utilized the vibrating copper loop principle which he discovered in Havana. Paper cones were replaced with tin cylinders to increase the resonant ring. He experimented with thin animal membranes, set into vibration by contact with the vibrating copper strip. This model begins to resemble the familiar form of the telephone as we know it.

Meucci wrapped fine electromagnetic bobbins around his copper electrodes, increasing vocal amplitudes considerably. In a second series, he explored the use of magnetic vibrators. A great variety of loops, coils, soft-iron bars, and iron horseshoes appear in Meucci's successive designs. These latter models gave amazingly loud results. In addition, Meucci's diagrams reveal experimentation with both separate and "in-line" copper diaphragms. These latter operated by the yet to be discovered "Hall Effect", where current-carrying conductors vibrate more strongly in magnetic fields produced by their own currents.

While power for his early teletrofonic system was derived from large wet cell batteries in the basement, Meucci made a pivotal discovery, discovered when he grounded his lines with large dissimilar metal plates. Suddenly, his system operated as if large batteries had been added to the line. Meucci disconnected the basement batteries and the system continued to operate, powered by ground currents alone.

This use of buried dissimilar plates repeatedly appears throughout early telegraphic patents. The actual devices by which this astounding electrification of lines was established were called "earth batteries". Several significant individuals made remarkable discoveries while developing earth batteries throughout the latter 1800's. They found that the earth batteries were not really generating the power at all.

Earth batteries tap into earth electricity and draw it out for use. Some telegraphic lines continue to operate well into the 1930's with no other batteries than their ground endplates. Certain systems continued using their original earth batteries without replacement in excess of 40 years!

Earth batteries are intriguing because they seem never to corrode in proportion to the amount of electrical power which they generate. In fact, they scarcely corrode at all. Exhumed earth batteries showed minimal corrosion.

A mysterious self-regenerative action takes place in these batteries, a phenomenon worthy of modern study.

Like Thomas Edison after him, Meucci was a master of practical chemistry. Numerous of his processes remain unused to this day. He developed strange chemical coatings; using saltwater, graphite, soapstone, wax, muriatic acid, asbestos, sulfur, and various bonding resins to treat wire conductors. Wire lines, specially treated by Meucci, had current rectifying abilities. These absorbed and directed both terrestrial and aerial electricity into the line, a one-way charge valve. Technically what he created is a large surface area diode.

When these specially coated wires were elevated, Meucci enhanced the absorption of "atmospheric electricity" into his system. Prevented from escape by chemical coatings, a steady stream of aerial charges were absorbed into the wire line. He succeeded in powerfully operating his system with "aerial electricity" alone.

Meucci now freely used aerial and earth electricity to power his teletrofonic system. In addition, he discovered that the latent power in strong permanent magnets could amplify speech with very great power. When coupled with energy derived from the ground, Meucci found that true amplifications could be effected. Meucci found that vocal force being sufficiently powerful to produced amplified reproductions at great distances in certain of his models which utilized magnetite "flour".

Sound-responsive soft iron cores were replaced with lodestone and surrounded by various powdered core composites developed in Meucci's laboratory. Lodestones, surrounded with cores of flour-fine iron powders, produced enormous outputs. Meucci used exceedingly fine copper windings. The vocal range of these magnetic responders was considerable when made in Meucci's own unique design.

Clear, velvety speech was communicated with great power in these fine-powder core designs. His use of flour-fine magnetite powders produced the world's first ferrites; composites of iron, zinc, and manganese later used in radiowave transformers.

His teletrofoni were now fully formed, handheld devices of some weight. Surviving models from his system resemble those much later manufactured by Bell telephone. They are cup-shaped, wooden casings...handheld transmitter-receivers. One speaks into the device, and then listens from the same for replies. Meucci's diagrams, notebooks, and models prove his priority over all the historically successive telephone designs.

In addition, Meucci used diaphragms which conducted the current which vocalizations could modulate. He developed remarkable graphite-salt coatings to enhance the electrical conductivity of his responder diaphragms, preceding Edison's carbon button microphone by a full 24 years!

TRANSOCEANIC

In addition to his existing system, Meucci conceived of entirely new directions in communication arts. His mind turned toward the sea...and to transoceanic teletrofonic communication. Meucci tested the idea that seawater could actually replace telegraph cables, bizarre as it must yet sound. His notion would be termed "subaqueous conduction wireless". Others had achieved moderate results across limited waterways. Sommering, Lindsay, and Morse each sent weak telegraph signals across streams. Meucci envisioned the whole Atlantic as a possible reservoir for the transmission of telephonic signals.

His experiments took him down to the Staten Island seashore with his teletrofono, batteries, and large plates of both copper and zinc. The dissimilar metal plates were submerged quite a distance from each other. Vocal messages spoken into the sea were electrically retrieved by a teletrofonic apparatus connected to an equivalent arrangement of widely separated, water-immersed plates on an opposed part of the distant shore. The signals were clearly heard.

Most engineers will object that these experiments could not sustain vocal communications across great distances. They will say this because transmitter power should be so dispersed that no intelligible signal could ever be retrieved. The experiment having been tried across short distances actually works. The most amazing rediscovery concerns the signal-regenerative ability of seawater. Seawater requires only an infinitesimal transmitter current in order to achieve strong signal exchanges.

The submerged plates themselves generate sufficient current to operate the teletrofonic system without batteries. Electrical signals do not diminish in seawater as theoretically expected. When Meucci spoke of transoceanic communications he was not exaggerating. Seawater seems to be a self-regenerative amplifier of sorts. The addition of a carrier frequency (an electrical buzzer) would pitch the signals toward a higher range, granting more signal focus.

Sir William Preece duplicated these experiments for telegraphy across the English Channel in the early 1900's. Their developing success was eclipsed by the appearance of aerial wireless. Some researchers have interpreted the work of G. Marconi to be a blend of Meucci conduction telegraphy and aerial wireless. While purists protest, it is intriguing that Marconi would later actually resort to mile-long submerged copper screens for transoceanic communications. The submerged copper screens acted as a "capacitative counterpoise", following his equally long aerials...out to sea.

Several segments of these Marconi aerial-screen systems have been located by investigators, both in New Brunswick (N. Jersey) and in Bolinas (California). The Marconi "bent-L" aerial system differs from Meucci's design only in that it utilized several hundred thousand watts of VLF currents. In effect, Marconi employed Meucci conduction wireless in his early transoceanic systems.

Meucci became prolific when designing these maritime inventions. It was told him that a certain deep-sea diver, having once distinctly heard a steamship engine while performing a salvage operation, was told (on resurfacing) that the ship was fully forty miles away! This phenomenon so impressed Meucci that his mind turned toward the use of his teletrofono in deep-sea communications and offshore ranging.

His notion was truly original, involving this submerged plate system for wireless vocal communication. The use of short aerial rods projecting from the diver's helmet formed the very first "aerials". Divers could maintain communications with their surface companions without interruption if such teletrofonic aerials and internally housed responders were installed in their helmets. Sealed aerial rods (one foot or less in length) would protrude out from the helmet, forming the wireless link; an invention truly worthy of Jules Verne! Transmissions and receptions would occur through the remarkable conductive-regenerative ability of seawater to conduct electro-vocal signals.

Of chief concern in Meucci's mind was the establishment of solid maritime wireless communications systems. He designed several systems intended to aid harbor approach and navigation during times of limited visibility. Clusters of tone-transmitters (positioned as fixed stations or anchored as buoys) could wirelessly communicate danger or safety to sea captains equipped with onboard listening devices. Both landmark stations and onboard responders would communicate through seawater with submerged metal plates. These plates would be fixed in position at some depth; much below each landmark and right under the ship hull.

Navigators would be guided into safe harbor by following a specific tonal signal, and avoiding the selected danger tones. These tones would be subaqueous transmissions...true tonal beacons. Navigators were to carefully listen for guide-tones while entering a harbor. Pilots could locate their offshore position with precision by simply listening for the designated subaqueous tonal beacons.

Position could be triangulated by comparing tones and their relative volumes. Tones could be determined by comparison with a small on-board receiver containing tuning forks. Maps could mark these tonal-stations and pilots could rely on their presence. Meucci wished to eradicate the blinding dangers of fog and storm for sailors. Meucci accurately foresaw that an entire corps of maintenance operators would find continual employment in such worthy service.

In all of this, Meucci actually anticipated the LORAN system by a full seventy-five years! In the years before radio pierced the night isolation of shipping, ships maintained tight commonly used sea-lanes when far from coastlands. Mid-oceanic collisions were not uncommon. Meucci conceived of systems by which ships could transmit warning beacons toward one another while out at sea. Helping to avoid such mid-ocean disasters, sensitive compass needles would detect passing ships. Plate-pairs would be poised beneath the ship's hull in the four cardinal directions. Relays could detect ships, responding with loud alarms.

In addition, ships could launch teletrofonic currents in the direction of specific approaching or passing ships, establish continual vocal contact. Meucci accurately foresaw the development of new maritime communications corps, anticipating those wireless operators who would later be called "sparks" by their crew mates.

EXPLOSIONS

Lack of funding alone prevented Meucci from making large scale demonstrations of his revolutionary systems. In addition, prejudices associated with his nationality prevented New York financiers from even knowing of his activities. Meucci turned to his own patriots for help.

Confident in the both the originality and diversity of his teletrofonic inventions, Meucci was now sure that he could convince Italian financiers to help commercialize the Teletrofonic System; not in America, but in Italy. Meucci (now fifty-two years old) set up a long distance demonstration of his system in 1860 in which a famous Italian operatic singer was featured. His songs being transmitted across several miles of line, Meucci attracted considerable attention. Featured in the Italian newspapers around New York City, he indeed attracted the attentions of financiers.

Sr. Bendelari, one such impresario, suggested that full scale production of the teletrofonic system begin in Italy. He travelled to Italy with drawings and explanations of what he had seen and heard. Contrary to the hopes of all, Sr. Bendelari found it impossible to interest financiers in the teletrofonic system. Civil wars distracted the ordinarily aggressive Italian development of all such new technology.

Italian production of the teletrofono having never begun, Meucci became extremely embittered over both the incident and his own circumstance in America. American financiers were no better. Most contemporary Americans who had any "practical financial sense" at all could not believe that any mechanical device could actually transmit the human voice. They were far less interested in investing their fortunes toward developing systems which they considered fraudulent.

On sound advice from sympathetic compatriots, Meucci was warned never to bring anything to the American industrial concerns without first protecting himself by legal means. Before Meucci could dare bring his models the short ferry trip to Lower Manhattan to the developers, he needed a patent. Patents have never been cheap to obtain, this the regulator's tool. Even in those days, a patent cost a full two-hundred and fifty dollars.

Exorbitant costs being established for the financier's benefit, no independent inventor-novice could ever become an independently successful competitor without "financial assistance".

Meucci settled the matter by obtaining a caveat, a legal document which was considerably cheaper than the patent. Antonio could now only afford a caveat, a legal declaration of a successfully developed invention.

The caveat describes an invention and shows the time-fixed priority of an inventor's work. Meucci had models as well as the legal caveat. His caveat would stand in court, bearing the official seal, a registry number, and the signatures of witnesses. The Meucci caveat was taken in 1871, when he was 63 years old.

While travelling from Manhattan to Staten Island, Meucci was nearly killed when the steam engine of the ferry exploded. He survived this explosion in some inexplicable miracle, severely burned and crippled. While he languished in a hospital bed, his wife sold his original teletrofono models for the small sum of six dollars in order to pay for his expenses.

These models were sold to one John Fleming of Clifton, a secondhand dealer. Attempting to repurchase these models, he was informed that a "young man" had secured the models. Unable to locate the purchaser, Meucci was devastated. He suddenly felt that his own creation was already taking on a life of its own...fleeing away from him, out of control.

Growing desperate with thoughts of his own growing age and poor condition, Meucci now pursued the issue of commercializing his invention without restraint. In 1874 Meucci met with a vice-president of the Western Union District Telegraph Company, a certain W.B. Grant. Meucci described his "talking telegraph" and the complete system which was now operational. Meucci requested a test of his teletrofoni on one of the Telegraph Lines and was promised assistance and cooperation.

Mr. Grant appeared in earnest, engaged Meucci for a long while, and requested Meucci to leave his models. Meucci did so, being encouraged that he would be contacted very shortly for the test run. Hours of waiting became days. At this point, Meucci attempted to contact Grant again. The vice president could "never be found". Meucci continued visiting Western Union in hopes of reaching Grant and performing the required long-distance tests as promised him originally.

Meucci became bitterly angry over this betrayal of trust. The duplicity involved in the act of such unprofessional denial so exposed the fundamental methodology of American business that he wondered why he had ever left Cuba. So infuriated was he that he maintained a vigil at the Union Office, becoming an annoying eyesore. White haired, bearded, and bowed over with age, Meucci was viewed as a harmless old fool by younger, more aggressive office workers.

Adamant to the last, Meucci finally and loudly demanded the return of his every model. He was then very curtly informed that they "had been lost". Grant had passed these devices onto Henry W. Pope for his professional opinion on the exact working of the devices, forgetting the issue completely in the course of a business day. The monopoly had beaten another victim. He stormed out.

The path which the Meucci models took inside Western Union has been traced. The models periodically kept appearing and disappearing in the electrical research labs of Western Union, revealed through the written studies of several curious individuals. The models were transferred among several engineers as successive new electrical directors were installed. Each examined the models in complete ignorance. Lacking introductory explanations, no one comprehended what the weighty wooden cups could do when electrified.

Franklin L. Pope, friend and partner with young Thomas Edison at the time, was given the models by his brother. Together Pope and George Prescott could not understand the nature of the devices, putting them into a storage area in Western Union. This seems to be the last mysterious repository of Meucci models. Given in trust years before, the models sat in the dustbins of Western Union. Lost science.

The true history of telephonics begins with Meucci. Others, far younger, were raised in an atmosphere which was enriched by Meucci's developments. Phillip Reis noted the telephonic abilities of loosely positioned carbon rods through which flowed electrical currents. His primitive carbon microphone was later stolen by a vengeful Edison, who was in search of some means for both "breaking" the Bell Company's hold on telephonics, and saving his own financial record with Western Union Telegraph.

Meucci led the way long before others. It must be mentioned that both Gray and Reis were independent and equally great discoverers who each, though antedating Meucci by some 20 years, actually predated Bell by at least 10 years. Some have suggested that, as Bell was encountering great difficulty in developing his own telephonic apparatus, these same models were given to him for the expressed purpose of speeding the race along.

Western Union would engage Edison to "bust" the Bell patent in later years. Edison's invention of the carbon button telephonic transmitter was an inadvertent infringement of Meucci's earliest responder designs. The industrialization of the telephone revealed the repetitious and convoluted infringement of Meucci's every system-related invention. Bell's own frantic rush to develop telephony had more to do with his need to "live up to" sizable investment monies given him for this research, and less with any true inventive abilities. The truth of this is borne out in considering Bell's later work, involved in his frivolous failed "kite developments". Indeed, without the fortunate "assistance" by friends at the Patent Office, Bell would have succeeded in neither defeating Meucci's caveat nor Gray's electro-harmonic patent.

TELEPHONE SYSTEMS

Those who wished the implementation of telephony for financial gain, chose more controllable and less passionate individuals. Neither Meucci, Gray, nor Reis fit this category of choice. The Bell designs are obvious and direct copies of those long previously made by Meucci. The dubious manner in which the Bell patents were "handled and secured" speak more of "financial sleight of hand" than true inventive genius. The all too obvious manipulations behind the patent office desk are revealed in the historically pale claim that Bell secured his patent "15 minutes" before Gray applied for his caveat. Today it is not doubted whether perpetrators of such an arrogance would not go as far as to claim "15 years priority".

Lastly, this fraudulent action denied the years-previous Caveat of Meucci, which "could never be found at all in the patent records" during later trial proceedings. No mind. Meucci is a legend. A name suffused by mysteries. The Meucci caveat remains to this day on public record. All subsequent telephone patents are invalid. Meucci bears legal first-right. No lawyer today will decline this recorded truth.

All other court actions taken against Meucci toward the end of his life was staged by both the corporate Telephone Companies and the Court itself for the expressed purpose of securing the communications monopoly. The complete and operational Meucci Telephonic System, witnessed and used by countless visitors and neighbors for equally numerous years before Bell, was well documented in both Italian and local papers of the day.

To read the transcript of the Meucci court battle waged around the now aged and infirm Meucci is to witness the fear which large megaliths sustain. Though Meucci was not able to afford the yearly renewal price of his caveat, his priority was damaging, otherwise they would not have taken such measures to examine him publicly. The Bell Company sought to minimize Meucci's system by calling it nothing more than an elaborate "string telephone" in court proceedings, exposing themselves on several counts of fraud. Scientifically, this line of defense was unfounded. The obviously slack lines made the Meucci System incapable of conducting merely elastic vibrations with such clarity and amplitude. Moreover, the velvety rich tones received through these devices were far too modified, clarified, and loud to be "mere mechanical transmissions".

It was then hoped that the elderly gentleman would desist the entire crude process and give up. Meucci was publicly and ethnically labelled by leading journalists as "that old Italian, that old...candlemaker". Meucci maintained his ground to the consternation of the prosecuting attorney. Priority of diagrams, witnesses, working models...nothing could satisfy the predetermined judgement of the court.

To add insult to injury, Meucci's character was vilified in the press. In numerous pro-corporate newspaper articles Meucci is referred to as "a villain...a liar...an old fool". Predetermined to satisfy the corporate megalith, a deliberate and shameful court examination had as its aim the eradication of Meucci and his claim of priority. This process would later become the normal mode of business operation when destroying competitive technologies. With no hope of financial reprise in sight, Meucci ceased the excessive court fees. This was precisely what the monopoly wished. The fact yet remains that Meucci was first to invent the system.

Throughout the years, Meucci's name was not even mentioned in the history of telephonics. Closer evaluation of this true social phenomenon in "information control" reveals that communications history sources were controlled and principally provided in later years by Bell Labs to school text companies. They would ensure that the otherwise complex story was "straightened out".

It is also obvious that Meucci and his countrymen were never truly "embraced" by the American establishment until they took deliberate action. To the very end of his life, Meucci simply and elegantly maintained his serene statements in absolute confidence of the truth which was his own. "The telephone, which I invented and which I first made known...was stolen from me".

The more important fact in these matters of intrigue is recognizing that discovery itself is no respecter of persons or indeed of nations. Discovery touches those who honor its revelations. Discovery is an inspiring ray whose tracings are never limited by laws, prejudices, unbelief, nation, ethnic group, or economic bracket.

LEGEND

Eager to maintain their ascendancy in the annals of corporate America, incredible odds were marshalled against the aged Meucci by The Bell Company. In this determined counsel, we see the singular insecurity which frightens all secure investments. In truth, no investment is ever secure, when once discovery is loosed on the earth. What corporations have always feared is discovery itself. It is an unknown. In attempts to capture discoveries before they have time to take root and grow, every corporate megalith employs patent researchers. Their job is to waylay new company-threatening inventions.

Inventors represent the true unknown. They are uncontrolled forces who truly hold the power of the economic system in their grasp. Were it not so, then corporate predators would not pursue them with such deliberate vehemence. No one can destroy an idea once it has made its appearance on earth. Discovery is neither controlled or eradicated by the powerful. Attempts at wiping out new technology mysteriously result in a thousand diversified echoes, moving in a thousand places simultaneously.

The biography of Antonio Meucci is suffused with the deepest of emotions. I have read the biographies of many great and forgotten science legends, yet have not found one whose pathos completely equals that of Meucci. Despite the manner in which the new world treated him, the dignity of this great inventor is silently mirrored in his every portrait. The face of Antonio Meucci is serene...the face of a saint.

CHAPTER 3

EARTH ENERGY AND VOCAL RADIO: NATHAN STUBBLEFIELD

MEANDERINGS

The scientific historian methodically searches out catalogues of forgotten phenomena by thorough examination of old periodicals, texts, and patent files. The retrieval of old and forgotten observations, discoveries, scientific anecdotal records, and rare natural phenomena provide the intellectual dimension desperately needed by modern researchers who work in a vacuum of dogma.

Those who are familiar with the lure of scientific archives understand very well that more potential technology lies dormant than is currently addressed, discussed, or implemented. Much of modern scientific research is the weak echo of work already completed within the last century. The notion of drawing up electrical power from the ground sounds incredibly fanciful to conventional scientists, but numerous patents support the claim. A number of retrieved patents list compact batteries which can operate small appliances by drawing up ground electricity. Others describe methods whereby enough usable electrical power may be drawn out of the ground itself for industrial use. The existence of these devices is concrete, documented in several unsuspected and unstudied patents.

"Earth batteries" have been detailed in a previous article. Their history can be traced back to experiments performed by Luigi Galvani on copper plates in deep stone water wells. Currents derived through these gave Galvani and his assistants "shivering thrills and joyous shocks". Thereafter, a certain Mr. Kemp in Edinburgh (1828) worked with earth batteries, so that we know these designs were already being seriously studied. They demonstrate the validity of very anciently held beliefs concerning the generative vitality of earth itself.

Several of these devices were employed to power telegraphic systems (Bain), clocks (Drawbaugh), doorbells (Snow), and telephones (Meucci, Strong, Brown, Tompkins, Lockwood). Earth batteries are an unusual lost scientific entry having immense significance. Developed extensively during Victorian times, the earth batteries evidenced a unique and forgotten phenomena by which it was possible to actually "draw out" electricity from the ground. The most notable earth battery patent, however, is one which operated arc lamps by drawing "a constant electromotive force of commercial value" directly from the ground. In addition to this remarkable claim, a vocal radio broadcast system...through the ground.

It all began one hundred and fifty years ago with the advent of telegraphy. Well before geologists and draftsmen were hired to mark telegraph line installations across a chosen territory, the linesmen selected the actual course and way. While major line directions were generally known, it was left to the linesmen to select the specific post-by-post pathway through the forests. Not necessarily the best geological trail, linesmen followed the path which seemed, to their aesthetic sense, to be the "right one".

The meandering telegraphic wire went through rich dark evergreen forests and around glades. Lush valleys, flowing with corn, languidly waved...as the linesmen drew their weaving trail. Across meadows where wild flowers covered the earth in fragrant bouquets, there went the line in its curious, twisting path.

Over rolling hills which soared into the hazy sunlight, the telegraph linesmen sang as they went. And the lines followed a mysteriously winding trail which few discern. There were no specific instructions for the orientations of these long systems. Linesmen chose paths which they felt best "secured" the elevated line. Early telegraph linesman "felt" their way through the woods, laying the paths for lines according to their peculiar intuitions. Theirs was a sense-determined path rather than a strictly mathematical one, carved through woods and vales in artistic meandering ways.

Older linesmen recalled the days when line installations took their characteristic winding routes through woods, across meadows, and sinuously along ridges, lakes, and streams in an expressive freedom which was otherwise difficult to explain. Old linesmen innately sensed the most favorable paths along which lines "should" be placed. One finds that a careless amble through the woods is not unlike the path which telegraph linesmen chose.

A surveyor might simply draw a straight line across a section of land, and engineers would then employ powerful means to cut that straight path despite all natural barriers. Much of modern housing development is based on this "draw and cut" method. The sharp paths of engineers is effective and direct, but the old meandering rural roads dotted with their naturally placed homes are...beautiful. Later concerns for conserving wire, insulators, and posts outweighed this aesthetic sense of direction. Telegraph lines then simply followed the rails, as trains cut right across the landscape with no concern for aesthetics or the tendencies of nature at all.

The ability is not completely lost however. There are a few groups of individuals who yet maintain this peculiar "sense of the land". One is the landscaper, who artfully designs gardens and grounds with an eye on maximizing beauty. This qualitative skill is based on a sense which is both visionary and visceral. Architects alike enjoin this special aesthetic skill when designing buildings which must be site-conformable. They must balance both alignment, form, material, and structure against the mysterious "urge" of the land upon which they are to build. Improperly placed, the building offends the delicate "forgotten sense"...which all art critics loudly exercise.

This sense-oriented technique for determining the best route through the countryside was most definitely based on a forgotten sensitivity which dowsers yet preserve. The ground "urge" was anciently noted and honored by ancient sensitives who understood this to be the mysterious intelligence of the earth itself. Selected ancient persons recognized that certain ground spots emanated a vitalizing power which energized the mind, emotions, and body to penetrating heights. Places where this energy was potent were deemed "sacred spots".

It was soon discovered that a very strange and meandering path system naturally connected sacred spots together. Seeing that these sacred spots held the key to their survival, these sensitives followed the energizing paths across the ground in search of new understanding. These lines were not "light-line" straight. They meandered, like veins across an outstretched arm. They were dendritic, like trees branches and nerveworks.

Sorciers and Templars alike called these sites sacred, the interconnective natural paths "woivres". These paths seemed to meander and waver across the countryside, as streams and currents of water meander across the earth. Soon, ancient societies developed technologies which employed the mysterious earth energies for agricultural and medical purposes. Stones were erected where these energies emerged. It was hoped that some means for concentrating and maintaining the vital flow would be secured in this manner. The collective name for such technology is "geomancy", the long-forgotten craft for raising the vital earth energy.

Works of the ancient geomancers remain. As the megaliths, a system of ground receivers and earth energy concentrators, we recognize an ancient and forgotten empirical wisdom. Near certain of these rock pillars we find that agricultural vitality remains maximum. Geomancers, exercised their heightened communion with the earth energy to find water, a gift prized by their societies.

Geomantic qualitative science exceeds geologic quantitative science. The gift persists among countless individuals. Holding metal rods or green twigs in their hands, these persons sense the presence of water by a peculiar reflex which is felt deeply in the abdomen. The reflex found validation in the research of Dr. George S. White and Dr. Albert Abrams, both of whom rediscovered the autonomic response of human physiology to the distant presence of certain substances.

Signalling to these sensitives its messages in peculiar runes and dream tokens, the geomantic energy pulsed and streamed over the countryside when properly cultivated. Geomancers maintained vigil at the sacred spots until satisfied with deep personal epiphanies. The sacred spots were known as sites where visions and dreams were exceptionally vivid. Their ability to heal the infirm is recorded in legend, forming the foundation and determining the altar stone placements of European Cathedrals. In fact, there are those who cite the Cathedral System as a most recent attempt at preserving the anciently heralded sacred spots of wood and glen.

Geomancy was the ancient qualitative science by which "holy spots" were discerned, and sacred edifices were properly founded. Intuitive discernment, rather than mathematical objectivity, governed the geomancer. Geomantic aesthetics preceded and ruled the building of ancient villages and towns. It is no wonder that most architects of any real artistic worth exercise these same aesthetics. Art-governed architects are natural geomancers.

The earth energy "sense" is found in all cultures, however separated in time and location. Empirically discovered by each society, we find repetitive examples of the geomantic art the world over. Geomancers each mapped the earth in their own vicinity, noting the presence or absence of ground energy. Vitality was the only energy of ancient survival. Geomancers were the priests. They were the sages, the gnostics. Geomancers were the architectural planners. They located every natural resource needed by their communities. Throughout the world we find their legacy, now largely forgotten.

Replaced by the quantitative skills of geologists and engineers, we now scour across the land along quadrants and grids of our own design. No longer do planners observe the "urge" of the land. Few can afford the fees which truly gifted architects demand when demonstrating the geomantic skill. None will disagree however, that the geomantic skill does shape both mood and vision when properly executed in architecture. The released power exceeds the modern ability of measurement, evaluation, or quantification. The geomantic energy...the earth energy...defies quantitative analysis. It is an entity whose presence links both sensual experience, dream, vision, thought, imagination, and place. Numerous societies called this mysterious power by their own names. Chinese geomancers called this energy "Qi". Anglo-Saxon geomancers called it "Vril". Each is an ethnic name for the one earth energy.

The organismic energy of earth which manifests as a mysterious black radiance is seen throughout natural settings. It is observed beneath evergreen trees at noonday. It shimmers above certain ground spots where it rises as a magnificent glowing crown at night. It combines sensation and consciousness, being simultaneously seen and felt. It unifies metaphysical and physical entities, being recognized only through personal contact and experience.

In addition, the earth energy freely saturates and modifies the operation of certain very specific technologies, where its presence creates quantitative anomalies. Any system whose primary elements require ground connection are sure to become hosts for the geomantic energy. When the telegraph system first appeared, it became flooded with this energy.

The old linesmen trekked across woods in a careful manner, turning aside from natural barricades. When maps of these first telegraphic lines are consulted, it is seen that these lines meandered with natural features common to the earth energy paths. As the telegraphic lines twisted and turned through the countryside and wilds in twisting vines of iron on tar-covered wooden poles, they directly intercepted ground energy.

Early telegraph lines intercepted earth energy with great regularity, often connected distant sacred spots together. We find all too numerous anecdotes and collections of reports in telegraphic trade journals which indicate that an anomalous ground energy was entering the system components at certain critical seasons. These reports affirm that an earth "electricity" was energizing telegraph systems without the need for battery power at all.

Other reports tell of strange, automatic telegraph signals which suddenly manifested during night hours. Still others report the peculiar ability of telegraph operators to "know who would call...why they were calling...and what the nature of the message would be". This phenomena would be repeated in later years, when wireless operators began experiencing the very same things. The heightened consciousness experienced near these large grounded systems had everything to do with the reappearance of phenomena anciently observed along the meandering paths between sacred spots. The rediscovery of these anciently known truths was again making its appearance during the industrial revolution.

VISIONARY

Who is Nathan B. Stubblefield, and why do most citizens in the state of Kentucky justifiably revere his name? A native of Murray, Kentucky, Nathan B. Stubblefield had a love for the lonely wooded areas on the outskirts of town. A self-educated experimenter and avid reader of every kind of scientific literature, Nathan Stubblefield supplemented his living with farming. He remained a practical inventor of some of the most unusual electrical devices ever developed in America.

What he discovered and demonstrated before hundreds of qualified observers in his day seems to challenge many basic axioms of electrical dynamics. It all began with his sensitivity to the "urge" of the land. Certain spots in the surrounding woods were mysterious, possessed of a strange magick all their own. Vitalizing and sense-provocative, Stubblefield instinctively knew that these locations might be unique natural energy sources.

Rock outcrops, evergreens, and flowing springs each registered as strong sensual attractants. Could it be that they were sensual attractants because they conducted and projected special ground currents? Was he enthralled and drawn into certain spots because of their projective energy? But...what energy? Did it contain or exceed the qualities of electricity?

He developed numerous "vibrating telephones" which were used by local residents in 1887. They were powered by an extraordinary receiver of ground electricity which produced great quantities of a strange "electricity". The telephonic devices were patented in 1888 and represent the first commercial wireless telephones, using the ground as the transmission medium. The years when telephonic lines were suddenly made available to the world betrayed the fact that the new medium was one which only the very rich could afford.

Common people could simply not be serviced with local telephones until prices were made cheaper. While telegraphy employed thrifty iron wire, telephony demanded the expensive and better conducting copper lines. Telephones designed by A.G. Bell did not give powerful enough signals through iron wire at any distance because of their additional high resistance. Among its numerous other telephonic problems, the Bell telephone could simply not transmit or receive a strong and clear vocal signal without very excessive battery power. The Bell System was thus not a truly "democratic" medium of communications.

A mysterious and unrecorded sequence of discoveries preceded Stubblefield's early developments, but he was able to dispense with wire connections entirely. His was not a "one-wire" system. Nathan Stubblefield performed the "impossible": he developed, tested, demonstrated, and established a small, democratic telephone service which did not require wire lines at all!

Mr. Stubblefield discovered that telephonic signals of exceptional clarity could be both transmitted and received through the ground medium alone. There was simply no precedent for this development. His system utilized the ground itself as the conductive medium, an inexplicable natural "articulation" connecting all ground placed telephones.

The first effect of this wonder was that common people could now have the much-needed communications which both great distance and poverty prevented. Farms could be interlinked by the Stubblefield exchange by simply plugging both terminals into the ground. Wire would not take up the expense which the telephone exchange would later charge to the customers in addition to service. Signals were loud...and clear. All those who experienced this kind of telephonic conversation declared that Stubblefield's telephone was "exceptionally clear". He had discovered a true wonder.

We have photographs of his telephone sets. These reveal small, ruggedly built wooden cases which are surmounted by conventional transmitter-receivers. Heavy insulated cables run to the outer ground from this apparatus. Stubblefield developed an "annunciator" (horn loudspeaker) which amplified the voice of distant callers. These telephone sets appeared in his numerous demonstrations on the east coast, from New York to Delaware.

The signals were so loud and clear that they defied commercial levels of excellence provided by the now growing monopolies of American Telephone and Western Telegraph. Thomas Edison broke the Bell telephone monopoly when he developed the carbon button microphone for Western Union. While sounds were indeed louder with the Edison carbon microphone, these carbon microphones needed excessive battery power...and batteries were not cheap.

Some telephone companies began utilizing dynamo systems to power their lines. The fuel needs of dynamos drive customer costs much too high, prohibiting the ordinary people from having their own service installed. But Stubblefield's devices defied all the known electrical laws. In the early Stubblefield system, twin terminals into the ground formed the initial bridge among telephones.

As system users were effectively joined together through the ground itself, the high cost of wire was eliminated! The signals were exceptional, and did not fade or intensify with rain. This fact was never considered in theoretical discussions of his work. Those who experienced speech through the Stubblefield system each reported similar impressions. While ordinary "soil conduction" telephonics require a certain degree of ground water for their operation, we know that his system did not operate on this principle.

The theoretical reasons explaining ground conduction telephony had later been established by researchers in England, notably Sir William Preece. Preece successfully attempted only telegraphic signals across great distances of land and sea. Stubblefield was telephoning through greater distances with the legendary clarity and strength which became equal to his other mysterious developments.

TERMINALS

The first telegraph lines of Morse were two-wire lines. The circular flow of electrical signals among station receivers, batteries, and keys was conserved with great efficiency. Double wire systems were very expensive however. It was quickly discovered that single lines, terminated in the ground with heavy metal plates, could exchange equally strong signals. The immense savings in wire, poles, insulators, and maintenance was an attractive feature of the single-wire method. Company owners were elated.

The problem in single-line telegraph systems was finding the right ground spot. It was quickly recognized that "good" and "bad" grounds could affect the behavior and operation of the line. Improperly placed ground plates could ruin a system by not conducting signals properly. In time improperly placed lines would actually fail. Spurious conductivity in a line could ruin critical transmissions and receptions. Telegraphic lore is filled with discussions about both "good ground" and "bad ground".

The linesmen, workers in a yet primarily agrarian society, had experience with soil and earth in general. Many of them were farmboys who had watched certain old-timers "divining" for water. Linesmen frequently discussed such natural means for discerning the "good ground" for terminating a telegraph line. Telegraph linesmen found to their delight that the dowser's skill always located "good ground". This is why so many of them guided the telegraph lines through those meandering paths across the countryside.

The true difference between the Stubblefield system and these early "conduction telegraphy" systems became obvious as soon as we delve into the record. Stubblefield developed a means by which calls could be individualized among customers. Later, his central telephone exchange included power-amplifying relays, set in the ground at specific distances. Calls were handled by an elaborate system of two-wire, ground connected automatic switches and relays which were placed in specific spots across the countryside. Telephone signal purity was remarkable for the time, using a single carbon button for both transmission and reception.

Furthermore, Stubblefield's telephones could be left on for days without weakening the power system at all. Now hundreds of ordinary people in widely separated places could afford the installation of telephone service. But, how were ground plugged relays acting as amplifiers in the Stubblefield system?

As telephony gradually replaced the telegraph service, lines were also accommodated to telephony. Before becoming entirely reclusive, Mr. Stubblefield befriended a few employees of the telephone service. These friends obtained cast-off telephone equipment and parts for his experiments. Older linesmen told Stubblefield much about their own empirical observations on the systems...phenomena which had no textbook explanations.

He became very familiar with the behavior of telephone exchange equipment in the natural environment. The telephonic systems of existing service companies were grounded systems. Each end of both telegraphic and telephonic lines were sunk into the ground, while the single expensive copper line formed the communications link.

Ground sites terminated specific lengths of these service lines in special, thick metal plates. Plates were well-buried in selected ground. These plates were composed either of zinc or copper, and required specific ground placement for their continued operation. Linesmen were taught to find "good ground" for these sites. Some later insensitivity among the growing numbers of hired crew members required the development of electric "ground location meters", none of which were to give the special and anomalous characteristics observed in early linework.

Certain telephonic patents reveal extremely "articulated" termination plates for these service lines. These were folded, stacked, coiled, and interleaved. Acting as accumulators of earth energies, these often became dangerously charged. It was found that signals would both self-clarify and self-amplify to unexpected degrees when these special terminations were employed.

Properly grounded telegraph lines were known to produce unexpected signal strengths...as well as unexpected signals. Night station operators were often "haunted" by spurious messages. These contained fragmentary words and sentences, and could not be traced to other station operators. It is not coincidental that the older lines demonstrated their remarkable, consistent operation throughout the years...requiring few or no batteries. This absolutely astounding fact is well documented in the telegraph and telephone literature of the day.

In these trade journals we find reports of lines in which current was everflowing! Company owners found this fascinating natural fact quite lucrative as well as surprising. The question was...where is the current coming from? The echo of the linesman resounded in the forest, the answer singing beneath his feet.

Another equally remarkable fact involved the engineer's methodically driven lines. Surveyed straight across land and through mountains, these lines did not manifest electrical self-excitations. Clearly, the difference of methods had produced completely opposed energetic results; the one active and the other inert.

As companies expanded across greater regions of ground, engineers replaced the oldtime lineman's sense of "proper placement" with surveyor's charts. It is not unusual for corporate expansions to bring about such a dramatic loss of quality...in exchange for a growth of quantity. What they derived from this obsession with thrift simply added to a continuing fund of ignorance which has swallowed up the more naturalistic empirical sciences.

In their movement toward economy and thrift, numerous companies wished to save on batter costs. The trouble with single-line telegraph systems was that the battery power was always "on". Usually lost to the ground in continual volumes, battery current simply drained off into the ground. This meant high battery costs. Owners insisted that a means for alleviating these high costs had to be found.

George Little found that this leakage could be reduced by employing carbon rheostats between the signal key, battery, and ground. One could control the actual leakage to the ground by carefully rasing the rheostat. By preventing unnecessary signal leakages into the ground, battery power could be conserved.

But, when these rheostats were installed, strange phenomena began taking place with great strength and repetition. The first phenomenon related rheostatic settings to actual line-developed power. Power kept appearing despite the battery status. Rheostats made by Little were sensitive enough to "valve" line signals and use the ground developed current.

Several of these terminal rheostat patents have been retrieved. One familiar model uses a thick cylinder of carbon with a slide spring contact. Another uses fixed resistive steps which are switched into or out of the circuit. Here were the very first control components of the telegraphic system, the first in a great series.

Thomas Edison dominated the method invented by George Little, including the use of terminal rheostats in order to control the amount of current flowing to and from the ground during signal time. Another amazing phenomenon was the great variation of rheostatic settings which each ground required before strong signalling could occur. There was no automatic means for determining these settings. No textbook formula could predict the settings. They seemed to obey unknown laws of behavior.

Some terminal rheostats needed to be closed completely. Others could be opened full until signals were of sufficient strength to operate the system. Seasonal rheostatic variations were always noted. It was thus seen that each ground site had its own "character". Each ground was possessed of activities which defied conventional quantitative description. In addition, the fact that these settings changed completely with the season, having little to do with ground water tables, was troublesome to most theoretical engineers.

Inventors, however, adopted whatever empirical experiment would offer. If a component worked, then it was employed. Empirical technology produced the most amazing devices ever seen, working with little-understood forces. The anomalous appearance of powerful currents in the end-grounded lines was one such marvel.

Telegraph line was not made of pure copper. Telegraph line was bare iron wire. Lines were not well insulated throughout their lengths. The line itself was supported on porcelain insulators and fixed to tarred wooden poles. Rain and corrosion caused the conductivity of the line to vary considerably with distance. Signal strength along such resistive wire would have theoretically been extremely poor.

The remarkable observation defied theoretical estimate. Signals were excessively strong at certain times, appearing in seasonal waves of strength. So great was the developed signal strength that operators could "remove battery cups" and work with almost no current at all. This was especially true of chemical telegraphs, which employed only earth battery energy for most of their operating time.

Where did this extra energy come from? From what mysterious depths did this strange power emerge? Was it electricity as we know it? It has been suggested that earth energy, the pre-electrical energy of the ground, was at work in all these systems. Called "vital energy" by Victorian Science, this presence exceeded the character and nature of ordinary electricity.

When later researchers began measuring and experimenting with the ground derived energy, they discovered several important distinctions between it and electricity. Where the force of electrical currents would radiate in a fan shaped radiance through grounds, earth energy evidenced a vegetative nature. Examination of the vegetative patterns taken by earth energy revealed discrete articulations...a thready nature which was unlike ordinary current.

Energetic threads of this energy could be measured between communication sites only along tightly confined trails. Also, while electrical force clearly dissipated through ground conduction, the thready earth energy actually evidenced growth characteristics in conductive lines. Electrical power grew to spark potentials in these lines when no exterior evidence allowed explanation of the energy levels.

Most recognized that electricity was simply a by-product or epiphenomenon of a more fundamental agency which entered the grounded lines. Rheostats somehow "tuned" the potentials of this earth energy. While Reichenbach discovered the fundamental permeating nature of "Od force", several others showed the essential unity of earth energy and the human aura. It was found possible to "match and tune" these energies through the use of rheostats and capacitors. Persons who were weak and infirm actually experienced vitalizing elevations when connected to the ground energy through these special rheostatic tuners.

Called "radionic" tuners by those who developed them, numerous investigations revealed the potentials of this ground energy for social use. Agricultural applications of radionic tuners produced greater crop yields. Moreover, large ground-connected radionic tuners produced extraordinary effects on the mind and emotions...relieving tensions and opening thought to new potentials. Taken from this viewpoint, telegraph systems behaved as radionic tuners on a vast scale. We would therefore expect them to produce anomalous energetic effects in several parameters of human experience.

Touching a well-grounded iron rod is a good first experiment to try in these regards. Try and find a place where power leakage into the ground is minimal...a park or wooded area. Take a yard-long solid iron rod whose surface is free of shellac or insulator coatings. Carefully drive the rod into the ground with a hammer. Wetted hands on the iron should produce a mild electrical sensation. These voltages may be measured. They "pin" sensitive galvanometers. The current does not cease after several weeks of activity when properly placed.

ELECTRICAL OCEAN

We find a good number of the earth battery designs in the Patent Registry. The earliest designs appeared in 1841 when Alexander Bain applied the phenomenon to telegraphy. While working a telegraphic line, he chanced to discover that his leads had become immersed in water. This short-circuit through earthed water did not stop the actions which resounded through his system.

Mr. Bain took the next step to a greater distance, burying copper plates and zinc plates with a mile of ground between them. These, when connected to a telegraphic line performed remarkably well without any other battery assistance. Bain obtained the patent for his earth battery years after his initial discovery (1841), using it to drive telegraph systems and clocks.

Stephen Vail (1837) observed the same effect, not knowing what caused it. The establishment of the first functional telegraph line seemed to require ever few batteries with time. Vail began with some twelve large battery cups, reducing them gradually until only two were needed. There came a point during certain operative seasons where he found it possible to remove all the batteries!

J.W. Wilkins in England (1845) corroborated findings made by Bain, developing a similar earth battery for use in telegraphic service. An early English Patent appears in 1864 by John Haworth, the first true composite earth battery. This battery is drum shaped, having numerous solid discs mounted on an insulative axis, end-braced, and buried. Their power was rated in terms of disc diameter and telegraph line distance: one foot diameter discs for seventy-five miles of line, two foot discs for up to four hundred and forty miles of line.

This mystery persisted for years. I have heard such an account by a close friend and electrical engineer who reported that local telegraph stations remained in operation despite the fact that their batteries had not been recharged for a great number of years. When the battery was examined it was actually dried out and physically corroded. Yet the signals continued (W. Lehr).

Patent Archives have revealed a great number of these devices including their remarkable operative descriptions. Earth batteries by Garratt (1868), Edard (1877) , Mellon (1889), and Hicks (1890) yield therapeutic powers. Earth batteries by Bryan (1875), Cerpaux (1876), Bear (1877), Dieckmann (1885), Drawbaugh (1879), Snow (1874), Spaulding (1885), and Stubblefield (1898) produce usable power.

In addition to these marvelous patents, there are those earth batteries which found their way into telephonic service. Designs by Strong (1880), Brown (1881), Tomkins (1881), Lockwood (1881) provided primary power as well as power boosts for telephonic systems throughout the countryside. The well reputed fame of "earth batteries" centers around their very anomalous electrical behavior. How they produce such volumes of current remains an embarrassing anomaly.

The central mystery about earth batteries is that they do not corrode to the degree in which their electrical production rate theoretically demands. Exhumed earth batteries reveal little surface corrosion. Earth batteries also varied in outputs when placed in different grounds. Some gave only weak and unusable outputs. Others continued to produce prodigious volumes of power for years unattended. Some researchers connected earth batteries in series (Dieckmann) to build a greater output, but Stubblefield was not interested in this arrangement.

It becomes apparent that Mr. Stubblefield had witnessed (or experienced) some natural occurrence of discharging electrical energy in a telephonic system, and had determined the mode of its manifestation with simple means. His ground energy receiver (Pat.600.457) remains a true electrician's mystery.

Nathan Stubblefield's observations of natural electrical manifestations led him to consider the taking of "free" electrical energy from the earth. His excessive study of theoretical literature taught that no such advantage could be obtained by using earth batteries. Writers contended that vast amounts of energy could never be used to drive the engineworks of industry by earth battery power. He saw that, unless a new breakthrough in the art could be found, the theoreticians would be correct.

Older linesmen taught Stubblefield about sensitive ground spots: how "uncommonly great" electrical activity had to be patiently searched out. These electrical hot spots, when compared with most adjoining ground, were like electrical oil wells. Finding the "right spot" would do more than simply insure good ground connection for telegraph lines. Certain ground powerpoints could actually power the lines! Motivated toward deeper research by his own natural observations and intuitive sensation, Stubblefield devised several earth batteries. His own peculiar ability to sense earth energies taught that it was vast in quantity, yet untapped by humanity. The means for drawing out the energy could be found!

Stubblefield knew that ground probes, placed into various spots, reveal an amazing degree of electrical activity. These currents varied across any chosen plot of ground. Wet soils often reverse the expected electrical strength; weakening rather than strengthening their magnitude. Stubblefield knew that a proper placement of metallic ground probes could produce stronger currents for use. But he did not anticipate what he then accidentally discovered.

His initial experiments involved the development and examination of simple earth batteries: buried metallic arrangements which produced weak electrolytic power. Mr. Stubblefield observed a strange "earth charging phenomenon", reporting that the burial of an "earth energy cell" required time to build up charge. During the first phase of this charge building process, the characteristic weak output was observed. This was usually a volt at half an ampere, the general electrolytic output of buried metals.

From his linesman mentors, Stubblefield knew that placement of any grounded metal was the key toward deriving power. If properly placed, the energetic output of his cell would be phenomenal. Finding such a powerpoint, he buried the cell. The process took a week or more to build strength. Once the cell was "saturated", however, it became (in his words) "a conduit of earth charge". This mysterious transition from weak battery to energy conduit required time.

Typical of his curt statements, Mr. Stubblefield simply stated that the fully saturated coil suddenly "manifested an electromotive force far greater than any known wetcell". This state being achieved, the cell flowed over in "commercial electrical volumes".

He did not claim complete knowledge of the phenomenon. He observed that the activity "reached into weeks and months of continuous work night and day".

Stubblefield envisioned the energy cell as a "plug", drawing out the electrical charge of the ground. The cell coils acted as a lumped conductor. Charge saturated this conductor and flowed up into it, powering any electrically connected appliance. After repeated exhumations, the copper element of these cells was found "not acted on in any perceptible degree...even after repeated renewals".

Mr. Stubblefield described means by which such cells could be connected in series at short distances from one another. "With these, acting as electrodes...you draw from the electrical energy of the earth a constant E.M.F. of commercial value". That phrase..."acting as electrodes..." is the heart of the Stubblefield energy cell. It is not a battery. It absorbs and flows over with the stupendous energy of the earth's charge.

This device, an earthed electrode, drew up enough natural electric charge from the earth to operate motors, pumps, arc lamps, and all the components of his ground telephone system. The implementation of his earth energy technology would have changed the nature of American Society, were it permitted free market expression in its day.

Mr. Stubblefield later stated in very plain language that the earth was filled with "an electrical ocean". This electrical ocean was surging with huge "electrical waves" which could be felt, brimming over in certain places. No doubt, he was one who felt the ground energy. Stubblefield sensed that the ground currents arrived in powerful electrical waves. In Stubblefield's visionary approach, the electrical waves permeated the ground. These electrical waves were like ocean waves: ceaselessly surging and cresting over.

As ocean waves crash against fixed shores and rocks, so electrical waves also surge and crash against underground geological features. Stubblefield reasoned that this electrical waving should be extreme in certain locales. The "rocky shores" of the electrical ocean were numerous but specific. Just as there are rocky shores, calm beaches, and surging ocean depths, Stubblefield clearly envisioned the mysterious dark waves of the vast and unsuspected subterranean electrical ocean.

Knowing these truths, Stubblefield arranged ground rods in very specific locales in order to intercept the electrical waves for power. He knew that these electrical waves would only appear in very specific places, so he did not expect to find them everywhere in abundance. Stubblefield constantly spoke of "working the ground" before power could be taken from it. Stubblefield observed the natural tides and boundaries of the electrical ocean in and around his lovely rural hometown.

Some researchers believed that the vast electrical ground reservoir finds its source in the enormous solar efflux. Certainly daytime grounds yield a remarkable amount of static. Ground terminal shortwave reception is excessively "choked" during the daylight hours on certain bands. Despite the supposed insulative qualities of the atmosphere the solar efflux finds its way through space, eventually permeating the ground. Some researchers have referred to the ground-permeating solar energy as the "slow solar discharge".

The "slow discharge" represents the enormous drift of aether through the entire body of ground. The earth evidences a constantly self regenerating charge. Tesla opposed the notion that this potent field was the result of decaying radioactive materials deep in the crust. Tesla charted and used the earth waves in their surging impulses for transmitting power across the earth.

Numerous other researchers would refer to this "electrical ocean" as the vast reservoir of untapped natural energy. Somehow this reservoir is regenerated in a constant swelling. Where did the energy come from? Earth static was presumed by Tesla to be a solar activity which manifested in and across the ground. The ever growing static of earth was problematic for physicists who could not see the source for such energetic growth potentials.

Tesla believed that ultrafine corpuscles from the sun permeated the entire earth, manifesting as static charge. Tesla further conjectured that these rays came primarily from the sun, since it was ejecting matter "at excessively high voltages". If this were so, reasoned Tesla, then sunlight contained something of this electro-active component...and it was certainly possible to derive electrical energy from sunlight.

Nikola Tesla announced these facts in 1894, finding only the silencing ridicule of academicians already hating his very name. When Tesla declared that "rays from space" were "bombarding the earth" he was absolutely rejected by the academic club who rejected these claims as "superstitious". Upbraiding his findings, they later claimed for themselves the very same discovery (Hess 1912, Millikan 1932).

Tesla stated that the electrical energy released by the sun is a far greater, more permeating supply than sunlight itself. He certainly believed it should be considered as a first rate natural electrical source of enormous potential for commercial applications. His assertion was based on experimentally verified facts when, measuring steadily growing charge states in vacuum tubes, it occurred to him that earth charge was sourced in solar activity.

Tesla also demonstrated the extraction of free electrical power from solar energy. A grounded mica capacitor is surmounted by a highly polished zinc plate. This plate may be poised in a highly evacuated glass container to best advantage, the zinc not exposed to corrosive influences. The tube is elevated and exposed to sunlight. The mica capacitor is connected in series with the vacuum tube. After only several minutes of exposure time, the stored electrical energy is formidable, producing a powerful white arc discharge. Tesla patented this device.

Samuel Morse originally planned the burial of telegraphic lines between cities. Having done so across some twenty miles (at great expense and through great labor) Morse found his system utterly incapable of operation. Static had so flooded his receivers that no signalling was possible at all. Receivers were paralyzed by the volume of ground-absorbed energy. This first bad experience with the static of ground presented such a discouragement, that he almost stopped the entire plan. The uneconomical task of elevating all his cables later became the normal format for telegraphic systems.

Early telegraphers observed a steady growth of static throughout night seasons. This growth continued despite the absence of winds or storm conditions anywhere along the line. Researchers have often referred to this kind of power as "free energy", meaning that the power source is extraterrestrial and natural in supply. Such an energy source would remain cost-free. The privatization of utility companies could conceivably be municipal and democratic. Municipal groups could share the cost of installing the ground-energy stations.

Since earth absorbs the permeating solar efflux, then these energies can be extracted for aeons. Others have viewed the generation of ground static as a natural "radiant process" from the ground itself. Static charge appears as the inert by-product of the mysterious earth energy, the self magnifying organismic ground energy. Solar effects mirror the ground states which absorb them, producing the static charge epiphenomenon. Vril, according to medieval mystical philosophers, is the ground of being from which all material manifestations emerge.

ENERGY RECEIVER

Mr. Stubblefield developed a peculiar bi-metallic induction coil which, when buried, draw up sufficient electrical power to operate lamps and other appliances which he designed and tested. A great length of both cotton-insulated copper and bare iron wires were wound together in a "bifilar" arrangement on a large iron stove bolt. The windings were held side by side throughout the coil. His patent specification describes the device as a "terminal which draws electricity out of the ground".

This successful operation of the device required very specific ground placement. It would not work with equal effectiveness in all locations. A very precise placement of the device required a precise knowledge which only dowsers have. Stubblefield shared this particular fact with only one person.

I spoke with an academician who had the extreme privilege of speaking with Mr. Stubblefield's son, Bernard Stubblefield. Bernard, by this time himself quite aged, told that his father's method in locating the "right spot" was deliberate and time consuming. His father referred to the device as a "receptive terminal" and not a battery. Despite the insistence of Patent Officers in calling the device a "battery", Stubblefield declared it to be an "energy receiver...a receptive cell for intercepting electrical ground waves". Its conductive ability somehow absorbs and directs the enormous volumes of earth energy.

Whether the current derived from this cell is electricity as we know it has been questioned. One indicator that it is not is found when considering his use of the energy in lighting lamps. With this energy Nathan Stubblefield operated a score of arc lamps at full brightness for twenty four hours a day. There was a definite trigger by which this energy was stimulated and maintained.

The induction coil which bears his name is equipped with three coils which are wrapped around upon a heavy iron core. Bare iron wire and cotton covered copper wire are wrapped side by side, comprising a primary coil body. Each layer of this primary coil body is covered by a band of cotton insulation, bringing four wire leads to the coil terminus. Two leads of iron and two of copper are external to the coil. Commercial electrical power is obtained through these connective terminals.

In addition to this bimetallic winding, there is a third winding: the "secondary". This third coil is insulated from the primary bimetallic coil, serving as a trigger device. Presumably, a stimulating impulse shock was introduced into the tertiary coil, after which the upwelling electrical ground response brought forth powerful currents in both iron and copper coils.

Electrolytically (as a battery in acid or saltwater) the Stubblefield coil is disappointing; producing less than one volt according to those who have duplicated its construction. Stubblefield's bimetallic coil was a "plug": a receiver which intercepts the vast and free electrical reservoir of the ground itself. His patent and subsequent company brochures define the manner in which his earth battery was to be activated.

Technically, the Stubblefield device is a modified thermocouple (a bimetal in tight surface contact) but could not supply the degree of power which he reported. While this arrangement could develop a few milliwatts of power in appropriately hot ground spots, the thermoelectric explanation of the device cannot explain the phenomenal output reported in news reports of Stubblefield's demonstrations.

Furthermore, though the Stubblefield power receiver is wound like an induction coil, it produces a steady direct current output. This poses additional problems for the conventional engineers. Electrical induction only occurs with electrical alternations, oscillations, and impulses.

Witnesses described ground-powered motors which ran unceasingly and unattended for months without need for replacing or replenishing the ground battery. Small machinery, clocks, and loud gongs were run by other ground-buried cells as reported by credible witnesses. Stubblefield may have discovered the auto-magnifying voltage effect of electrostatic induction in coils before Tesla, who later utilized the effect in his special electrostatic Transformers.

These buried coils may have become saturated with earth electrostatic energy, which travelled from subterranean depths. In such a case, the mere battery power of the coil was replaced by the electrostatic flow, the coil acting as an electrode. This seemed obvious when considering the fact that its ordinary battery current (1 watt) was gradually replaced by a continually growing electrical current of far greater proportion.

TREE ROOTS

Experimenters have observed the "slow accumulation and creep" of current up through vertically buried coils and large solid rods. This current has growth characteristics which gains strength with lengthened burial time. Buried coils and rods do not give their full output until they have "developed" power over a few hours of time.

This behavior resembles nothing like a true electrical current. The best model to explain the phenomenon is vegetative growth...a biological expression. Only a full scale test of the reconstructed Stubblefield device in proper grounds will give conclusive and satisfying answers.

Witnesses convey that Mr. Stubblefield's batteries were usually buried at the roots of certain very old oak trees. From these sites it was possible for him to bring small arc lamps to their full candlepower. Tremendous amounts of energy are required for this expenditure of power. Not only was he remarkably able to draw such volumes of current from the ground reservoir for lamp lighting, but the power was available to him throughout the day.

Arc lamps were hung in the trees themselves with their receiver coils buried in the roots. Such was the nature of this current that the lamps did not heat excessively, and seemed to burn on forever in a brilliant white light. Nathan was not replacing his lamps with the frequency demanded by such continuous operation. Obtained through his employment with the telephone company, he was able to recharge old wetcell batteries with energy from these buried receivers for other experiments.

Certain conventional thinkers claimed that the Stubblefield simply used wetcell power for his telephones. Later demonstrations indicate the fundamental error of this conventional view. Stubblefield ran most of his apparatus nonstop for days; without turning off the power. It is more than likely that charged wetcells were used to "jump start" the ground electrode during certain seasons, since the patent reveals that an outer third coil could be added to the copper-iron bimetal.

We do not know the secrets of the earth charge as Nathan determined. Others since this time have observed fluctuations at certain times of the year in ground energy. It may be that a sudden induction is required before the excess ground charge surges to the surface...like priming a pump. The arc lamps could have been low pressure gas arclamps of the kind demonstrated by Daniel MacFarland-Moore; but these required high voltages. Nathan did not utilize such excessive voltages.

Another Stubblefield paradox deals with the erroneous notion that he simply connected hundreds of his small-wattage batteries together, producing a large and commercial output. Nowhere is this evidenced. Nathan showed that one or two such batteries were sufficient to draw off "the charge of the earth"; a very different kind of energy.

When properly placed, the weak power of the Stubblefield "battery" becomes an electrode for the powerful earth charge. But arc lighting and battery charging was not his only specialty; there were other marvels which he began developing in methodical succession. His bimetallic coil receiver intercepted electrical waves and produced enormous power outputs which could be modulated: superimposed with additional signals, sounds...and voices.

GROUND RADIO

Salva (1795) suggested several electrical schemes for long-distance, and even transaquatic telegraphy. He suggested that physiophonic telegraphy be the communications mode; where human recipients would receive the mild shocks of a distant signal station, and so convey messages.

Salva also believed that earthquakes had subterranean electric origins. Working on the hypothesis that subterranean electricity caused violent communications under vast earth strata, Salva suggested that ground and water be used to replace wires for electrical signalling.

Sommerring (1811) first attempted telegraphic transmissions through water-filled wooden tubs. The signals were effectively passed as if through wire conductors, the thought of wireless ground resulting. James Lindsay (1830) first developed the notion of utilizing artificially generated electricity for special modes of lighting, motor-power, and communication. Mr. Lindsay suggested that submarine cables might be laid between land masses while using "earth batteries and bare wires" as the means for power transfer.

Steinheil (1838) demonstrated the remarkable passage of signals along one-wire to the ground. When trying to use earth as the "second line" he measured large currents. This complete success proved the great conductivity of ground; and so the "earth circuit" was born, liberating telegraph systems from the expense of using the two-wire system. Morse (1842) sent telegraph signals across a river. Antonio Meucci (1852) had already demonstrated the transmission of vocal signals through seawater, but traversing the ground represents a different thing altogether.

Mr. Stubblefield reasoned that, since electrical waves traverse the whole earth, it might be possible to send signals to distant places. These ground-permeating natural electrical waves might serve as carriers for the human voice. The ground would act as both power generator and signal conductor. Like a gale carrying messages downwind, these electrical waves could bring wireless communications instantly to any part of the world.

To this end, Mr. Stubblefield experimented with the buried power receiver and a system of telephone sets. He found it possible to send vocal signals through the ground to a distant receiver, referring to this system as a "ground telephone". Telephoning through the ground became routine for this remarkable man.

Signals sent through the Stubblefield method were notable for their reported "great clarity". What is strange about this system is its elegant simplicity. Stubblefield's transmitting system evidences an almost crude minimalism which offends some researchers, while surprising others.

Numerous private and public demonstrations of this first system were made in Murray, Kentucky (1886-1892), where his mysterious "black boxes" were seen. Two metal rods were stuck into the ground a few feet apart from each distantly placed set. Speech between the two sets was loud and clear despite distances of 3500 to 6000 feet.

These transmission were made through the ground itself and used the Stubblefield cell for power. In several photographs we see special loudspeaking telephones outfitted with long (1 foot) horns, designed to act as annunciators. Calls from these annunciators brought his son Bernard to the telephone transmitter. The system was never switched off. Power was limitless and did not diminish with time of day or length of use.

While Marconi and others were barely managing the transmission of telegraph signals for equivalent distances, Nathan Stubblefield was transmitting vocal dialogue. The clarity of these signals and their sheer volume was the most widely recognized feature of the Stubblefield system. He was developing the system to operate through far greater distances, using automatic relays to boost signals for very great distances.

He published an extraordinary brochure in 1898 to attract investors who had expressed interest in consolidating a small corporation around his work. In this brochure, Stubblefield insisted that power for his device was not generated in the cell. He calmly stated that the cell received its surplus energy from the earth. In a less discussed portion of this brochure, Stubblefield stated that "electrotherapeutic" devices had been developed from his earth battery. Other researchers made similar claims for their earth batteries (Hicks, Mellon).

STATIONS

In 1902, Stubblefield set up one of his sets in a "Mainstreet" upper office... in a hardware shop. From that point to his farm (some 6000 feet distant) he conducted continuous conversations with his son Bernard. Tapping with a pencil on his one-piece transceiver, Bernard was quickly heard in a loud, very clear voice. This transceiver was a carbon button placed in a tin snuff box. Speech and response were transacted through the self-same device, which acted as both microphone and loudspeaker. Cells were placed downstairs from the office in the ground. They were never removed and never wore out, though operating twenty four hours around the clock.

Nathan Stubblefield offered to construct a large scale power station for the town of Murray. His quoted initial installation cost were estimated at five thousand dollars. The town politicians declined the offer. Now, the technique of drawing up electricity from the earth remains a mystery.

The Stubblefield ground radio system was demonstrated for approximately one thousand Murray residents (January 1902). Photographs of Stubblefield and his family, and a good crowd of witnesses from town show the cell lying on the ground among all his assembled inventions; a flower-pot sized coil of good volume. Other devices show motors and large capacitor stacks for aerial voice transmission experiments.

After the successful completion of these preliminary tests, Stubblefield travelled to Washington, D.C. for a public demonstration which was to be one of his crowning public achievements (March 1902). Stubblefield sent wireless messages from a steamship to stations on the shores of Georgetown. In this successful test, Stubblefield trailed long wires in the river water. Signals were engaged from ship to shore in a remarkable demonstration. Witnesses later acknowledged that Stubblefield's ground telephony sounded louder and came through with greater clarity than the subaqueous tests. Photographs of this event are all available.

During this time, Stubblefield declared that news, weather, and other announcements could be broadcast through the ground across a great territory for private reception. He also added that simultaneous messages and news of all kinds would soon be transmitted through the ground from a central distribution station.

Nathan also stated that, while such broadcasts required district-wide transmissions, he was developing a means by which privacy of telephonic messages could be maintained among callers. This "method of individuation" would also take place through the ground, insuring that no one could eavesdrop on conversations. Stubblefield had conceived and demonstrated these systems some twenty years before, anticipating statements made by Nikola Tesla, when referring to his Wardenclyffe Station.

The Washington D.C. demonstrations were followed by a trip further north. Mr. Stubblefield took his apparatus to New York City for additional tests, preparing for a public demonstration in Manhattan's Central Park. The demonstration was to take place in less than twenty four hours after his arrival. To his very great shock, Stubblefield found that the ground was not conducive to easy ground telephony, there being no "powerpoints" available. He requested more time to discover the powerpoints before setting up the stations properly. Time to "work the stony earth" of the Park left a few investors foolishly wary of the system's worth. This demonstration was immediately withdrawn.

His next public expositions were given in Philadelphia's Fairmont Park with greater success (May 1902). He now recognized, more than ever before, the role of geologic formations in determining and establishing his stations. Natural powerpoints would determine the location of each such station central. Stubblefield published a prospectus for his WTCA (Wireless Telephone Company of America), stating that "I can telephone without wires a mile or more now, and when the more powerful apparatus I am working on is finished and combined with further developments, the distance will be unlimited".

Despite each of these remarkably extensive demonstrations, Stubblefield sold only one telephonic system to another corporation, the Gordon Telephone Company of Charleston. This system was used to communicate with offshore islands. It would be interesting to retrieve this system and examine its contents.

He entered these commercial aspects with some trepidation. By June of the same year he withdrew from the project completely. A few persons managed to discover the reason for his quiet, sudden retreat. Because of his difficulty in instantly stationing his system in New York City, it was suggested that he adopt the method of burying lines to "fake" the operation...if just for the purpose of making a good show. Nathan declined.

Technology is lost at the market place, where inventors meet with astute businessmen of shrewd and cautious intent. There, the confrontation determines future world expressions. Certain businesses simply do not want to revolutionize any thriving technology for financial reasons. The individuals who fill such historical episodes are often incapable of seizing the new opportunity because they are simply not venture capitalists. New technologies produce far greater profits and energetically stimulate the economy to positive productive states.

In many such confrontations, the investors are merely heirs and custodians of fortunes they did not make. Zealous of maintaining the family fortune, they find the easiest and most infantile means at their ability level. By eradicating competitive technology they imagine themselves in possession of security. Some have retreated so into their own reclusive worlds that they imagine themselves in full control of national economy.

Others more aggressively attempt duplicating any competitive technology. Patent stealing is not a new phenomenon. After witnessing the public Stubblefield demonstrations, another inventor (A.F. Collins) duplicated some of Nathan's early inventions. Filing a counter-patent (patent 814.942 for "Wireless Telephony", 1906) for a ground telephonic system, Collins thought to seize Stubblefield's market outright. One of the signing witnesses on the Collins patent was, one Walton Harrison. Harrison, himself a WTCA member, later infringed on another Stubblefield experiment with his "Transmitter for Wireless Communication". This inferior telegraphic-telephonic system (patent 1.119.952, 1914) did not achieve the groundpower status by which Stubblefield is known.

It became apparent that certain WTCA members were trying to oust Stubblefield himself. The WTCA now took on a life of its own. Stubblefield was thoroughly disgusted at the display of human greed and ambition, and left them to their own devising. Collins, Harrison, and their co-conspirators were later accused of petty crimes having to do with mail fraud. The WTCA failed in time. Internal disputes over money, rather than technological progress and implementation, was their own death knell. Marconi arrived with an inferior (though highly publicized) system. When Marconi began his work, the effective signal transmission distance was equal to that achieved by Stubblefield.

Stubblefield was experimenting with ground radio since 1888, but did not patent his developments until much later. Credible witnesses saw his ground radio experiments in action during this time frame, establishing the historical priority of Stubblefield, a true and original American genius.

While Marconi could barely send telegraphic "dot and dash" signals with great difficulty through a static-filled medium, Nathan had already transmitted the human voice with loud, velvet clarity. Others would adopt and implement the Collins system (Fessenden, DeForest, Bethenod, Braun), but none could duplicate the Stubblefield System.

Nikola Tesla performed double ground experiments with impulses as early as 1892, reporting these in lectures and patenting some embodiments in 1901. No one of these later systems ever achieved the same results of clarity, tone, and volume of Stubblefield ground telephony. Tesla never discovered the true powerpoints which powered Stubblefield's devices.

Priority in all these arts belongs to Nathan Stubblefield alone. In addition, his was the only system in which natural energies were obtained, magnified, and entirely employed as the empowering source. All the other inventors used "artificial" sources (batteries, alternators, dynamos).

Following all these ground radio demonstrations, Stubblefield researched "magnetic waves" and developed several systems which did not use ground terminals for exchanging signals. Long distance wireless telephone communications were his aim. Many imagined this to be radio as we know it, but several features of the Stubblefield aerial system are distinctive and different.

First, his transmitters and receivers were telephonic, not telegraphic. In his preliminary experiments, the earth battery was used to energize an apparatus to which was connected a long horizontal aerial line. Marconi later adopted this "bent L" symmetry in conjunction with a grounded copper conduction screen. We do not have photographs of these arrays, but have handwritten manuscript copies of certain diary notes in which a progressively greater telephonic distance is reported. Nathan made steady progress in this form of telephonic transmission, but used neither alternators or spark discharge.

A second series of experiments reveal the development of stacked capacitors. Photographs reveal two large capacitor stacks, presumably for inductive transmission purposes. Some researchers induced ground oscillations of electrical current, while absorbing each "flyback" into large capacitors. This system evidenced the "hydraulic" model of electricity, popular during the latter Victorian Epoch.

Photographs reveal a final form of Stubblefield's aerial telephone which utilizes a two foot in diameter single turn copper band. This outer copper band is spaced from a second inner copper band, and is mounted on wooden pedestals. A telephone is connected to this array. This compact apparatus transmitted inductive rays, not waves, for great distances when earth energy was modulated by the human voice. In some strange manner, he had found a more potent means for activating, resonating, and projecting ground powerpoint energy. His was no ordinary radio transmitter.

A truly honest and humble man, he justly considered the ambitious and aggressive northern investors as "scalawags and damned rascals". He became suspicious of others. Considering the time frame in which these events took place, we may understand his reaction. Rejecting their tempting swindle, he was compelled to leave for home in order to continue his beloved experiments in privacy. He became mysteriously compulsive about his privacy after this.

In the words of several persons with which I have had the good fortune to speak, "Nathan was honest to a fault". He, disappointed again in human behavior, packed away his equipment and went home. After this unfortunate time period, Mr. Stubblefield preferred to be alone. Some say he became increasingly intolerable to live with. These patterns mark the disappointed genius, the broken-hearted dreamer. His hurtful nature began to hurt others. Friends forsook him, and he continued to allow them to leave. Finally, his wife left him with their children. Bernard was the only child who seemed to maintain contact with his father.

HOMESTEAD

As visitors approached the Stubblefield farm, yet a good way off, Stubblefield would appear at the door to wave them away. This often occurred when they were simply too far away to be visually located. He refused to speak to anyone for long periods of time. Many of these occurrences were reported during the night, when visibility from the cabin to the distant parts of his fields would be impossible. Nathan would always appear at the door, somehow knowingly, waving would-be visitors away.

Pranking schoolboys, intent on stealing vegetables or fruits, would ever so secretly crawl onto his farm, quite out of possible sight. Nathan would always be right next to them laughing in no time, somehow mysteriously detecting their presence. In a later embodiment, bells would sound when anyone approached so much as a half-mile from his cabin. It has been suggested that he had developed a device which could actually indicate the positions of any intruder across a space of ground.

Some declared that Nathan, jealous of his privacy, rigged the whole farm with delicate trip wires in order to locate and surprise pranksters. Sometimes the intruders would be met by Stubblefield, waiting at the very spot where they were stealthily heading. No intruder ever managed to feel or find these supposed wires. This tantalizing mystery has never been fully explored.

Others would say that Nathan buried sound-sensors all over the farm. These, when pressed, could model a trace across a map of the farm inside the cabin. Each sensor, tied to an indicator could show up on the map. Studying this map, he could see where intruders were on the fields. Nathan could then gleefully sneak up on them and chase them away.

Methods of distant ranging and location were devised by Antonio Meucci, employing tone signals. These required receivers, however, at the distant end. But Nathan knew where the intruders were coming from and where they were going as well. Nathan may have developed ground-wireless relays which responded to ground-buried sensors. These may have transmitted a tonal signal to the cabin, where a receiver would be triggered. This receiver may have been the bell-sounding mechanism. How did he locate people with pin point accuracy however? No complex array of detectors was ever found in his cabin when he died.

In light of all his experimentation with earth energy and wireless, we will assume that his last two mysterious inventions speak of utterly new and unknown (though related) ground energy phenomena. But, what natural phenomenon permitted him to achieve this feat?

Ocean waves often contour the shoreline, evidencing something of the shore outlines to distant places. Electrical waves might conceivably do this. But how would Nathan model this inside his cabin? No such map was ever really found. Also, if he were using some kind of ground impulse Doppler radar or sonar (electrical ground impulses outward) then what feature beneath the approaching intruders would signal an echo back to the receiver?

Some have even suggested that Stubblefield was utilizing distant variable ground conductivity. Intruders would alter this by their weight and step. But how would such a signal be transferred back to the measuring station? Such reciprocation in ground currents would require that the energy used is somehow...irritable and sensitive. This would evidence an unsuspected permeating biological nature in geology...a song, a personality with which the old linesmen-dowsers were intimate.

MOTORS

A motor, designed by Stubblefield to operate entirely by fluctuations in ground static, has been stored in a local museum. The device features several mobile pithballs around a compass-like perimeter, resembling the equally mysterious electrostatic hoop telegraphs of the 1700's. Students of Stubblefield's work have examined the pithball pendulum device and ignorantly concluded it to be a useless piece of junk.

Pithball (static) telegraphs of the early 1700's reveal this Stubblefield design to be a very special "find". Pithball telegraphs utilized a grounded metal hoop, an underlying dial, and a pendulum on which a pithball (cork) was hung. A single line (sometimes of silk) connected two such arrangements.

Signals were made and received in a very curious manner with pithball hoops, an equally historic mystery. Moving the pithball to a particular letter on the dial resulted in identical displacements in the receiver: an anomaly. These arcane devices managed the articulate transaction of messages by earth energies.

Through unknown phenomena characteristic of earth energy, these devices approached true intelligent transfer by a single wire connection. One simply swung the pithball toward a letter or word, indicated on a dialette. At a great distance away, an equivalent set registered identical swings. Electricity does not produce such responses.

Witnesses of these signalling devices were credible persons in the scientific community. No one questioned how it was possible to articulate such a transfer with static electricity. In any event, any researcher not familiar with the designs would pass over Stubblefield's "pithball table" without counting it as worthy of study.

The device found in Nathan's cabin after he passed away is of singular mystery. One person actually thought that Nathan built it just because it "looked really strange"...like some science artform made to baffle the unwary. It sat upon a trunk off to the side of his cabin room. Bernard Stubblefield, his son, did not recognize the device. Nathan must have built it after Bernard was taken away with his mother. Too young to independently pursue his father's developments, Bernard did not remember seeing the device before. It was taken to a local museum, where it now resides unheralded.

This device is a square arrangement, having several insulator-mounted pithballs in each quadrant of the central square table. It is quite likely that this was the means by which Nathan detected movements and positions in his field. If this analysis proves true, then it represents a major leap in his earth power technology.

I have surmised that this device is the Stubblefield long-range detector. Motions in a specific pithball pendulum gave the direction and position of the intruder. Such a device relies on phenomena which are unknown in conventional electric science.

Natural observations in systems lead to unexpected, theory-busting discoveries. Such an effect demonstrates that an articulate quasi-intelligent energy permeates the natural environment...of which electricity is a minor part. The natural phenomenon which is responsible for this ability is truly remarkable...nothing short of the miraculous. In its realm, we see that nature is suffused with an almost biological organization which includes the supposed inert world of geology. This would be equivalent to acknowledging that geological structure is suffused with a neurological sensitivity; a thing which academic science is neither prepared nor equipped to endorse.

Nevertheless, different aspects of this ground sensitivity were discovered and differently implemented throughout the following years. T.H. Moray (1935) also discovered long-range articulate tuning through the ground from a fixed single site. His "radiant energy listening device" permitted him to scan a tract of land and actually eavesdrop on distant conversations and sounds through earphones. This device did not implement a microphone.

The Moray Listening Device used a grounded rod and special large germanium detector. How does a stationary tuner sweep across land and pinpoint sound sources? Stanley Rogers (1932) discovered the same long-range scanning effect when, using a radionic tuner for mineral detection, he found it possible to sweep a field or meadow with a variable capacitor. Adjustments on these grounded tuners could sweep across land, revealing and mapping every mineral contour. Dr. R. Drown (1951) independently developed a compact device which could sweep, scan, and delve through subterranean grounds for the specific purpose of ore detection. This device permitted photographic detection of ores swept through the ground, isolating specifically sought mineral deposits.

The Stubblefield pithball pendulum represents a leap in ground power technology. It is an engine which operates without electrical transformations at all: a ground powered "auric" engine.

SUNLIGHT

Two more mysteries have lingered from this latter period of invention in the Stubblefield biography. The nature of each reveals the extent to which he had developed and advanced his new earth power technology. Nathan continued to pursue his experiments, but little was seen of him for long time periods. Alone and tired, Nathan stopped working his farm completely.

Later visitors felt sorry for Nathan, now aged and abandoned by his wife and children. Several of the town's many charitable ladies decided to take him some food. On one occasion, they arrived at his farm to find the ground "ablaze with light...like pure sunlight was coming right up out of the hillside".

Later investigators entered his land area and found heavy wires leading from the roots of trees. To these wires were attached small arc lamps, hung in the trees. These were long extinguished. They imagined this to be the explanation of his hillside sunlight. Their hasty analysis proved problematic from stories which witnesses report.

The warm and diffuse sunlight which came from the ground itself around his house was not localized in specific lamps. The light came from the ground, not from the trees as before..."a whole hillside that would blossom with light"..."lit up like daytime". These observations indicate that Stubblefield had managed indeed the direct conversion of earth energy to light and warmth.

This would be acceptable, were Mr. Stubblefield simply working on a newer means of drawing electricity from the ground to light small arc-lamps; a feat which he had accomplished earlier. But these kind persons could never find any evidence of arc-lighting or any other form of known lighting anywhere near the area. In their own words "the light seemed to come out from the ground itself".

In addition to the ground sunlight effect, many heard very loud and unfamiliar noises coming from the whole area surrounding his cabin. What could this be? Had he managed to directly transduce the natural impulses of the ground energy into audio?

His own last claim, made two weeks before he passed away was made to a kind neighbor: "The past is nothing. I have perfected now the greatest invention the world has ever known...I have taken light from the air and earth...as I did sound".

SUNSET

I was the quite fortunate recipient of an unexpected personal letter while writing my original treatise on Nathan Stubblefield. It was told by a gentleman who received the account through a man who witnessed the following.

Neighbors had not seen Nathan for several days. As they were worried about his health, they attempted to call on him. The lock was secured from the inside. It was a lonely, cold, and rainy March day when old friends and neighbors broke the lock on Nathan's cabin and entered. He had passed away in his bed, the probable victim of malnutrition and fatigue. They all noticed that the interior of the cabin was "toasty warm", as if heated by a strong fire. Moved to locate the source of this heat, town officials found "two highly polished metal mirrors which faced each other, radiating a very great heat in rippling waves". Now this, I must say, is a truly great discovery and last mystery. It fulfills what Nathan reported in his last testimony.

Nathan's deepest confidence was in those kind and compassionate people who continued to seek him out with love and concern to his last days. Abandoned by all, he wished one of his dearest neighbors to write a biography. Perhaps he wished to explain his life, an apology for all his ways. He said "I have lived fifty years ahead of everybody else". While often sounding inspirational, these are words of deepest sorrow.

To live with a vision of the future is to experience the surprising, often disappointing rejection and resistance of all who surround. Some said he was incapable of loving others. But...it was love, his love, which coaxed the living sunshine out of hard, rocky ground...the resounding waves of an eternal subterranean sea of energy.

CHAPTER 4

BROADCAST POWER: NIKOLA TESLA

POLYPHASE

The drama of Twentieth Century Science and its intriguing relationship with financiers and governments unfold together in the remarkable life of Nikola Tesla. His is a biography replete with all the elements of tragedy. Tesla, a great discoverer of unsurpassed force, became the focal point of old insidious forces intent on destroying the future for the selfish sake of the status quo. Tesla remains a focal point of wonderment, of dream, and of worlds which yet should be to those who are familiar with his biography. For them, Tesla stands astride the quaint past century and the gleaming future. He is a technological Colossus, pointing the way to a new dawn.

The biography of Nikola Tesla should be the very first chapter in every child's science text. Yet, we find his name stricken from the record in every avenue of which he alone holds priority. This conspicuous absence prompts wonderment. What the world does with discoverers determines the world course. In the life of Nikola Tesla we see the portrayal of our own future, the fate of the world. The achievements of this researcher were lofty. The world has not yet implemented his greatest works. For a time, all the world's dramatis personae focussed on Tesla. He remains the legend, the theme, the archetype of all Twentieth Century scientists.

But who was Nikola Tesla, and where was he from? How did he reach such a mighty stature, and what did he actually invent? Tesla was born in 1856, the son of an illustrious Serbian family. His father, an Orthodox priest, his uncles noteworthy military heroes of highest rank. He was educated in Graz, and later moved to Budapest. Throughout his life he was blessed, or haunted, by vivid visions. In the terminology of Reichenbach he would be termed an extreme sensitive. It was through these remarkable visions that Nikola Tesla invented devices which the Victorian world had never seen. Indeed, his visionary experiences produced the modern world as we know it.

He attended various Universities in Eastern Europe during his early adulthood. While delving into his studies, he became aware by the new and insidious scientific trends which questioned the validity of human sense and reason. An impassioned soul, Tesla felt the pain of modern humanity in its intellectual search for a soul. Finding no solace in any of his classes, he sought refuge in a more romantic treatment of science and nature. None could be found. Professors dutifully promoted the "new view" by which it was declared that the natural world was "inert...dead...a mere collection of forces".

This quantitative regime was mounting force among academes, who were then attempting the total conversion of scientific method. Those who would not accept the new order were compelled to depart from academic pursuits. Tesla totally rejected these notions on the strongest of inner intuitions. Most of his instructors would have said that he was not University material. Tesla, sensitive to every such dogmatic wind, rejected their thesis and sought some better means for knowing nature. If he was to excel in engineering, there could only be cooperation with natural force, never violence. It was clear to him that the new scientific world-attack would ultimately lead to violent responses from nature itself.

His inner conflict expressed itself openly and candidly, bringing young Tesla into certain disrepute among rigid University authorities. Universities were more like military academies than places where original thinking was conducted in open forum. Tesla challenged too many persons of esteemed rank with probing questions for which he was given rebuke but no real answers.

A gifted researcher and voracious reader, he chanced upon some forgotten volumes of natural science written by Goethe. He had not been aware that Goethe, long before he chose poetry for the vehicle of his scientific themes, had written several magnificent tomes on the natural world. Tesla found to his wonder that Goethe had experienced the very same emotions. When the new scientific dogma was just in its infancy, Goethe caught wind of it and reacted violently, even as one who stands watch in the night.

Goethe was well aware of the new scientific trend and its implications. The reduction of nature to forces and mechanisms was utterly revolting to Goethe. Now, Tesla found a notable compatriot in his experience. He secured a thorough collection of Goethe's scientific texts and read these to the exclusion of all other philosophies. It was through this window that we may comprehend all of Tesla's scientific methods and later statements. For in Tesla we see the quest for communion with nature, one based on the faith that mind, sensation, consciousness, and ordained structure form the world-foundations.

The sense-validating Qualitative Theme again appears in Nikola Tesla. Armed with this foundation, he was able to filter and qualify every other new study with which he was presented. In addition, he was irresistibly drawn into the study of electricity, the "new magick". In the following months, he absorbed the electrical engineering courses so rapidly that he no longer attended classes. He had taken a technical position in Budapest. Several new intuitions had seized him. Tesla became fascinated, obsessed with alternating current electricity. The problem he faced was considered insurmountable. Tesla was sure that he could devise an engine which was turned, not by contact-currents, but by magnetic field actions alone.

The struggle toward designing such a device, begun as a puzzling amusement, was now completely consuming his strength. The answer, tantalizing and near, seemed elusive. Undergirding all these efforts was the strongest desire to achieve something original, and by this, to attain financial independence for the sake of pure research. His only dream was to have a laboratory facility of his own.

The excessive labors and mental exertions nearly drove him to the brink of madness. He was, for as time, seized with strange maladies and sensitivities which physicians could not address. Reichenbach accurately describes these symptoms, characteristic of extreme sensitives. There come times when the neurological sensitivity of these individuals literally transforms and processes through their being. The emergence of these rare sensitivities affects such persons for the remainder of their lives.

Tesla found that his senses were amplified beyond reason. He was terribly frightened at first, nervous exhaustion permeating his frail being. Eventually learning to manage these rare faculties, he again resumed his life. But the visions which began in his youth were now more vivid and solid than ever before. When they came, unbidden, he could literally touch and walk around them. Now also, he was equal to receiving them. He was waiting for the revelation by which his alternating current motor would appear.

Tesla's life came into a new focus while walking in a park with some friends, the year 1881. It was late afternoon, and Tesla became entranced with the sight of a glorious sunset. Moved to indescribable emotions, he began quoting a verse from Goethe's "Faust":

"The glow retreats, done is our day of toil;
it yonder hastes, new fields of life exploring,
ah, can no wing lift me from this soil...
upon his track to follow, follow soaring?"

As he reached this last line of verse, Tesla was suddenly seized by an overwhelming vision. In it, he beheld a great vortex, whirling eternally in the sun and driving across the earth with its infinite power. Completely absorbed in this glory, he became catatonic and irresponsive...to the great fear of his companions. His mind and body buzzing with the power of the vision, he suddenly blurted out, "see my motor here...watch me reverse it". They shook him, believing he had lost his mind completely.

Rigid and resisting all of their efforts, he would not move until the vision subsided. When he was finally led to a bench, he seemed completely transformed. The remainder of the day was spent in a grand and joyous celebration, Tesla's remaining funds supplying the feast. Throughout the long hours of that night he shared with his friends the great sight he had beheld. They spoke of the sure implications portended for the world's future, and departed with very great expectations.

Moving to Strassburg, he was employed as an engineer in a telephone subsidiary of the Continental Edison Company. It was in a small machine shop that he constructed the world's first brushless motors. He called them "magnetic vortex motors". Their whirling magnetic fields baffled electrical engineers. Now, Tesla's professors were studying his work. Goethe was absolute in his judgement of science and human nature: nature leads humanity to "follow, follow soaring".

Tesla's strange whirling devices worked on their very first trial. There were no connections between the rotors and stators, no sparking, lossy brushes. The motion was smooth and efficient. Numerous alternating current generators, transformers, and "brushless" motors, all were developed by Tesla in quick succession. The vision in material form. Himself a professional draftsman, he mapped out his entire Polyphase System. Tesla emigrated to America with a full portfolio of plans. America would be the place where his dreams would find fulfillment.

Continually attracted to engineering problems which none could master, his sudden visualization of the solutions became his normal mode of operation. In this respect, as well as others, he remained the wonder of all his technical assistants. He worked for Thomas Edison in New Jersey for a very short time period until securing a laboratory and financial supporters of his own.

In his first independent venture he developed arc lamps and lighting systems. When his financial supporters betrayed his trust, they left him bankrupt overnight. He became a ditch digger, suffering all the indignities which immigrants faced in America during the 1880's. He learned the value of publicity after his incessant mention of polyphase and alternating current managed to attract the attention of certain new financial supporters. They drew him out of the ditch, but not before he demanded his own laboratory, a machine shop, and a sizable personal percentage "up front". The result was our present day electrical distribution system.

Tesla did not invent alternating current. Tesla reinvented alternating current in the form of Polyphase Current. His Polyphase System was a novel means for blending three identical alternating currents together simultaneously, but "out of step". The idea was similar to having three pistons on a crankshaft rather than one. Tesla's method had wonderful advantages, especially when motors were to be operated. Formally, no one could make an alternating current motor turn at all simply because no net motion could be derived from a current which just "shuttled" to and fro.

Polyphase applied a continuous series of separate "pushes" to rotors. Tesla's Polyphase System made brushless motors and brilliant lighting methods possible. Polyphase made it also possible to send electrical power to very great distances with little loss. Alternating electrical currents vibrated in the line. Current did not flow continuously from end to end, as in Edison's flawed system. Edison's direct current system could not supply electricity beyond a few city blocks before current virtually disappeared.

In efforts to discover a more efficient kind of polyphase, Tesla explored higher frequency alternating currents. During this research, he built and patented several remarkable generators. Higher frequency polyphase was found by Tesla to perform with far greater efficacy than the common sixty-cycle variety which we still use. He fully intended on implementing these special generators in the system which his patron and friend, George Westinghouse, had proliferated. The business arrangement rendered Tesla fabulously wealthy at a young age.

Tesla extended his generator frequencies in multiples of sixty until reaching some thirty thousand cycles per second. These very high frequency alternating current generators became the marvel of all the academic and engineering world. They were copied and modified by several other subsequent inventors including Alexanderson. Remarkably driven at excessive speeds, they constituted Tesla's first belief that high frequency alternating current generators would supply the world's power.

High frequency current phenomena were new and exceedingly curious. A line of experimental research was conducted in order to evaluate new safe and possibly more efficient ways for transmitting power along long elevated lines. Tesla stated that the transmission of such safe currents across very long powerline distances in the future would be a certainty, seeing their wonderful new qualities.

Tesla found that high frequency currents were harmless when contacted by the human body. Discharges from these generators traversed the outer surface of materials, never penetrating matter with depth. There was no danger when working with high frequency currents. He also observed their very curious and beautiful spark effects. They hissed and fizzled all over wire conductors, could stimulate luminescence in low pressure gas bulbs, seemed to traverse insulative barriers with ease, and made little pinwheels spin like delicate little fireworks displays.

Though curious, the effects were weak and furtive. They seemed to intimate some future technology which he was yet unable to penetrate. Tesla learned that his intuitions and visions were infallible. What he guessed usually proved true. This very personal revelation, he later claimed, was his greatest discovery.

As the safety of all personnel was his main concern, he was consumed with the idea of making his High Frequency Polyphase System completely safe for human operators and consumers alike. An extensive examination of each System component was undertaken with this aim in mind. Tesla was thorough and relentless in his quest for safety and efficiency.

But, his involvement with alternating currents would come to an abrupt and unexpected end. During a series of experiments which followed these high frequency tests, an amazing seldom-mentioned accident occurred in which Tesla observed a phenomenon which forever altered his view of electricity and technology.

SHOCKING DISCOVERY

Tesla was an avid and professional experimenter throughout his life. His curiosity was of such an intense nature that he was able to plumb the mysteries of an electrical peculiarity with no regard for his own comfort. Whereas Edison would work and sleep for a few hours on the floor, Tesla would never sleep until he had achieved success in an experimental venture. This marathon could last for days. He was once observed to work through a seventy two hour period without fatigue. His technicians were in awe of him.

The Victorian Era was flooding over with new electrical discoveries by the day. Keeping up with the sheer volume of strange electrical discoveries and curiosities was a task which Tesla thoroughly enjoyed...and preferred. His Polyphase System in perfect working order, the pleasurable occupation of studying new gazettes and scientific journals often fascinated his mind to the exclusion of all other responsibilities. A millionaire and world-heralded genius before the age of thirty, Tesla sought the pure kind of research he had so long craved.

Whenever he observed any intriguing electrical effect he immediately launched into experimental study with a hundred variations. Each study brought him such a wealth of new knowledge that, based on phenomena which he observed, he was immediately able to formulate new inventions and acquire new patents.

Tesla's New York laboratories had several sections. This complex was arranged as a multi-level gallery, providing a complete research and production facility. Tesla fabricated several of his large transformers and generators in the lower floors, where the machine shops of this building were housed. The upper floors contained his private research laboratories. He had attracted a loyal staff of technicians. Of all these, Kolman Czito was a trusted friend who would stand by Tesla for the remainder of his life. Czito was the machine shop foreman in each of Tesla's New York laboratories.

Tesla observed that instantaneous applications of either direct or alternating current to lines often caused explosive effects. While these had obvious practical applications in improvement and safety, Tesla was seized by certain peculiar aspects of the phenomenon. He had observed these powerful blasts when knife-switches were quickly closed and opened in his Polyphase System. Switch terminals were often blasted to pieces when the speed of the switchman matched the current phase.

Tesla assessed the situation very accurately. Suddenly applied currents will stress conductors both electrically and mechanically. When the speed of the switch-action is brief enough, and the power reaches a sufficiently high crescendo, the effects are not unlike a miniature lightning stroke. Electricity initially heats the wire, bringing it to vapor point. The continual application of current then blasts the wire apart by electrostatic repulsion. But was this mechanistic explanation responsible for every part of the phenomenon?

The most refractory metals were said to be vaporized by such electrical blasts. Others had used this phenomenon to generate tiny granular diamonds. Yes, there were other aspects about this violent impulse phenomenon which tantalized him. Sufficiently intrigued, he developed a small lightning "generator" consisting of a high voltage dynamo and small capacitor storage bank. His idea was to blast sections of wire with lightning-like currents. He wanted to observe the mechanically explosive effects which wires sustain under sudden high-powered electrifications.

Instantaneous applications of high current and high voltage could literally convert thin wires into vapor. Charged to high direct current potentials, his capacitors were allowed to discharge across a section of thin wire. Tesla configured his test apparatus to eliminate all possible current alternations. The application of a single switch contact would here produce a single, explosive electrical surge: a direct current impulse resembling lightning. At first Tesla hand-operated the system, manually snapping a heavy knife switch on and off. This became less favorable as the dynamo voltages were deliberately increased.

He quickly closed the large knife switch held in his gloved hand. Bang! The wire exploded. But as it did so, Tesla was stung by a pressure blast of needle-like penetrations. Closing the dynamo down, he rubbed his face, neck, arms, chest, and hands. The irritation was distinct. He thought while the dynamo whirred down to a slow spin. The blast was powerful. He must have been sprayed by hot metal droplets as small as smoke particles. Though he examined his person, he fortunately found no wounds. No evidence of the stinging blast which he so powerful felt.

Placing a large glass plate between himself and the exploding wire, he performed the test again. Bang! The wire again turned to vapor...but the pressured stinging effect was still felt. But, what was this? How were these stinging effects able to penetrate the glass plate? Now he was not sure whether he was experiencing a pressure effect or an electrical one. The glass would have screened any mechanical shrapnel, but would not appreciably shield any electrical effects.

Through careful isolation of each experimental component, Tesla gradually realized that he was observing a very rare electrical phenomenon. Each "bang" produced the same unexpected shock response in Tesla, while exploding small wire sections into vapor. The instantaneous burst produced strange effects never observed with alternating currents. The painful shocking sensation appeared each time he closed or opened the switch. These sudden shock currents were IMPULSES, not alternations. What surprised him was the fact that these needle-like shocks were able to reach him from a distance: he was standing almost ten feet from the discharge site!

These electrical irritations expanded out of the wire in all directions and filled the room in a mystifying manner. He had never before observed such an effect. He thought that the hot metal vapor might be acting as a "carrier" for the electrical charges. This would explain the strong pressure wave accompanied by the sensation of electrical shock. He utilized longer wires. When the discharge wire was resistive enough, no explosion could occur.

Wire in place, the dynamo whirred at a slower speed. He threw the switch for a brief instant, and was again caught off guard by the stinging pressure wave! The effect persisted despite the absence of an explosive conductor. Here was a genuine mystery. Hot vapor was not available to "carry" high voltage charges throughout the room. No charge carriers could be cited in this instance to explain the stinging nature of the pressure wave. So what was happening here?

The pressure wave was sharp and strong, like a miniature thunderclap. It felt strangely "electrical" when the dynamo voltage was sufficiently high. In fact, it was uncomfortably penetrating when the dynamo voltage was raised beyond certain thresholds. It became clear that these pressure waves might be electrified. Electrified soundwaves. Such a phenomenon would not be unexpected when high voltages were used. Perhaps he was fortunate enough to observe the rare phenomenon for the first time.

He asked questions. How and why did the charge jump out of the line in this strange manner? Here was a phenomenon which was not described in any of the texts with which he was familiar. And he knew every written thing on electricity. Thinking that he was the victim of some subtle, and possibly deadly short circuit, he rigorously examined the circuit design. Though he searched, he could find no electrical leakages. There were simply no paths for any possible corona effects to find their way back into the switching terminal which he held.

Deciding to better insulate the arrangement in order that all possible line leakages could be eradicated, he again attempted the experiment. The knife switch rapidly closed and opened, he again felt the unpleasant shock just as painfully as before. Right through the glass shield! Now he was perplexed. Desiring total distance from the apparatus, he modified the system once more by making it "automatic".

He could freely walk around the room during the test. He could hold the shield or simply walk without it. A small rotary spark switch was arranged in place of the hand-held knife switch. The rotary switch was arranged to interrupt the dynamo current in slow, successive intervals. The system was actuated, the motor switch cranked it contacts slowly. Snap...snap...snap...each contact produced the very same room-filling irritation.

This time it was most intense. Tesla could not get away from the shocks, regardless of his distance from the apparatus across his considerably large gallery hall. He scarcely could get near enough to deactivate the rotating switch. From what he was able to painfully observe, thin sparks of a bright blue-white color stood straight out of the line with each electrical contact.

The shock effects were felt far beyond the visible spark terminations. This seemed to indicate that their potential was far greater than the voltage applied to the line. A paradox! The dynamo charge was supplied at a tension of fifteen thousand volts, yet the stinging sparks were characteristics of electrostatic discharges exceeding some two hundred fifty thousand volts. Somehow this input current was being transformed into a much higher voltage by an unknown process. No natural explanation could be found. No scientific explanation sufficed. There was simply not enough data on the phenomenon for an answer. And Tesla knew that this was no ordinary phenomenon. Somewhere in the heart of this activity was a deep natural secret. Secrets of this kind always opened humanity into new revolutions.

Tesla considered this strange voltage multiplying effect from several viewpoints. The problem centered around the fact that there was no magnetic induction taking place. Transformers raise or lower voltage when current is changing. Here were impulses. Change was happening during the impulse. But there was no transformer in the circuit. No wires were close enough for magnetic inductions to take place. Without magnetic induction, there could theoretically be no transformation effect. No conversion from low to high voltage at all. Yet, each switch snap brought both the radiating blue-white sparks and their painful sting.

IMPULSES

Tesla noted that the strange sparks were more like electrostatic discharges. If the sparks had been direct current arcs reaching from the test line, he would surely have been killed with the very first close of the switch. The physical pressure and stinging pain of these sparks across such distances could not be explained. This phenomenon had never been reported by those who should have seen and felt its activities.

Tesla gradually came to the conclusion that the shock effect was something new, something never before observed. He further concluded that the effect was never seen before because no one had ever constructed such a powerful impulse generator. No one had ever reported the phenomenon because no one had ever generated the phenomenon.

Tesla once envisioned a vortex of pure energy while looking into a sunset. The result of this great Providential vision was polyphase current. A true revelation. But this, this was an original discovery found through an accident. It was an empirical discovery of enormous significance. Here was a new electrical force, an utterly new species of electrical force which should have been incorporated into the electrical equations of James Clerk Maxwell. Surprisingly, it was not.

Tesla now questioned his own knowledge. He questioned the foundations on which he had placed so much confidence in the last several years. Maxwell was the "rule and measure" by which all of Tesla's polyphase generators had been constructed. Tesla penetrated the validity of Maxwell's mathematical method. It was well known that Maxwell had derived his mathematical descriptions of electromagnetic induction from a great collection of available electrical phenomena. Perhaps he had not studied enough of the phenomena while doing so.

Perhaps newer phenomena had not been discovered, and were therefore unavailable to Maxwell for consideration. How was Maxwell justified in stating his equations as "final"? In deriving the laws of electromagnetic induction, Maxwell had imposed his own "selection process" when deciding which electrical effects were the "basic ones". There were innumerable electrical phenomena which had been observed since the eighteenth century. Maxwell had difficulty selecting what he considered to be "the most fundamental" induction effects from the start. The selection process was purely arbitrary. After having "decided" which induction effects were "the most fundamental", Maxwell then reduced these selected cases and described them mathematically. His hope was to simplify matters for engineers who were designing new electrical machines. The results were producing "prejudicial" responses in engineers who could not bear the thought of any variations from the "standard". Tesla had experienced this kind of thematic propaganda before, when he was a student. The quantitative wave of blindness was catching up with him.

Tesla and others knew very well that there were strange and anomalous forms of electromagnetic induction which were constantly and accidentally being observed. These seemed to vary as the experimental apparatus varied. New electrical force discoveries were a regular feature of every Nature Magazine issue. Adamant in the confidence that all electrical phenomena had been both observed and mathematically described, academicians would be very slow to accept Tesla's claims.

But this academic sloth is not what bothered Tesla. He had already found adequate compensation for his superior knowledge in the world of industry. Tesla, now in possession of an effect which was not predicted by Maxwell, began to question his own knowledge. Had he become a "mechanist", the very thing which he reviled when a student? Empirical fact contradicted what that upon he based his whole life's work. Goethe taught that nature leads humanity.

The choice was clear: accept the empirical evidence and reject the conventional theory. For a time he struggled with a way to "derive" the shock effect phenomenon by mathematically wrestling "validity" from Maxwell's equations...but could not. A new electrical principle had been revealed. Tesla would take this, as he did the magnetic vortex, and from it weave a new world.

What had historically taken place was indeed unfortunate. Had Maxwell lived after Tesla's accidental discovery, then the effect might have been included in the laws. Of course, we have to assume that Maxwell would have "chosen" the phenomenon among those which he considered "fundamental".

There was no other way to see his new discovery now. Empirical fact contradicted theoretical base. Tesla was compelled to follow. The result was an epiphany which changed Tesla's inventive course. For the remainder of his life he would make scientific assertions which few could believe, and fewer yet would reproduce. There yet exist several reproducible electrical phenomena which cannot be predicted by Maxwell. They continually appear whenever adventuresome experimenters make accidental observations.

FOCUS

High voltage impulse currents produced a hitherto unknown radiant effect. In fact, here was an electrical "broadcast" effect whose implementation in a myriad of bizarre designs would set Tesla apart from all other inventors. This new electrical force effect was a pre-eminent discovery of great historical significance. Despite this fact, few academicians grasped its significance as such. Focussed now on dogmatizing Maxwell's work, they could not accept Tesla's excited announcements. Academes argued that Tesla's effect could not exist. They insisted that Tesla revise his statements.

Tesla's mysterious effect could not have been predicted by Maxwell because Maxwell did not incorporate it when formulating his equations. How could he have done so, when the phenomenon was just discovered? Tesla now pondered the academic ramifications of this new effect. What then of his own and possibly other electrical phenomena which were not incorporated into Maxwell's force laws? Would academes now ignore their existence? Would they now even dare to reject the possibility of such phenomena on the basis of an incomplete mathematical description?

Seeing that the effect could grant humanity enormous possibilities when once tamed, Tesla wished to study and implement the radiant electrical action under much safer conditions. The very first step which he took before proceeding with this experimental line was the construction of special grounded copper barriers: shields to block the electrical emanations from reaching him.

They were large, body sized mantles of relatively thick copper. He grounded these to insure his own complete safety. In electrical terms, they formed a "Faraday Cage" around him. This assembly would block out all static discharges from ever reaching Tesla during the tests. Now he could both observe and write what he saw with confidence.

Positioned behind his copper mantle, Tesla initiated the action. ZZZZZZ...the motorized switch whirring, dynamo voltage interrupted several hundred times per second, the shock action was now continuous. He felt a steady rhythm of electrostatic irritations right through the barrier accompanied by a pressure wave which kept expanding. An impossibility. No electrical influence should have passed through the amount of copper which composed the shield. Yet this energetic effect was penetrating, electrically shocking, and pressured. He had no words to describe this aspect of the new phenomenon. The shocks really stung.

Tesla was sure that this new discovery would produce a completely new breed of inventions, once tamed and regulated. Its effects differed completely from those observed in high frequency alternating current. These special radiant sparks were the result of non-reversing impulses. In fact, this effect relied on the non-reversing nature of each applied burst for its appearance. A quick contact charge by a powerful high voltage dynamo was performing a feat of which no alternating generator was capable. Here was a demonstration of "broadcast electricity".

Most researchers and engineers are fixed in their view of Nikola Tesla and his discoveries. They seem curiously rigidified in the thought that his only realm of experimental developments lay in alternating current electricity. This is an erroneous conception which careful patent study reveals. Few recognize the documented facts that, after his work with alternating currents was completed, Tesla switched over completely to the study of impulse currents. His patents from this period to the end of his career are filled with the terminology equated with electrical impulses alone.

The secret lay principally in the direct current application in a small time interval. Tesla studied this time increment, believing that it might be possible to eliminate the pain field by shortening the length of time during which the switch contact is made. In a daring series of experiments, he developed rapid mechanical rotary switches which handled very high direct voltage potentials. Each contact lasted an average of one ten-thousandth second.

Exposing himself to such impulses of very low power, he discovered to his joy and amazement that the pain field was nearly absent. In its place was a strange pressure effect which could be felt right through the copper barriers. Increasing the power levels of this device produced no pain increase, but did produce an intriguing increased pressure field. The result of simple interrupted high voltage DC, the phenomenon was never before reported except by witnesses of close lightning strokes. This was erroneously attributed however to pressure effects in air.

Not able to properly comprehend their nature at first, Tesla also conservatively approached the pressure phenomenon as due to air pressure. He had first stated that the pressure field effect was due to sharp soundwaves which proceeded outward from the suddenly charged line. In fact, he reported this in a little-known publication where he first announced the discovery. Calling the pressure effects "electrified soundwaves", he described their penetrating nature in acoustic terms.

Further experimentation however, gradually brought the new awareness that both the observed pressure effect and electrical shock fields were not taking place in air at all. He demonstrated that these actions could take place in oil immersions. Impulse charged lines were placed in mineral oil and carefully watched. Strong pressure projections emerged from sharp wire ends in the oil, as if air were streaming out under high pressure.

Tesla first believed that this stream was wire-absorbed air driven off by electrical pressure. Continual operation of the phenomenon convinced him that the projected stream was not air at all. Furthermore, he was not at a loss to explain the effect, but was reluctant to mention his own theory of what had been generated by high voltage direct current impulses.

Tesla made electrical measurements of this projective stream. One lead of a galvanometer was connected to a copper plate, the other grounded. When impulses were applied to wire line, the unattached and distant meter registered a continual direct current. Current through space without wires! Now here was something which impulses achieved, never observed with alternating currents of any frequency.

Analysis of this situation proved that electrical energy or electrically productive energies were being projected from the impulse device as rays, not waves. Tesla was amazed to find these rays absolutely longitudinal in their action through space, describing them in a patent as "light-like rays". These observations conformed with theoretical expectations described in 1854 by Kelvin.

In another article Tesla calls them "dark-rays", and "rays which are more light-like in character". The rays neither diminished with the inverse square of the distance nor the inverse of the distance from their source. They seemed to stretch out in a progressive shock-shell to great distances without any apparent loss.

MAGNETIC ARCS

Nikola Tesla now required greater power levels than those provided by his mechanical rotary switch system. He also saw the need for controlling ultra-rapid current interruptions of high repetition ("succession") rates. No mechanical switch could perform in this manner. He had to envision and devise some new means by which ultra-rapid interruptions could be obtained. In his best and most efficient system, highly charged capacitors were allowed to impulsively discharge across special heavy duty magnetic arcs.

The magnetic arc gap was capable of handling the large currents required by Tesla. In achieving powerful, sudden impulses of one polarity, these were the most durable. Horn shaped electrodes were positioned with a powerful permanent magnetic field. Placed at right angles to the arc itself, the currents which suddenly formed in this magnetic space were accelerated along the horns until they were extinguished. Rapidly extinguished!

Arcs were thus completely extinguished within a specified time increment. Tesla configured the circuit parameters so as to prevent capacitor alternations from occurring through the arc space. Each arc discharge represented a pure unidirectional impulse of very great power. No "contaminating current reversals" were possible or permissible.

Reversals...alternations...would ruin the "shock broadcast". The effect was never observed when alternating currents were engaged. High voltage was supplied by a large dynamo. Tesla could speed or slow this dynamo with a hand operated rheostat. Power was applied in parallel across the capacitor. The magnetic arc was linked almost directly to one side of this capacitor, a long and thick copper strap connecting the magnetic arc and the far capacitor plate.

This simple asymmetric positioning of the magnetic arc discharger to one side of the dynamo supply produced pure unidirectional electropositive or electronegative impulses as desired. Tesla designed this very simple and powerfully effective automatic switching system for achieving ultra-rapid impulses of a single polarity. Capacitor values, arc distances, magnetic fields and dynamo voltages were all balanced and adjusted to yield a repetitive train of ultrashort singular impulses without "flyback" effects.

The system is not really well understood by engineers, the exceptional activities of the arc plasma introducing numerous additional features to the overall system. While the effects which Tesla claimed can be reproduced with electron tube impulse circuitry, these produce decidedly inferior effects. The overall power of the basic arc discharge is difficult to equal. Tesla eventually enclosed the magnetic arc, immersing the gap space in mineral oil. This blocked premature arcing, while very greatly increasing the system output.

Most imagine that the Tesla impulse system is merely a "very high frequency alternator". This is a completely erroneous notion, resulting in effects which can never equal those to which Tesla referred. The magnetic discharge device was a true stroke of genius. It rapidly extinguishes capacitor charge in a single disruptive blast. This rapid current rise and decline formed an impulse of extraordinary power. Tesla called this form of automatic arc switching a "disruptive discharge" circuit, distinguishing it from numerous other kinds of arc discharge systems. It is very simply a means for interrupting a high voltage direct current without allowing any backward current alternations. When these conditions are satisfied, the Tesla Effect is then observed.

The asymmetrical positioning of the capacitor and the magnetic arc determines the polarity of the impulse train. If the magnetic arc device is placed near the positive charging side, then the strap is charged negative and the resultant current discharge is decidedly negative.

Tesla approached the testing of his more powerful systems with certain fear. Each step of the testing process was necessarily a dangerous one. But he discovered that when the discharges exceeded ten thousand per second, the painful shock effect was absent. Nerves of the body were obviously incapable of registering the separate impulses. But this insensitivity could lead to a most seductive death. The deadly aspects of electricity might remain. Tesla was therefore all the more wary of the experiments.

He noticed that, though the pain field was gone, the familiar pressure effect remained. In its place came a defined and penetrating heat. Tesla was well aware that such heat could signal internal electrocution. He had already made a thorough study of these processes, recognizing that such heating precedes the formation of electrical arcs through the body. Nevertheless, he applied power to the dynamo in small but steady intervals.

Each increase brought increase in the internal heating effects. He remained poised at each power level, sensing and scoping his own physiology for danger signs. He continued raising the power level until the magnetic arc reached its full buzzing roar. Tesla found that this heat could be adjusted and, when not extreme, was completely enjoyable. So soothing, relaxing, and comfortable was this manifestation that Tesla daily exposed himself to the energies. An electrical "sauna".

He later reported these findings in medical journals, freely offering the discovery to the medical world for its therapeutic benefits. Tesla was a notorious user of all such therapies from this time on, often falling into a deep sleep in the warm and penetrating influences. Once, having overindulged the electro-sauna therapy, he fell into a profoundly deep sleep from which he emerged a day later! He reported that this experience was not unpleasant, but realized that proper "electro-dosages" would necessarily have to be determined by medical personnel.

During this time, Tesla found shorter impulse lengths where the heating effect disappeared altogether, rendering the radiance absolutely harmless. These impulse trains were so very high that the deepest nerves of one's body could not sense the permeating radiant energy field. Now he could pursue his vision of broadcast energy systems without fear of rendering to humanity a technological curse, rather than a true blessing.

TRANSFORMERS

Tesla operated the magnetic arc system at higher power levels, experimenting with various impulse lengths and repetition rates. He measured the mysterious electrical current which apparently flowed through space from this system. These radiant fields operated at far greater power than before. Strange effects were suddenly appearing at certain distances from the magnetic impulser.

For one thing, Tesla noticed that metallic surfaces near the impulser became covered with white brush-like corona discharges. While the sparks played in trails across the metal surfaces, Tesla observed physical movement among the metal objects. Tensions and rocking motions. Both phenomena occurring simultaneously, he was utterly fascinated. The sparks themselves seemed alive. The moving metal objects seemed to suggest new motor effects. What was this strange coalition, this synchronicity of phenomena?

Brilliant white coronas came forth with a gaseous "hissing" sound from metal points and edges. Metal plates were soon poised all around the device for observation. Tesla recognized at once that these effects were not identical with those obtained earlier while using high frequency alternating currents. These new discharges were white, energetic, and strong.

The electrical behavior of copper plates, rods, cylinders, and spheres near his primary impulser brought forth a great variety of white fluidic discharges. Strong discharge brushes appeared from the ends of copper plates. These came in prodigious volumes, hissing and arcing wildly in all directions, especially from sharp points. Tesla tried copper discs. These seemed to produce more stable discharges. He observed the curious manner in which these white discharges seemed to "race" around the disc edge at times, blending and separating with all the other sparks. Here was a greatly magnified example of Reichenbach's Od force perhaps!

He noted the manner in which white brush discharges appeared from copper conductors of different shapes. Each form, poised near his impulser, gave a characteristic corona distribution. This coronal correspondence with specific geometric form greatly impressed him. With certain metal forms the discharges were very fluidic in appearance. Smooth, fluidic sheaths covered copper cylinders of specific size. This absolutely fascinated Tesla. There was an aerodynamic nature inherent in radiant electricity.

Copper cylinders produced remarkable volumes of white discharges. The discharges from certain sized cylinders were actually larger than those being applied. This inferred that an energy transformation effect was taking place within the cylinder. This reminded him of his initial observation with the shock-excited wires. Those which did not explode gave forth far greater voltages than were initially used. He had never understood why this was occurring. Here was another instance in which applied energy was seemingly magnified by a conductor. Why was this happening?

The key to understanding this bizarre phenomenon might be found here, he thought. He observed the discharges from copper cylinders of various diameters. Each became edged with white brush discharges when held near or actually placed within the conductive copper strap of the impulser. The discharge effect was most pronounced when cylinders were placed within the periphery of the copper strap.

Tesla noticed that white corona sheaths were actually covering the outer cylinder wall at times. These would appear, build in strength, and disappear on sudden discharge with a surprising length. The sheathing action was repetitive when the cylinder had a critically small volume. Very small cylinders behaved like rods, where discharges only appeared at their edges. The stability of these strange sheath discharges varied with cylinder diameter and length.

Tesla noticed that not every cylinder performed well near the impulser. Only cylinders of specific volume produced stable and continuous white electrical sheaths. If the cylinders were too small, then the sheaths were intermittent and unstable. There was an obvious connection between the supplied impulse train and the cylinder volume. But what was it?

Tesla surveyed the entire range of his recent discoveries. Impulses produced a radiant electrical effect. Radiant electricity, was mysteriously flowing through space. As it flowed, it focussed over metal conductors as a white fluidic corona. When the shape and volume of the metal conductors were just right, the energy appeared as a stable white corona of far greater voltage than the impulse generator supplied. More questions. More discoveries.

Rods produced sparks from their edges, but not as long as copper cylinders did. Tesla selected a cylinder which worked very well, and placed several horizontal "cuts" all around its surface. He was totally surprised when, on testing, the spark discharge from the cut cylinder was notably larger than before. Increased spark length means increased voltage. But why did this diminished conductivity force the voltage up?

The cuts diminished conductivity in the cylinder by forcing the energy into a tighter "squeeze". He had noted that electrical impulses displayed a tendency to traverse the outer surface of metal conductors. Certain cylinders were often ensheathed in a fluidic white discharge which smoothly travelled between coil ends in a tightly constricted layer. Here was something truly notable. His input voltage was far less than that produced from the upper coil terminal. But why from end to end?

The essential reason why current preferred outer surface conduction was precisely because they were impulsing. The sudden shock which any conductor experienced produced an expansive effect, where the electrical charge was rejected by the conductive interior. This "skin effect" was a function of impulse time and conductor resistance. Highly resistant objects forced all of the impulse energy to the surface.

Now he was getting somewhere. Frustrated radiant electricity constricted into a tighter surface volume when encountering metal surfaces. This intense surface focussing effect brought the voltage up to tremendous values. Here was a new transformer effect! He believed it was an electrostatic transformation. Impulse currents each possessed an electrostatic nature. The bunching of charge in the impulser brings this electrostatic field to a peak in a small instant of time.

Constricting this field volume produces a greatly magnified voltage. Placement of any conductor in the field space alters the field by constricting its shape. When symmetrical conductors of special shape, volume, and resistance are placed in this space, the field is greatly constricted. Because the impulsing electrostatic field is very abrupt, it "snaps" over the conductor from end to end.

Tesla knew that here is where the secret lies. If resistance in the conductor is great enough, the snapping electrostatic force cannot move any charges. It is forced to "grow" over the conductor surface until it discharges at the end point, where greatly magnified voltages are obtained. When the wire diameter is small enough, the wire explodes under electrostatic pressures which exceed those seen in dynamite.

In effect, Tesla had managed to interrupt a high voltage direct current several thousand times per second. In doing so, he had discovered a way to completely separate electrostatic energy from current impulses. Tesla pondered these facts, wondering if it was possible to force the magnification effect beyond the limits of standard electromagnetic transformers. In other words, how high could voltage be raised? Was there a limit to the process?

In order to achieve such enormous voltage levels, he needed a conductive shape which offered so much resistance to charge movement, that all the applied energy would become electrostatic. In effect, Tesla wanted to convert a quantity of supply power into a pure electrostatic voltage. This phenomena suggested that his goal was not impossible.

Tesla extended his idea of the cut copper cylinder to coils. From the viewpoint of electrostatic impulses, flat copper coils appear to be "continuously cut" cylinders. The electrostatic field fucuses over the coil as it did with the cylinders, from end to end. A simple magnet coil of specific volume would offer so much resistance that it would be difficult to predict the actual resultant voltage which results without an empirical test.

WHITEFIRE

Constructing several of these, he was ready for the test. When each copper magnet coil was impulsed, Tesla saw tremendous white brushes leaping from their free ends: discharges approaching one million volts! But his supply power was nowhere near these voltages, and the coil was not wrapped in thousands of windings. These previously unexpected voltage magnifications were the result of an energy transformation, one which took electrical power and converted it completely into pressure. Watts into Volts, an unheard thing. It was the key to a new and explosive technology.

Tesla also found that such coils required very thin coil forms. He ceased using cellulose and cardboard forms, preferring "squirrel cage" type forms made of thin end-braced wooden rods. Wire was wound about these cylindrically disposed rods, producing the very best effects. Spacings were also tried between successive coil windings with excellent results. Spaced windings reduced sparking to a minimum.

Tesla remarked that the electrostatic potentials along the coil surface (from end to end) could be as much as ten thousand volts per inch of winding! A ten inch coil of proper volume could produce one hundred thousand volt discharges. In addition, and in confirmation of his suspicions, no current was ever measured at the free terminals of these coils. A "zero coil current" condition! It was simply another paradox which would occupy the academicians for several more argumentative decades.

Tesla suddenly realized that coils represented a truly special and valuable component in his quest. The instantaneous resistance which any coil offered to an applied impulse was so immense that current could not flow through the wire length. As a phenomenal consequence, no current flowed through the coil windings at all! But sparking was observed, travelling from coil end to end. Here was yet another anomaly!

He began placing these "secondary" coils within his "primary" impulser circuit. The strap which connected his magnetic arc to the capacitors formed the "primary". He made necessary distinctions among his Transformer components. Few engineers actually appreciate these distinctions. The "primary" and "secondary" of Tesla Transformers are not magnetic inductors. They are resistive capacitors. Coil-shaped capacitors! Tesla Transformer action is electrostatic induction.

There were conditions for the most efficient manifestation of the effect. Maxwell could not predict these values. Tesla empirically discovered most of the rules for impulse behavior. He found that the transformative abilities of these smooth copper coils were maximum when the coil mass equalled the mass of the impulser's conductive copper strap. It did not matter how thin the coil windings were. The equality of copper masses brought maximum transformative effects. When this equal mass condition was fulfilled, Tesla said that the coil-capacitors were "in resonance". Electrostatic resonance.

Tesla found it possible to produce millions of electrostatic volts by this method. His first Transformers were horizontal in orientation, both free ends of the secondary coil-capacitor producing unidirectional impulses of great power. White discharges from each of these free ends had very different characteristics, indicating the unidirectional flow. Electropositive terminals always appeared brushlike and broad. Electronegative terminals always appeared constricted and dartlike.

His next Transformer series employed vertical cylinders with the base connected directly to ground. Free terminals stood quite a distance above the primary capacitor strap, spouting a brilliant white crown. These marked a turning point in his theories concerning electricity, since it was possible for him to develop well over one million volts impulse power in a device scarcely taller than a child.

These discharges were of an intense white coloration. Whitefire. Very sudden impulses color discharge channels with the brilliant whitefire because Tesla Transformers separate the effusive aether from electrons. Tesla Transformer conduct aether, not electrons. The whitefire brilliance is the distinctive aetheric trademark of Tesla Transformers.

During this time, Tesla discovered the peculiar necessity for streamlining his Transformers. Cylindrical secondary capacitors suddenly became conical forms. These presented the most bizarre appearance of all. Tesla used cone-shaped secondaries to focus the impulses. Whitefire discharges from these forms evidenced real focussing effects, the discharges themselves assuming inverted conical shapes. Their greatly intensified nature is seen in photographs which were taken under his own intrigued supervision. The magnified voltages were reaching those thresholds in which his laboratory enclosures were far too small to continue making industrial scale progress on radiant energy systems.

The fact that whitefire discharges pass through all matter, notably insulators, revealed the aetheric nature. Tesla saw that whitefire discharges could permeate all materials in a strangely gaseous manner. This penetration scarcely heated matter. In fact, the whitefire brushes often had a cooling effect. The sparks themselves, though violent in appearance, were "soft" when compared to all other forms of electricity. He had successfully removed the hazard from electricity. In blocking the slow and dense charges, he had freed the mysterious effusive aether streams inherent in electricity. Because of this, new and intensified radiant effects were constantly making their appearance across his laboratory space.

Tesla found that as these new "Impulse Transformers" greatly magnified power supplied to them, so also their radiant electric effects were equally magnified. He found it possible to wirelessly project electrostatic power to very great distances, lighting special lamps to full candlepower at hundreds of feet. In these experiments, he also conceived of signalling systems. It would be possible to switch radiant effects in telegraphic fashion. Distant vacuum tube receivers would then light or dim in corresponding manner. Tesla experimented with a special breed of telegraphic wireless in 1890.

He also found it possible to wirelessly operate specially constructed motors by properly intercepting this space-flowing energy stream. He had made his own polyphase system obsolete! The new vision was vastly more enthralling. The world would be transformed. He discovered ways to beam the energy out to any focus, even to the zenith. His plan to illuminate the night sky with a radiant energy beacon captured the minds of all who listened.

Tesla now possessed the means by which the radiant electricity could be greatly magnified and transmitted. He could now transform the very nature of the radiance so that it could carry increasingly greater power. Now he could begin developing a new technology which would completely revitalize the world order. Power could be broadcast to any location without wire connections. Radiant electricity could be utilized in completely new appliances. A new world was about to be released!

SPACE FLOWING CURRENT

Understanding the analogue between these electrical impulse effects and the behavior of high pressure gases was of paramount importance. This gaseous aspect of impulse electrical radiance was perhaps the most mystifying aspect of these new-found energies. Those who sought out Tesla's every lecture were very aware that a new electrical species had been discovered.

While yet a student, Tesla had became aware of certain scientific imperatives enunciated by Johann von Goethe. One of these was the preservation and extension of all activities-natural. Goethe implied that when natural conditions were preserved during experimentation, then nature itself was in the best configuration to reveal more unified phenomenal exhibitions to qualitative observers.

Tesla recognized that his new discovery of impulse, the result of an accident, was a total departure from polyphase alternating current. While his original vision of the vortex was applied by him to the designing of motors and generators, Tesla now realized that this was not its primary message. In fact, taken from the viewpoint which Goethe expressed, polyphase was a most unnatural form of energy.

Natural activity is suffused with impulses, not alternations. Natural activity is initiated as a primary impulse. Nature is flooded with impulses of all kinds. From lightning to nervous activities, all natural energy movements occur as impulses. Impulses were now seen by Tesla to fill the natural world. But, more fundamentally, Tesla saw that impulses flood the metaphysical world.

The mysterious flow of meanings during conversation occurs as a sequence of directed impulses in space. Though inert air vibrates in alternations with sounds uttered, the flow of meaning remains unidirectional. Intentions are also impulses. The unidirectional flow of intentions appear as impulses. Motivations proceed from the manifestation of sudden desires. Overtly expressed as actions, the initiating impulses are then fulfilled.

Tesla wished to comprehend where this "motivating force" came from, and where it went during the expressed actions. In all of this, he was very much the wonderful stereotype of the Victorian natural philosopher. His scientific pursuits followed these considerations until the last. Those who study his announcements recognize his metaphysical foundations, the basis of all his subsequent scientific quests.

Tesla observed the amazing "coordination" of new phenomena which daily seemed to bring new technological potentials before him. This wonderful synchronicity, this vortex, revealed his new and fortunate position in nature. Having somehow "broken" his fixation with the unnatural...with polyphase...he re-entered the natural once again. Impulses. Could it be that the induction of electrical impulses summoned the other impulse characteristics of nature? Was he producing a metaphysical vortex, into which all the impulse phenomena of nature would now flow? Was this the real sunset message which seized him in Budapest, so many years ago? Was electricity the fundamental natural energy...the motivator?

Victorian Science was not exactly sure what electricity was, there being so very many attributes associated with the term. Seventeenth and Eighteenth Century natural philosophers conjectured on the nature of both electric and magnetic forces. Gilbert and Descartes shared the belief that these forces were a special kind of "flowing charge", a space radiant stream which took place in tightly constricted lines. Some equated the electromagnetic forces with a "dark light", which Karl von Reichenbach later proved in part.

Faraday adopted and modified the view that electromagnetic forces acted through space because they were a special flow of charge. This effusive charge movement changed when travelling through conductors, becoming more densified and retarded in velocity. Faraday's "lines of force" were not conceived by him to be mere static tensions as modernists view them. Faraday envisioned these force lines as radiant, streaming lines. They were mobile, moving longitudinally into space.

Others would change the names, referring to electric force lines as "diaelectric" or dielectric flux, but the view remained essentially as conceived by

Faraday. Young James Clerk Maxwell also believed that force lines were dynamic, longitudinal lines of flow. But flowlines of what substance? Here lay the principle problem which occupied physicists throughout the Victorian Era.

Victorian researchers and natural philosophers wished to discover the exact nature of the "flowing charge" of which force lines were composed. Most agreed that the mysterious flowing "substance" had to be an effusive, ultra-gaseous flux. This flux was composed of infinitesimal energy particles which effected the various pressures and inductions observed.

Henry and Faraday struggled with the idea of deriving usable electric power from static charges. The notions was that, since forcelines were made of a "flowing charge substance", then fixed contacts placed on charged masses would supply electrical power forever. No one was able, however, to derive this flowing charge. Lossy discharges preceded every contact. Most researchers, whose attempts with highly charged Leyden Jars failed, sought a more benign source of concentrated charge. The quest shifted to magnets, but the attempt remained as futile as ever. There remained no available way to derive power from the individual flowing charges of a forceline.

J.J. Thomson discovered electrons in vacuum discharges, assuming that these "electric particles" operated in all instances where electrical activity was observed. Victorian researchers did not accept this view completely. Thomson's "electrons" were viewed as the result of violent collisions across a vacuum acceleration space. It was not possible to ascertain whether these same "Thomson currents" were active within electrical conductors operating at small voltages.

Very reputable experimenters besides Tesla continued claiming that "space flowing electricity" is the real electricity. Tesla's classic demonstrations proved that rapid electrical impulses actually exceed the ability of fixed charges to transmit the applied forces. Charges lag where electrostatic forces continue propagating. One is compelled to see that electrostatic forces precede the movement of charges.

Tesla saw that electrostatic impulses could flow without line charges. His "zero current coils" operated simply because the charges themselves were immobilized. Electricity was shown to be more in the nature of a flowing force rather than a stream of massive particles. But what then was this "flowing current"?

In Tesla's view, radiant electricity is a space flowing current which is NOT made of electrons. Later Victorians believed that there was a substance which both filled all space and permeated all matter. Several serious researchers claimed to have identified this gas. Notables, such as Mendeleev predicted the existence of several ultra-rare gases which preceded hydrogen. These, he claimed, were inert gases. This is why they were rarely detected. The inert gases which Mendeleev predicted formed an atmosphere which flooded all of space. These gaseous mixtures composed the aether.

Tesla and others believed that both electrical and magnetic forces were actually streams of aether gas which had been fixated in matter. Materials were somehow "polarized" by various "frictive" treatments by which an aether gas flow was induced in them. Most materials could maintain the flow indefinitely, since no work was required on their part. Matter had only to remain polarized, transducing the aether flow. The aether gas contained all the power. Unlimited power.

This aether gas power manifested as the electromagnetic forces themselves, adequate reason to pursue the development of an aether gas engine. Such an engine could run forever on the eternal kinetic energies of the aether itself, it being both generated and driven by the stars.

Tesla believed that radiant electricity is composed of aether gas. He based this belief on the fact that his zero current coils were not conducting the "slow and dense" charges usually observed in ordinary electrical circuits. Abrupt impulses produced distinctive and different effects...fluidic effects. The qualities ascribed by Tesla to "electricity" or things "electrical" in his numerous patent texts and press interviews are those which refer to the aether gas. Tesla did not refer to electron currents as "electricity". He did not equate "electricity" with electron flow. Whenever Tesla spoke of "electrical" effects he always described their effusive, gaseous quality.

Tesla referred to space as the "ambient or natural medium". Space, he claimed, was that which "conducts electricity". He had found a means by which this gaseous electrical flow could be greatly concentrated, magnified, and directed. He saw that this radiant electricity was, in reality, a gaseous emanation. An aetheric emanation. This is why he made constant reference to fluidic terminology throughout his lectures.

Resistance, volume, capacity, reservoir, surface area, tension, pressure, pressure release: these were the terms upon which Tesla relied throughout his presentations. The terminology of hydraulics. Tesla also recognized that because aether was a gas, it had aerodynamic requirements.

Aether, in Tesla's lexicon, was space flowing electricity: a gas of superlative and transcendent qualities. Aether was the electricity which filled all of space, a vast reservoir of unsurpassable power. Motive, dynamic, and free for the taking. Aether gas technology would revolutionize the world. Aether gas engines would provide an eternal power source for the world. Science, industry, corporations, financial alignments, social orders, nations...everything would change.

INTRIGUES

Completing a tour of the major scientific institutes in America, Tesla expected to retire for a season of rest in New York once again. News of his advancements however, flooded every technical trade journal. The name Tesla was everywhere once again. First polyphase and now radiant electricity. He was the "darling" of the press. Tesla captured the public eye once again. People everywhere were thrilled with the projected future visions which Tesla freely provided. He was a model European immigrant, suave and debonair. These are probably the qualities which first attracted Anne Morgan. Irresistible, wealthy, unattached, and warm. Tesla was her obsession.

Despite his great personal charm and magnetic personality, he maintained his serious tone and poise wherever he went. The vision of the future was far more important than the attentions of a young and flirtatious lady. In anticipation of these forthcoming events, Tesla often invited other socially esteemed guests to his laboratory for special demonstrations. In this manner, it was noised abroad that what he claimed was in fact real. Anne often attended these gatherings, breathing silently in the shadows of his large loft laboratory.

There were others who, although not attending these demonstrations, were equally watchful of Tesla's newest radiant energy developments. Several of these persons, shall we say, were interested in his new discovery and its implications...because their fortunes were threatened. Tesla had swept the world once with polyphase. He wiped out Edison's Direct Current System overnight.

J.P. Morgan, Edison's recent "patron", had lost a considerable sum during that fiasco. It was certain that Tesla would soon sweep the world again with broadcast electricity. This destabilizing influence would not be tolerated. Anne complicated the affair considerably. She was in love with Tesla. Obsessed in fact. Too obsessed and desperate to let go.

ROYAL SOCIETY LECTURES

In the very midst of all these national attentions, Tesla received an invitation from Lord Kelvin. He was formally requested to address the Royal Society, his latest findings were earnestly desired. The English, usually extremely conservative, were sure that Tesla would change the course of world history.

Tesla, adjourning from his daily researches now prepared himself for the lectures which would start the world-change. He packed nearly every piece of delicate equipment one can imagine. Vacuum tubes, Transformers, strange motors, and equally strange wireless apparatus. All were carefully crated and personally brought to Europe by Tesla himself. His beloved elder and personal mentor, Sir William Crookes, greeted him.

In the opening portions of his Royal Society lectures Tesla first described his preliminary work with high voltage high frequency alternating currents in some length. He explained that these devices embodied the very last investigations and improvements of his Polyphase System. He demonstrated several of the first small high frequency alternators and iron-core induction coils in order to prepare his audience for a final announcement.

In this very last dramatic demonstration Tesla revealed to British Academia the disruptive electric discharge and the properties of electric rays. Tesla made a rare and complete "full disclosure" of the electric ray effect at the very end of his lecture. It was the very last time he would ever do so again in academic circles.

Tesla showed that the new radiant electricity was distinctive, having been openly proclaimed during the London Royal Society lectures. Tesla deliberately compared and contrasted the potent impulse radiance to his previous weak effects produced by alternating currents (February 1892). Fluorescent lamps and other luminous wonders held his audience spellbound. All the while his voice, tenor-like by excitement, rang throughout the silent awestruck hall.

He demonstrated wireless lamps, lit to full brilliance by radiant electricity. He ran small motors at sizable distances for his audiences to see. This last lecture represents the only recorded instance in which Tesla openly announced his discovery of the electro-radiant impulse. He tells the personally revolutionizing aspect of his discovery and how it virtually eradicates his previous work. He went to great detail verbally describing and disclosing the exact means for eliciting the phenomenon.

In his closing time Tesla quickly demonstrates special "electrostatic" motors and lamps made to utilize the radiant effect. Examination of these first lamp and vane-motor devices reveals their primitive and initial state. Tesla modelled the motor after the Crookes radiometer, stating this fact publicly for the benefit of his revered mentor. Tesla finally stated the vast implications of the discovery. He pointed their minds toward the establishment of true power transmission.

He prophetically announced the new civilization which would emerge from these first devices and systems. The world would be completely revolutionized by this new principle. Tesla described beam-transmission of electrical energy, and the possibility of harnessing the radiant energies of space itself.

Those who had witnessed Tesla's entire demonstration were completely enthralled at his results, but misunderstood his new announcement completely. This became apparent to Tesla a short while after he, highly decorated and honored, departed for his Parisian tour. British Science was yet delving into Teslian high frequency alternations. Tesla had already disposed of these discoveries as mere preparatory introductions to impulses.

Tesla showed by way of comparison that disruptive field impulse transcendently exceed all other electro-inductive effects by several orders. He expressed difficulty in discerning whether the effects were electrostatic or electrodynamic in nature, preferring to associate them more with electrostatic effects. We deduce that he had only recently begun developing the electric impulse effect because of his hesitance in identifying the phenomena properly.

Tesla was stringently exact in all his statements. This seems uncharacteristic of his scientific nature. But he did this in true scientific openness. Tesla did not know exactly what was occurring in the electric impulse at that time, desiring only to share the discovery openly and candidly. Academic disapproval of his personal semantics came swiftly in journal after journal.

It is clear that Sir William Crookes completely grasped the significance of Tesla's entire demonstration and realized the closing formal announcement of the new electric force. Crookes could not contain the thrilling implications. He was also sure that the new force would completely revolutionize the scientific world.

Crookes upheld Tesla thereafter as the true discoverer of an unrecognized electrical force. Tesla continued correspondence with his mentor after his departure from England. He had hoped that his dramatic announcement and demonstration would produce a new regime of electrical engineering, and that others would now reproduce the radiant electric effects as described. His hopes would be strangely dashed to pieces in the coming years when the derisive academic attacks began.

To European academes, the lecture series was astounding. It was a glimpse of the future, so clear that few could find time to argue with Tesla at all. Tesla concluded his tour of England and France, everywhere heralded in typical Victorian heroic style. One night, while in Paris, a telegram informed him that his mother was on point of death. Rushing to her bedside, he managed a few hours of final conversation.

He always referred to her as the one who completely understood his strange abilities. Was she not the woman who had encouraged him when he first remarked about his childhood visions? When siblings and friends derided him, she was his support. Early the next morning, in an adjacent house, he was abruptly awaken by a vision. What he beheld changed his life. A seraphic host surrounded his mother. She was ascending into bright clouds. Several minutes after that, the announcement came. His mother had quietly passed away. He spent a torturous week in his native land for her funeral, and fled back again to New York.

REVERSALS

When English engineers wrote, asking the means for generating his impulse effects, Tesla gave them very strict descriptive parameters. He never failed to openly disclose the secret by which his spectacular effects were obtained. He had learned to freely share what he knew with all. He was surprised to discover that the academic societies who so warmly addressed him in Europe, were gradually losing interest in his discovery. Being utterly incapable of duplicating his specified parameters, most believed the effects to be "dubious".

The impulse effect had very stringent requirements before its manifestation. Care in constructing impulse generators was the basic requirement. Engineers wanted equations. Tesla gave them descriptions. A few experimenters succeeded in later duplicating Tesla's broadcast electricity effects. But these systems were direct descendants of Tesla's earliest and less efficient designs.

It is often in the nature of academes to forgo empirically evident facts and argue personal differences, especially when foreign personalities are given excessive adulation. Fixated on issues having to do with words and personal poise, Tesla's audiences found several acrid voices whose equally vile publications dared tamper with Tesla's character.

New critics were everywhere, even at home. Dolbear, Thomson, and even Pupin found time to criticize and deride Tesla. Because most younger academes relied entirely on schooling and less on empirical method, they were easily swayed by academic opinion. Tesla underestimated the power of media and of opinions in underrating his abilities. He quickly found that public opinion could actually sway scientific opinion. He failed to see who was behind the media campaign.

Tesla disregarded his antagonistic colleagues. Crookes always deferred to Tesla, whom he admired and loved as a younger protege. Tesla revered the aged Crookes, upon whose confidence he came to rely during more difficult years. Crookes had been given a true Tesla Transformer when Tesla had given his lectures. The small device was potent, giving the uncharacteristic effects which Tesla had always claimed. This single piece of evidence was left in England for all to see. Remarkably, this evidence did not silence the critics.

Tesla could see no reason in all of this. Something did not quite "add up". Even Tesla could see that there was a missing part of the "equation". Discovering this part would explain his own reversals. As if these personally devastating events were not enough for him, the insolent young Anne continued haunting him at his every turn. He continued being "polite" to her, but never more than this.

Crookes wrote many times to the Royal Society and to Tesla concerning this fact. Sure that Tesla was a modern Faraday, Crookes continued espousing the belief that Tesla had discovered the next historically important electrical advancement. He was encouraged to continue research despite his protagonists. Few academes trusted Tesla's methods now. Fewer yet listened any longer to his statements.

Losing credibility as quickly as he had found it, financiers were slow to trust investing in his new systems. His inventions continued their steady march into electrical history. Each new device chronicles a new step in the technology which should have changed the world. He plunged himself headlong into work. Only work would vindicate him. Opinion would fade when others gradually saw the astounding developments which he would produce. In these actions, Tesla revealed his noble and naive nature. The world had changed, but changed toward a more brutish rule.

BROADCAST POWER

He set to work developing more powerful embodiments of his initial Transformers. In order to make a Broadcast Electrical System possible it would be necessary to devise more efficient transformers. He set to work on this very task, examining and dissecting every fundamental part of his existing Transformers.

Tesla discovered that excessive sparking, though impressive to observers, were actually "lossy instabilities". The distant radiant effects he desired were interrupted and distorted whenever sparking occurred. Both sparking and brush discharges actually ruined the distant broadcast effects of radiant electricity, a situation which had to be remedied. Tesla sought elimination of the discharges now. Tesla had already found that metals could focus radiant electrical effects. Additional stability in his Transformers could be achieved with the addition of large copper spheres to the active terminals. Tesla considered copper spheres to be "aether gas reservoirs", providing his transmitters with an additional aether gas supply.

Copper spheres attached to Transformer terminals reduced the required electrical levels for an efficient electric radiance. Copper spheres significantly reduced the injurious instabilities of visually spectacular brush discharges, but did not eliminate them entirely. What Tesla required was a new means for transmitting the radiant electricity without loss.

Tests with elevated copper spheres facilitated efficient transfer of radiant power between the Transformer and surrounding space. Now, Tesla Transformers became true Tesla Transmitters. Tesla found it possible to broadcast harmless radiant electricity with great power to very great distances. Numerous subsequent patents recorded his progressive conquest of the broadcast power principle.

He succeeded in making radiant electricity safe for human use. It would simply travel around conductors if made to impulse quickly enough. Only specially entuned receivers could properly intercept the radiant power for utility. Not three years before he had accidentally discovered the radiant electrical effect. He dreamt of safely sending electrical power without wires in 1892. Now, in 1895, he had realized his dream. Would the system work across the vast distances which he envisioned?

He took his more portable Transmitters outdoors, away from the confines of his South Fifth Street laboratory. Both in northern Manhattan and Long Island, Tesla tested his radiant broadcast systems without restriction. He measured the distant radiant electric effects of these designs in electrostatic volts. Broadcast power could be converted back into current electricity if so desired, the harmless high voltage becoming current in appropriate low resistance transformer coils.

He found to his very great surprise that very distantly positioned vacuum tubes could be lit to great white brilliance when the primary system was operating. The requirement for this action was twofold. First both the system and the receivers had to be grounded. Second, specific volumes of copper had to be connected to the receivers. When these two requirements were satisfied, lamps maximized their brilliance, and motors operated with power.

Copper in the receiver had to "match" the copper mass of the transmitter in a very special equivalence, otherwise radiant transfer would not be efficient. The requirements differed very much from those of ordinary radio antennas. He also found that elevated copper spheres more powerfully enhanced the broadcast radiant power from his transmitters. This was Tesla's means by which his transmitters and receivers could be better "connected" despite their distance.

Tesla believed that these electrical beams invisibly linked both his transmitter and receivers together. He considered each as "disconnected terminals" to ground. Electrical radiance spread out in all directions from the elevated copper sphere of his transmitter. The secret in receiving a maximum signal was to match the transmitter's copper mass with the receiver mass. Then, the aether streams would actually focus into the matched receiver. This affinity would take time, the transmitter energy "searching" for better ground sites. Radiant electricity evidenced curiously vegetative "growth characteristics".

Receivers now were outfitted with small copper spheres. These provided a more efficient affinity and absorption for the radiated power. The additional copper spheres which surmounted Tesla transmitters effectively lowered the input electrical power for the production of focussed aether discharges.

Tesla took the gas dynamic analogy to another level when he found that both low pressure gaseous and vacuum tubes could replace copper. Electroradiant effects from gas-filled globes were projected with less electrical loss and even greater power. Large low pressure argon gas filled globes were empirically found to broadcast tremendous radiance when used atop his transmitters. Additionally, he found that argon gas at low pressures could serve as an equivalent receiver as pure copper spheres.

The gas filled globes would be less costly than copper spheres to disseminate in public use. He was approaching a totally efficient system. Numerous personages were invited to observe these historic tests. J.H. Hammond Jr. was one such individual. Enthralled with Tesla's developments, he and his wife invited Tesla repeatedly to their home in later years. Tesla was their honored guest for months at a time. Later in years, after World War I, both Tesla and Hammond worked on robotics and remote control.

Tesla envisioned small power units for both home and industrial use. The installation and maintenance of these units would require a small monthly fee. Through these wireless units one could draw sufficient power to operate factories and homes alike. Electrical usage could be metered. The superiority of this new broadcast power system was obvious to all who observed it in operation.

Tesla also described the use of these power units for transportation. Transatlantic ships could simply draw their motive power from continental power broadcast stations. Trains and automobiles could be operated by drawing their power. The potential fortunes would soon stimulate financiers to invest heavily in the "coming activity".

In keeping with his publicity-mindedness, several investors were always invited to Tesla's private demonstrations. Tesla knew that their urge to support his new world-shaking venture would become irresistible when once each had beheld his small broadcast power system. The demonstrations were deemed by these individuals as "entertaining", in their typical dry tone. But, he rarely heard from these people again.

Here was a new change. Shy moneymen. A true contradiction. Their reticence left Tesla in a state of bewilderment. Once, in a ditch, his conversation alone was sufficient perfume to attract the bees. Now? None would dare leap into the new world sea. Why? What sharks were there besides themselves? Tesla could simply not understand this new "dearth", this incredulous conservatism and lack of imagination on the part of New York investors.

Eager to begin, Tesla patiently waited for the messengers to call. Had he known more of the world around him, however, he would have stopped waiting. Shortly after Tesla's private demonstrations were concluded, Morgan's agent approached Tesla with a "business proposition". The bribe being sizable, contracts would have placed Morgan in control of Tesla's new system. Tesla laughed at the pale little Mr. Brown in his pinching-tight tails, informing him that he himself was already a millionaire. Why should he need such an affiliation at all? He was escorted very graciously by the amused Tesla.

While dining in the Waldorf several hours later, a rude interruption informed him that his laboratories were ablaze. The connection between his refusal to bow and the flames which now reached skyward was not made until all was consumed. That night, the world changed completely for Nikola Tesla. He lost everything of his past. Everything. The totality of his technological achievements were burned into vapor. Books, priceless souvenirs, delicate equipment, patents, models, drawings, new pieces of apparatus. Everything was burned. He read the message well.

There was a two week period where he simply vanished. No one could find him. Kolman Czito, his trusted technical foreman and machinist feared for Tesla's life. Katherine Underwood Johnson was beside herself with anguish. She was the wife of a close friend, the only real love of Tesla's life. The fire was meant to kill. It was a message as clear as anyone would need. The assassination attempt failed to kill the intended victim. It certainly did not kill his dreams.

Wherever he was for those two weeks, the dreams were with him. But a part of Nikola Tesla died in the fire. It was the part which was tied to the past. His eyes on the future, Tesla developed his discovery into a major technology which the world seems to have forgotten. Of all those who prayed and wept over Tesla's disappearance, one person was no longer concerned. Never again would Anne need to be troubled by the thought of Nikola Tesla. His love was already sealed. Tesla recovered from the flames.

His subsequent discoveries and inventions surpassed his former works for forty more years; special radiation projectors, self-acting heat engines, power transmitters, remote control and robotics, the "World Broadcast System", Beam Broadcast transmitters, "aetheric reactors and aetheric engines", cosmic ray motors, psychotronic television...the list of astounding inventions is truly awe-inspiring. Tesla demonstrated each of these systems for a select group of witnesses.

Furthermore, despite rumors of his public and scientific demise, Tesla maintained two penthouse suites atop the Hotel New Yorker in a time when such extravagance was otherwise unobtainable. One of these suites was converted into a complete radio laboratory, several accoutrements of which having been retrieved by antique radio enthusiasts. Tesla was an indefatigable researcher. The biography of Nikola Tesla is replete with truly mysterious designs and developments. But these are parts of his biography which must be told in other volumes.

CHAPTER 5

ULTRA MICROSCOPES AND CURE RAYS: DR. R. RAYMOND RIFE

LIGHT

There is a constant war being waged which most prefer to ignore. Living out our days in the joyous sunshine, we rarely choose to glimpse full-faced into the horrid visage of disease as physicians so often do. Perhaps it is pain, perhaps fear. Despite our willful ignorance, hideous armadas of pathogens march through all nations unhindered. These insidious enemies wage their continual war against the human condition, with a cruel and merciless deliberation.

Pride and wealth cannot keep these legions away. They are deadly, having no conscience or allegiance. They are the universal enemy of humankind, a relentless foe. It is a wonder that nations have not surrendered their petty personal feuds long enough to recognize the common specter. Joining our best forces to defeat this dread army long ago would have secured major victories for all of humanity.

It has therefore fallen to the sensitive and impassioned few who seek alone, armed with vision and swords of light. The independent medical crusaders enter the battle alone. Their names are seldom seen in major journals any longer. Their private research forever dangles on gossamer threads of grants and endless bureaucratic labyrinths. Yet these are the ones, the men and women who make the discoveries from which cures are woven. Real cures.

They often live on shamefully minuscule budgets, preferring to pour out their personal funds into the work. They are the seekers. They are always close on the brink of a possible new development. One never knows when such will come. The important thing is that they are prepared, and wait in prepared chambers for the gracious and providential revelations upon which humanity depends. Theirs is the excitement of the chase. Their quest is "the breakthrough". They are the ones who fill little lab rooms, closet spaces which line university hallways. Their intuitive vision has guided them into research alleys which are too small for the big concerns of profiteering medical agendas. If these researchers are fortunate, they find an impassioned patron. Perhaps the patron is a sensitive one, whose life has been touched with the sting of tragedy. Perhaps a loved one was lost. Perhaps also in the heat of that pain, the recognition came that gold must be transmuted by passion and devotion before it can cure. These quiet ones who go about their work daily, so many devoted hearts, are driven on behalf of all who bear such sorrow over what has been lost.

There was once such a man. His discovery gave eyes to the blind. He perfected a means by which humanity's enemies could be detected. His microscope could optically sight viruses, and sight them in their active state. And he developed a means by which viruses, any virus, could be eradicated with the flick of a switch. His medical developments won him no reward because his research did not fit the desired agenda.

The Microscope. The Super Microscope. There were predecessors to the prismatic marvel of Dr. R. Raymond Rife, but no equals. Others had designed and used oil immersion lenses, dark-field illuminations, and deep ultraviolet light, each holding part of the secret for optically magnifying infinitesimal objects. But the design which Dr. Rife developed outpaced all of these.

It is doubtful whether you have ever heard his name. Reasons for mass-forgetfulness run deep. Truths have been kept from you. Only a careful and relentless study of the past will relinquish secrets purposefully and cunningly buried. The information is safely nestled in dust laden libraries which few now venture to search. Perhaps you will recognize why his name has been blotted out of the historical records before we reach the end of this amazing biography.

Dr. Rife began as a research pathologist. A medical crusader of the very highest qualifications, is was a heart filled with but one goal: the eradication of disease. Dr. Rife recognized first and foremost that successful medicine relies on vision, on light. What we cannot see we cannot battle. An unseen foe is impossible to destroy. Therefore his first quest was to secure a vastly improved system of microscopy. Once he could see, once all could see, then the pursuit of medical knowledge could again move forward. An armada of equipped seers could assail the foe on every shore. Looking for more light.

Dr. Rife's study of microscopy detailed every component and premise which tradition had presented to our century. The creation of a super microscope would run counter to every physical law and restriction which the previous two centuries had accumulated. Academes began again to love the writing of papers. Without the exercise of experiment, however, all these papers were so much tinder.

Dr. Rife wanted to develop super microscopes capable of seeing viruses. His aim was to chart and catalogue them, understanding that these represented a deadly foe which exceeded bacilli in their destructive assault on humanity. His quest now began. He reduced the fundamental premise by which microscope design had developed, analyzing each separate component and premise.

FOCUS

Optical designers had been adding ever more complex components to the design which began with van Leeuwenhoek. Lenses were compounded to lenses, crowns were added to compounds, crowns were added to crowns...the complexity was frightful. Simplifying the problem necessarily led to Rife back to the study of optical geometry and the comprehension of simple ray divergence.

Rife thought on these ancient principles. An ideal magnifying system is a geometric construction of extreme simplicity. Diverging light rays can magnify any object to any magnification. Given a strongly divergent light source and a great enough distance, one can theoretically magnify the indivisible! This is the principle which underlies projection microscopy. Dr. Rife realized that the projection microscope represented the best and simplest means for magnifying infinitesimal objects. One simply needed to discover a means by which a vanishingly small, brilliant radiant point could project divergent rays to the surface of any material speck. No virus, however indistinct and cunning, could hide from such an optical magnifier.

The theoretical design of microscopes relies purely on geometric principles. Actual materialization of these principles requires material manipulation, since geometric rays and light rays are significantly distinct. What is a microscope essentially? What is achieved in a microscope with light rays? The notion is quite simple. Take divergent rays from a vanishingly small point of brilliant emanations and allow them to pass through any specimen which is to be viewed. Light from this encounter is then made to diverge as far apart as possible in a given space. Geometrically it is possible to divert rays from a vanishingly small point out to an infinite distance. This geometric construction would produce unlimited (ideal) magnifications. Provisions toward this ideal goal would require that the point radiant source be tiny and brilliant enough, the specimen be close enough to the radiant point, and the image diverging space be very long. The geometric divergence of the point light source is the magnification factor. But geometry is an idealized reality. And ideal geometry encounters significant frustrations when implementing light in inertial space.

The most basic type of microscope is the projection microscope. It is the most simple system which is employed to greatly magnify the most infinitesimal objects. In the more common version, light is made to pass through a tiny specimen. Light from the specimen is forced to diverge across a long space by means of a very small focussing lens. Rays from this lens cross, diverging and expanding across a long space. This widely divergent beam is then projected onto frosted glass. The viewing of images derived through these means is indirect, but provides superior magnifications with ultrahigh resolutions.

Formerly, laboratories required compact units capable of close personal manipulations. The development of fine optical microscopes became frightfully complex when more powerful but compact models were required. The notion of a compound microscope is to physically compress the long projection space into a compact tube, delivering the shrunken design to customers who wish to conserve space. The "problem" with compact optical microscopes was bending the necessary wide beam through a small space. The "trick" in a compound microscope was to keep the image rays from prematurely diverging between lens stages.

The long expansion space required for divergent beam magnification had to be "folded" and "convoluted" within imaging tubes. Large numbers of lenses performed this duty. Being thus "convoluted" by lenses to achieve magnification, images produced by most expensive bench microscopes were inherently limited. Since the diverging image in these microscopes is "interrupted" within a greatly shortened space by means of several optical stages, it cannot produce great magnifications with either clarity or brilliance.

Each optical stage continually bends the image until a tremendous effective divergence is achieved. The effects are dramatic, but the necessary stages introduce optical resistances by which magnifications are inherently limited. Fundamental problems with white light alone complicated the problems which designers faced. Breaking into spectral components, each color refused to focus in exactly the same point. As a result, chromatic aberration blurred every image.

The light-crossing action of each lens brought widely diverging light beams into the ocular lens. It was pivotal that these rays be parallel. Images lost most of their radiant power against the tube walls before arriving in the final ocular. Therefore more corrective lenses were added in the beam path to bend the light back from the tube walls. Differences when light travelled between lenses and air introduced more aberrations. Batteries of corrective lenses, crowns and compounds, so loaded the light path with crystal that images lost their original brightness. These horrendous optical problems were never completely solved, despite the high cost of these instruments.

All of these optical horrors were the result of an old tradition which yet compels designers to maintain familiar outward forms. The projection microscope is so simple and potent, one wonders why newer designs had not been developed with as much dedication and zeal. It was the outward form which compelled the convolution of projection microscope simplicity, detracting from the excellence of magnified images. What was really lacking in optical microscopy was the development of true, tiny radiant points of monochromatic light. These diverging ray sources could produce novel and economical projection microscopes.

The numerous optical components of most excellent laboratory microscopes are configured to prevent image splitting, image incoherence, and other optical aberrations. All the differences between geometric ideals are suddenly and severely limited when using light and glass. The optical ideals come short of the geometric ideal.

Geometric rays do not fade with infinite distance. Light rays do. Geometric rays do not blur at their edges with increasing divergence, Light rays do. Geometrically magnified lines do not diminish in their intensity. Light images do. A successful optical approximation to the geometric ideal would produce a super microscope. Dr. Rife decided to manipulate all the possible variables in order to approach, as closely as possible, each part of the ideal geometric construction. If such a feat could be accomplished, he would have successfully bridged the gap between optical and electron microscopy.

POINTS

To be sure, numerous individuals had accidentally discovered enormous magnification effects while experimenting in completely different fields of study. A magnifying system which magnified much smaller infinitesimals than viruses appeared in 1891. Nikola Tesla developed a remarkable carborundum point vacuum lamp and made an accidental observation which opened a new world of vision to science.

Tesla began inventing single wire vacuum lamps for purposes of illumination. These were large glass globes powered by very rapidly impulsed currents. The impulse currents made the single supported wires glow to white brilliance, melting them. Impractical for public use, he sought to alleviate this condition by using special crystals. High melting points were required. An assortment of such materials were poised at the single wire termination. When electrified, they suddenly became radiant.

His experiments included using diamond, ruby, zircon, zirconia, carbon, and carborundum. He found it possible to blast the natural gems after a few seconds' electrification time. But before exploding, each of these crystalline terminations released puzzling patterns of light across the globe surface. This symmetrical pattern of points attracted Tesla's attention. They appeared when the current was turned on for just an instant.

Moreover, Tesla noticed that the brilliant fixed points of light remained in fixed positions each time he applied the current. Equally astounding was the fact that each material portrayed distinctive point symmetries upon the glass enclosure. The most resilient and successful crystalline material was carborundum, which he ultimately adopted for practical use. This too gave its characteristic point symmetry across the globe.

Tesla was not sure what he had discovered. He intuitively surmised that these point patterns of light somehow revealed the crystalline structure of the excited material. He also utilized the geometrical construction to obtain his deduction. His thoughts turned to the internal crystal conditions. As electrically charged particles were propelled and ejected through the carborundum, they were deflected by infinitesimal points. Diverging from such infinitesimal points, they impinged upon the inside spherical globe which housed the carborundum point. These brilliant points of light were always of the same symmetry because the ejected particles were passing through a fixed grating: a crystalline grating.

He theorized that this fixed pattern represented the greatly magnified crystalline symmetry. This simple apparatus was the world's first point-electron microscope. The phenomenon responsible for the defined projection of crystalline spaces is referred to as "field emission". Later, others would duplicate these same results with other crystalline specks. The remarkable X-Ray photography of Max von Laue already permitted the sighting of crystalline atoms. In this scheme a thin crystal point was placed at a critical distance from an X-Ray source. Entering and passing through the crystal slice, divergent X-Rays produced a greatly magnified image of crystal atoms on photographic negatives.

The result of von Laue's experiment was astounding, but was a purely geometric consequence. Divergent rays from a vanishingly small radiant point can theoretically magnify equally small specks to immense size. But while both Tesla and von Laue produced wonderful results with particle-like emissions, the practical achievement of these ideals were diminished when using optical light rays.

Emile Demoyens (1911) claimed to have seen extremely tiny mobile specks under a powerful optical microscope...but only at noon during the months of May, June, and July! Colleagues thought him quite mad, but Dr. Gaston Naessens has comprehended why these specific time periods permitted such extreme viewing. During these seasonal times the noonday sunlight contains great amounts of deep ultraviolet light. The shortened wavelengths provide a sudden optical boost, permitting the observation of specks which are normally invisible.

Progress in optical science seemed limitless and free. It was anticipated that no limit could bar humanity from viewing the very smallest constituents of matter. But when the physicist Ernst Abbe challenged the high hopes of optical science by imposing certain theoretical limits on optical resolution, all these hopes seemed to dissolve. Abbe claimed that optical resolution depended entirely upon incident light wavelengths, the limit being one-third of the light wavelength used to illuminate the specimen. According to Abbe, the extreme ultraviolet light of 0.4 microns wavelength could not be used to resolve the details of objects smaller than .15 microns.

This theoretical "death-knell" discouraged most optical designers of the time. Since, he claimed, resolution of optical microscopes was restricted from 1600 to 2500 diameters, developing newer optical microscopes was a futile pursuit. Since resolution is the ability of a magnifying instrument to identify details and ultrafine levels of internal structure, the Abbe limit imposed a serious halt on the development of newer optical microscopes.

Continual medical progress rides entirely on the excellence of its instrumentalities. In the absence of new and excellent optical instruments of greater precision, medical progress grinds to a screaming halt. When this happens, academes write papers in the absence of true vision. True knowledge, reliant on vision and experiment, is replaced by unfounded speculation.

Others conceived of electron microscope designs, taking advantage of the Abbe restriction for lucrative purposes. These developers were not good planners, failing to recognize that electron microscopy would place equally grave limitations on biological researchers. Electron beams kill living matter. Magnifying images only after killing them, no living thing could ever be observed in natural stages of activity through electron microscopy. But, if money was to be made, then "all was possible". Despite the protests of qualified medical personnel, RCA continued its development with Zworykin at their helm.

Electron microscopy, rationally impelled by the Abbe limit, became the new quest of young financiers. Despite the protests of major researchers, RCA continued its propaganda campaigns. This technological imposition, were it developed into a marketable product line, would severely handicap the work of every medical researcher. Pathologists would be literally forced to accept the limitations of the anticipated electron microscope.

Bracing themselves for the announcement of mass-produced electron microscopes, corporate researchers prepared themselves for the laboratory adaptations they would be forced to adopt. Manuals were already being distributed. They would be unable to watch progressive activities in the boasted "highest magnifications ever achieved". Before RCA reached the goal however, others had already challenged the capabilities of electron microscopy. The unexpected development temporarily threw RCA off balance. The competitors had challenged the Abbe limit, and seemed to be optically working their way into realms in which RCA had claimed "exclusive" rights.

DEEP VIOLET

Vibrating above the deep ultraviolet range were the X-Rays of von Laue's projection microscope. But this realm was not good for pathologists since X-Rays would only reveal the structure of crystalline substances. Some designers went ahead and built soft X-Ray microscopes. These devices placed heavy requirements on the preparation of specimens. X-Rays passed right through specimens and would kill them if they were alive to begin with. The very best X-Ray images of tiny specimens required organism-killing metallic stains. Biologists needed to see their specimens in the living state.

While engineers at RCA were yet scrambling to take the competitive edge and seize the new market, several designers of ultra-microscopes began to successfully challenge the Abbe limit. Abbe stated that the maximum resolving power of any ultraviolet ray microscope would be restricted from 2500 to 5000 diameters and no further. But ultra-microscopes constructed by Graton and Dane (Harvard University) succeeded in developing resolutions of 6000 diameters with magnifications of 50,000 diameters.

Dr. Francis Lucas of Bell Telephone Labs developed a modified version of this system in which a maximum magnification of 60,000 diameters was developed. Not only did this work significantly reduce the theoretical limits set by Abbe, but the ultra-microscope which Dr. Lucas designed actually empowered Bell Labs to compete with RCA in the microscope field. Dr. Rife had previously achieved resolutions of 6000 diameters with resolutions of 50,000 diameters. And now, Dr. Rife believed he had a means by which these preliminary feats could be greatly outperformed. The Abbe Limit, a theoretically perfect expression, was dissolving before the new empirical evidence.

Of course, RCA ultimately outdid the propaganda campaign for their own electron microscope system, wiping out the optical systems of both Bell Labs and Harvard University. Nevertheless, independent researchers preferred these ultraviolet microscopes to any system which RCA could market. Attractive because the ultraviolet microscopes permitted life-active observations, pathologists were not impressed by the extra magnifications of electron microscopy.

OBJECTIVE

Ultraviolet light for ultra-microscopes is an absolute necessity. The successful operation of any such device depends on deep UV rays. Monochromatic ultraviolet sources prevented many of the familiar optical aberrations common to optical microscopy. Blurring and fringe degeneration when passing through the optical resistance of lenses would be minimized. The ultraviolet source would also need to be of the shortest possible wavelength in order to approach the geometric ray ideal.

All optical components in the ultra-microscope would then have to be composed of pure quartz crystal in order to flawlessly transduce the deep ultraviolet rays. Even the specimen slides were made of thin quartz glass. The ultra-microscopes of Dane, Graton, and Lucas used as few lenses as possible, being virtually pure projection microscopes.

According to Dr. Lucas, resolution one-tenth of the illuminating light wavelength was obtained. This broke the so-called Abbe optical restrictions by an order of 300 percent; the resolution being brought up to .05 microns. How was this possible? Drs. Dane and Graton further stated that far greater resolution could be obtained through lenses than claimed by their manufacturers. The reason for this? So long as the manufacturers had accepted the theoretical limits there was no incentive toward progress in the field. No one bothered to find out!

The ultra-microscopes demonstrated beyond question that lenses do in fact surpass theoretical limits. The manufacturers, eager to maintain credibility in the academies, had simply endorsed whatever the physicists wrote. Equally as significant was the fact that each of these ultra-microscopes did not require the fixation of specimens before viewing. The embodiment of each ultra-microscope gave new drive to researchers who wished to see live pathological stages in tissue cultures. The systems were immediately demanded and obtained by numerous serious research institutes on both sides of the Atlantic.

Certain highly respected researchers came to believe that the most basic laws concerning physical light were fundamentally flawed. Perhaps light was of an entirely different nature than supposed. This, they mentioned, was why the Abbe limit was such a distorted mathematical expression. Light was not what the physicists declared it to be. This is why Abbe's assessment was so obviously flawed. But what other assumed truths were holding back fresh discovery? Empirical observations now replaced the theoretical piles with discoveries which were once termed "unlikely" by qualified authorities.

When researchers realized the great cost which the Abbe limit had so long imposed on microscope designers, they began challenging every known theoretical limit which pertained to their fields of study. Every scientific premise was questioned during the astounding decade of the 1930's. Every applicable optical rule was again subject to fresh questioning, the epitome of renewed scientific mind. New vision filled the researchers, challenging the inertial world again. The most significant effect of these new ultra microscopes was a renewed questioning process. Now also pathologists and biologists alike were given instruments with which to peer into the most infinitesimal natural recesses.

With the ability of medical researchers to peer into the deepest pathogenic lairs, new cures for ancient maladies could be effected. The war was on, and fresh crusades came to the battlefield armed with light. Curiously, the lines of battle brought two distinct groups to fight the same foe. Unfortunately, one group desired all the glory and crushed its more sensitive brother.

Rockefeller Institute extended their campaign by highlighting the efficacy of electron microscopy, securing the sale of their new units. The RCA cash flow was unrestricted now. Electron microscopy coupled its forces with the pharmacological industry, producing its line of allopathic medicines. Those who took upon themselves the inquisitorial profession, rather than the profession of truth, found themselves drowning in seas of new developments which their business-minded patrons wished to eradicate. Independent university researchers maintained their poise as the prime recipients of fresh and astounding discoveries which shook the medical world. This would not long be tolerated by the growing pharmaceutical monopolies and trusts who wanted total domination of the field.

MAGNIFICAT

The encroaching economic depression of the time period had crushed the general populace. Dr. Rife had been designing and assembling ultraviolet projection microscopes of superior quality from 1920 onward. He had planned to build a far superior instrument. The super microscope. The design was based on theoretical considerations developed during his preliminary experimentation in optics. Now this work was abruptly terminated. Finding himself out of employment, Dr. Rife sought the ordinary work of those who are in need. Humbled and not proud, he sought a salary in less intellectual venues for the time being.

Hired as private chauffeur by H.H. Timkin, a wealthy and philanthropic motor magnate, he gradually won both the respect and willing ear of his adventurous employer. He could not keep his wonderful dream to himself. On long journeys to boring board room meetings, Timkin engaged Dr. Rife in detailed discussions on his medical work. Dr. Rife eagerly entered these discussions with an enthralled candor which caught his employer quite by surprise. The seriousness and integrity of the man did not catch Timkin by surprise. He recognized quality when he saw it, and listened.

The man's great stature was not hidden, despite his humbled position. But when he spoke of these designs and research goals, the very air began to brighten around him! He mentioned regret at having to postpone his work, but was very sure that all would turn out well. What he had shared was enormous. Inspiration of the purest kind. When Timkin and his business partner Bridges realized exactly what Dr. Rife had hoped to achieve, they made a resolute decision to arrange financial support for the work at hand.

Timkin and Bridges created a endowment fund to finance Dr. Rife and his astounding research. Rife was delighted. Delighted to tears. An emotional man, he promised that no one would be disappointed. He would work until success. A laboratory was constructed on the Timkin estate grounds, (Point Loma, California) and Dr. Rife set to work with a fury which surprised those who lovingly surrounded him. Timkin and Bridges were taken aback by the rapidity with which Dr. Rife completed each design which he began. His efforts were relentless, a true inspiration to the equally loving people who supported his research. It was very apparent that the pensive and gentle doctor was serious in the extreme.

Dr. Rife aggressively pursued and achieved what had not been done in the field of ultra-microscopy. His mind had turned over the method which he had conceived so many years before. The dream which Dr. Rife originally received was now in view. Looking for more light. He decided to try filling the entire objective with cylindrically cut quartz prisms. There would be no difference in refractive index from start to finish along the optical path. Quartz prisms would "open out" each ray convergence, maintaining strictly parallel ray cadence. An increased ray content being thus returned to the ocular, the image would be brilliant in appearance and of high resolution.

This configuration of quartz prisms caused the rays to "zig-zag" in 22 light bends. The internal optical path was now entirely composed of 22 quartz blocks, fitted snugly to lenses. It was as if the entire device were one solid crystal of diverse surfaces. Now, specimen emergent light would launch out in parallel paths through quartz prisms, being magnified only when they reached each quartz lens. This optical tracking method would insure the brilliance of the emergent image.

A second optical innovation was added to this brilliant configuration. Dr. Rife decided to use a phenomenon by which strong specimen-entrant light stimulates internal fluorescence in the specimen. Pumping the specimen with brilliant ultraviolet-rich light would shift the divergence point into the very heart of the specimen rather than beneath, forcing the specimen to radiate its own brilliant ultraviolet rays.

Here was a true, vanishingly small radiant source with which to illuminate the specimens: they themselves would become the radiant source! This concept was truly sublime, since the very infinitesimal particles themselves were now made to radiate brilliant and divergent rays. This scheme was truly original from the very start. Dr. Rife then designed a system by which selected portions of the ultraviolet spectrum could be split and directed into the specimen using a polarizer. Turning this component of the system would allow each specimen to brightly fluoresce in its own absorption spectrum, the infinitesimal specks radiating their own maximum brilliance.

Theoretically, it was possible to magnify these brilliant specular rays to any degree. But a secondary monochromatic ultraviolet ray would perform an unheard wonder. When combined with the brilliant internal fluorescence of the specimen, this secondary ultraviolet addition would heterodyne the light. This meant that light pitches from the specimen would be raised far above its original values. At such shorter wavelengths, the resolving power of this device would be incredible.

An additional monochromatic deep ultraviolet beam was mixed with the fluorescent radiance of specimens, producing an astounding visual sharpness of otherwise invisible objects. The illumination scheme and the tube-filled cluster of quartz prisms (designed to maintain the specimen emergent rays in absolute parallels) were now brought together. Dr. Rife claimed that these parallel lines were within one wavelength of accuracy, an astounding claim.

He soon created a small ultra-microscope whose fundamental mode of operation violated the supposed laws of optics. This design outperformed all previous ultramicroscopes. So astonishing was this feat that the Franklin Institute, in rare form, published a long and detailed series of articles concerning the developments of Dr. Rife. They were also given several of these units for preservation, where they remain to this day.

This microscope was different, totally different. This microscope revealed not just viruses in their dormancy. This microscopes could see viruses in their active stages with magnified clarity. Dr. Rife's Prismatic Microscope surpassed the theoretical limits which were possible for optical microscopy in 1930, giving unheard resolutions of 17,000 diameters; three times the resolution developed by Dr. Lucas.

The first Prismatic Microscope was a horizontal optical bench assembly, mounted on a massive pier. Fitted with the finest photographic instruments, Dr. Rife took breathtaking photographs at unheard magnifications. The resolution was so staggering that research institutes rushed to watch Dr. Rife's demonstrations.

His accomplishments were extolled by the entire medical establishment on both sides of the Atlantic. An incredible amount of professional research publications devoted lengthy articles to his achievements. His findings were duplicated and reported by leading medical institutes whose names are well known. Therefore our general lack of knowledge concerning his life story is equally conspicuous.

Dr. Rife, humble enough to have worked as a chauffeur, had been raised from obscurity to fame...from shadows to light. The man's genius was only equalled by the upstanding character by which he was loved. The Timkin family adored him. The Rife laboratory was completely equipped with the very finest apparatus money could buy. Dr. Rife methodically designed new research ventures. Incredible new biological discoveries followed him in every direction. Now, with this "ultra vision", he was able to peer with his colleagues into unheard dimensions. These discoveries quite often challenged accepted biological and medical notions.

Dr. Rife had in mind the creation of an Institute, in which he could train younger specialists in the operation of these wonderful ultra-microscopes. Mass production of the devices would be insured. They would become fixtures in every professional laboratory. Money was not the aim of this research. Monies were already secured. Dr. Rife had a singular goal, and demonstrated the passion associated with his quest.

He developed seven different models of this initial projection-type prismatic microscope in quick succession. The horizontal projection format was converted to a more compact vertical orientation, best serving the needs of pathologists and biologists in practical laboratory settings. Several of these wonderful Prismatic models may be seen in the various archival films and photographs taken in Dr. Rife's laboratories.

If the Rife Prismatic Microscopes outperformed every standard laboratory microscope, being able to discern and photograph virus particles in their active state, the Universal Microscope outdid all the former records. In 1933 the creation of the Universal Microscope afforded resolutions in an astounding excess of 31,000 diameters, with magnifications in excess of 60,000 diameters.

Using technically precise photographic enlargement techniques, he was able to provide 300,000 diameter magnifications. His calculations indicated that a ultra-optical projection microscope giving clarified magnifications of 250,000 diameters would be possible. After photographic enlargement, there would be no limit to the optical viewing power unleashed for researchers.

The Rule of Abbe was mentioned as a failed byword in Dr. Rife's laboratory. He had succeeded in breaking the "vision barrier". There are those whose familiarity with optics and attainable optical precisions state that the claimed magnification effects cannot be obtained with ordinary principles of light. Beyond simple optical parameters, other light energies become the more active in such devices. The focussing process is radionic in effect, utilizing the penetrating Od luminescence. The stimulation of special retinal modes releases the anomalous perception with its reported ultra-optical magnifications. Careful examination of the Rife Ultramicroscope reveals tubes filled with quartz prisms, identical in basic use as the patented radionic analyzers of T.G. Hieronymus (Lehr).

Viruses remained absolutely invisible to the eye when cultures were searched with the then-standard Zeiss darkfield (oil-immersion) microscope. Dr. Rife's Prismatic Microscopes were immediately obtained by Northwestern University Medical School, the Mayo Foundation, the British Laboratory of Tropical Medicine, and other equally prestigious research groups. These models produced magnification and resolution up to 18,000 diameters.

A space composed of brilliant light, where mind illuminating light merged with light in the eyes was now opened before him. Fields, all of light. The new vision would be unstoppable. No cloak of invisibility could protect the foe now. Soon everyone would see, and the armadas of death and shadow would be vanquished. The spoils of this war would flood humanity with indescribable treasure. Life and light would again be unleashed in a world where shadow and death had reigned far too long. The immense task of cataloguing viral pathogens had begun.

QUEST

With the new Prismatic Microscope models, both he and Dr. A.I. Kendall (Northwestern University Medical School) were able to observe, demonstrate, and photograph "filterable" pathogens (viruses) in 1931. Moreover, they were perhaps first to discern the transition of these bacilli from dormancy to activity over a specific period of time. Freshly made cultures were sampled at specific stages, revealing fixed periods of quiescence and activation.

An initial tissue substrate was prepared in which bacillus typhosus was cultured. After several days' growth, samples of this lethal culture were filtered through a fine triple zero Berkefeld "W" filter. This filtration process was repeated ten times. When viewed under the best available laboratory microscopes a turbidity was seen, but there appeared no organisms whatsoever.

Under the Rife Prismatic Microscope, polarizer adjusted, the bacilli in this sample fluoresced with a bright turquoise blue coloration. Two forms were observed, taking the researchers by surprise. Long, relatively clear and non-motile bacilli were found alongside a great population of free swimming ovoids, granules of high motility. The motile granules glowed in a self-fluorescent turquoise light at a magnification of 5000 diameters.

These motile forms were transferred to a second fresh substrate, and allowed to grow for days. The same filtration process was performed. When sampled randomly before the four day period, the filtered specimen revealed something remarkable. Dr. Rife and Kendall observed relatively quiescent clear containing bright turquoise ovoids at one end. The implication was enormous. Exact transition periods were thereafter determined with precision, the entire process photographed through special attachments designed by Dr. Rife. At specific intervals of activation, the clear bacilli were discharging the turquoise motile forms into the culture. These blue ovoids were the real cause of the disease. The long and clear bacilli were only hosts. Transitions back and forth (between clear host-dormancy and motile turquoise granules) were observed and reported in the professional journals. These findings were corroborated first by Dr. A. Foord, chief pathologist at Pasadena Hospital, and later confirmed and reported by Dr. E.C. Rosenow at the Mayo Foundation (1932). The Rife Prismatic Microscope was quickly earning its reputation.

Soon, other specimens were obtained and studied by the team. Active poliomyelitis cultures were studied, the virus successfully isolated, identified, and photographed in 1932 by Rife and Kendall. In these cultures the team recognized streptococcus and motile blue forms resembling typhosus. These last reports were immediately transmitted to the Mayo Foundation and duplicated by Dr. E. Rosenow. Dr. Karl Meyer (Director of the Hooper Foundation for Medical Research, University of California) came to the Rife Research Laboratories with Dr. Milbank Johnson, examining and corroborating the stated results. The impossible and anomalous became fact. Bacilli could act as virus carriers. Furthermore, poliomyelitis victims evidenced a startling degree of typhosus-like associated virus.

Frightening implications came when comparisons between the Prismatic Microscope and the Zeiss scopes were made. All of the previous studies made with Zeiss scopes returned negative results. Such reports flooded the literature. The filtrates had been maintaining their cloak of invisibility for years. Professionals, bereft of this clarified vision, were concocting numerous speculative explanations for the appearance of these disease states. The vacuum produced by lack of visible evidence was producing erroneous theories. Many highly qualified persons, in absence of the sight required to know better, steadfastly maintained that victims of certain diseases were suffering from internally developed conditions.

The Rife ultra-microscope was about to trigger a war on viruses. Because of the self-fluorescent "staining" method, Dr. Rife observed live specimens exclusively; a distinguishing feature of his technology. The fluorescent coloration of each pathogen was catalogued, an historic endeavor. Tuberculosis bacilli appeared emerald green, leprosy was ruby red, E.Coli were mahogany colored...each wickedly deceptive in their pretty colors. The degree of precision demonstrated in Dr. Rife's catalogues bears the unmistakable mark of genius. We can view him at work in the archival movies.

Photographic arrays of all kinds may be seen in this footage, including the professional Scandia 35 mm movie camera with which he made stop-action films of viral incubation periods. Dr. Rife made sure to document every discovery. It was novel at the time to document every image on movie film as well as in still shots. He methodically went through every possible pathogenic specimen, photographing the deadly families. Suddenly new viral species began appearing: non-catalogued species.

The prismatic microscope was piercing into new shadows. Dr. Rife recognized unknown virus species everywhere. And then he turned his vision into the deepest shadow. He looked at the dreaded disease. To this very day the very utterance of the disease is foul. It carries the nimbus of finality. Cancer. It is an arrogant boast, a victory over cringing humanity. All who speak its name whisper in fear, afraid that it will hear and come for them. Immigrants refused to even mention the name, fearfully crossing themselves...calling it "the evil sickness".

In the absence of fact, in the absence of vision, researchers developed contradictory theories concerning cancer and its development. These contradictory theories were eventually consolidated in the professional literature, a self-neutralizing amalgam of conjecture. Researchers were forced to examine the biochemical effects, and not the cause, of cancer. Most could not imagine what would drive cells into the bizarre and abnormal cycles common to cancer tissues. There was certainly "no visible cause".

Dr. Rife began obtaining a wide variety of malignant tissues in 1931. The full power range of the first Prismatic Microscope was turned on these tissue samples with a vengeance. Dr. Rife was a master pathologist. His techniques can be observed in his cinematic presentations. Was he seeing correctly? What were those motile forms, glowing with a beautiful violet-red coloration? He watched them for a long while. They moved swiftly through the field of view. Clocking their motility in triple distilled water, he watched them darting across the grating. These stretching ovoids moved with startling speed.

Dr. Rife obtained yet more and diverse tumors from wider and more diverse clinical sources. An amazing 20,000 of these tissue samples were obtained and cultured. Incubating and culturing each of these required care and time. Absolute sterile conditions were maintained. He employed several groups of large high pressure steam autoclaves. No question of contamination could exist in this setting. His methods may be surveyed in the archival films which show every room of his facility. Specimens, removed from these cultures were always filtered through unused triple zero Berkefeld porcelain, mixed with triple distilled water.

Examination of each separate sample under the Prismatic Microscope revealed a consistent truth. There they were again! Always the same violet-red presence. He called it the BX virus, finding it present in every case of cancer in humans. Were these same violet-red motile forms the very cause of cancer? They were always found in every sample, a deceptive beauty. Could this be the cursed stream? Had he alone been brought to see these first? Colleagues were able to verify these findings only when using his microscopes. Both Dr. Rife and Dr. Kendall successfully demonstrated the isolation of the BX virus to more than fifty research pathologists associated with the most noteworthy institutions.

Many writers of medical theory had already postulated that there were some cancer cases which were viral in origin, but they never cited these agencies as the universal cause of cancer. The speculation, the papers, the lectures, the theories. Talk and more talk. Rife saw the universal cause of cancer. There, in full sight lay the proof positive. In case after unmistakable case, Dr. Rife found the very same agency at work. Always the same violet-red motile forms. It mattered not where the tissue materials came from. There could be no mistake. There was no citing possible contaminations. Independent acquisition of tissue samples were obtained by others who then verified these findings in distant laboratories. They were using the Rife prismatic microscopes.

He succeeded in isolating the BX virus in 1931, filming the process so that posterity would hopefully learn of its enemy. He cultured this evil spawn and proceeded to demonstrate its incubation and activation periods. Transferring BX virus from culture to host, and from host to culture, all became routine. One hundred and four separate transfers were successfully made with various BX strains. Dr. Rife witnessed the appearance of another related viral cancer-causing strain, the BY virus, found to be a much larger strain of the sarcoma group. Demonstration of the infection and incubation process was subsequently affirmed by other professionals.

The same virus appeared in every case of cancer in humans. He assembled high speed movie cameras in order to clock the periods of BX virus activity. When the film ran out and was developed, he and all his colleagues could watch the deadly dance. He stepped back for a moment and surveyed the photographic evidence, flickering on the wall. Those wicked damned wriggling specks! From how many souls had they drawn away the life?

Dr. Rife now watched in horror as the malignant act was revealed before his eyes at high speed. BX virus infection required special "weakened" physiological states. Contracted as a flu-type infection, the virus incubates in host physiology for a time. When specific detrimental physiochemical states are compounded, the virus stirs into activity.

Stimulating the rapid proliferation of cell division, the BX virus forces the host body to manufacture needed nuclear material on which to further its survival. Tumors were found to be sites where BX viral colonies were rampant. Occasionally there were persons who demonstrated spontaneous remissions. These were exceedingly rare cases where antibodies actually drove off the attacking virus. Most persons could not summon this degree of response. Once the virus took control of cellular integrity, death was imminent. Shadows sweeping over humanity. There had to be a means for destroying this enemy. There had to be...a light.

SPEARS

Others, working in distant laboratories, did not claim the same success. Why had they not seen them? Because, using the over-celebrated electron microscopes, they could not see. The frightful truth concerning the BX virus was that electron microscopy could not image them at all. What had occurred in the other research labs became clear again to the man with eyes to see. The others overlooked this obvious pathogenic presence simply because their microscopes could never reveal it. This hideous specter exalted itself in what cover it could find. Unfortunately, it found cover among those who claimed to be professional seers.

Otherwise excellent researchers became completely blind when searching for the BX virus because electron microscopy was itself the blinding agent. How could so obvious a pathogen not been imaged in a technology which boasted greatest visual resolving powers? In preparing specimens for an electron micrograph, technicians "kill" the tissue specimens. The process involves placement of the specimen in a high vacuum chamber. Bombardment of the specimen with metal ions is the "staining" procedure. The thin metal film gives highly projective electrons a detailed surface upon which to impinge. The electron spray is directed into this prepared specimen and is then magnified by successive intense magnetic field coils. Images are then watched on a phosphorescent screen, or photographed directly.

Electron microscopy mishandles frail viruses. It mishandled the frail BX virus, destroying it during each preparation process. Destroyed evidence. The same ritual was repeated a hundred times with the same negative results. Unable to think clearly, few of these technicians could surmount the situation and comprehend why this virus did not appear on their viewing screens. Overconfidence in the RCA system blocked common reason. Electron microscopy does not resolve frail viruses because they are shattered and dissolved during the preparation stage.

Well then, there was the flaw. Why would no qualified person see this simple truth? Why was the light eluding those who claimed to have all of it? The technological marvel, designed to replace all competitive microscopes had brought a secure sleep on those supposed to resolve such obvious dilemmas. Medical technicians had forgotten how to think. Its newly adopted methods actually destroyed frail pathogens intended for study. Quite recently the search for the HIV virus evidenced frustration again because of these inherent limitations of electron microscopy.

The BX viruses cavorted and wriggled boldly before his eyes. But...how to destroy them? To find an immunological tool for each of these would represent an enormous task, a project which would take centuries. Humanity did not have that much time to wait. No, some other more universal means had to be developed by which this, and all pathogenic forms could be dissolved.

Protozoa and bacteria of all kinds could be destroyed by exposing them to special ultraviolet spectra. Perhaps the BX virus would succumb to such exposures. He had to know. He had the tool with which to see. So he began a long and arduous search, looking for spectra which could destroy virus cultures.

Dr. Rife discovered that deadly viruses actually thrived in the radiations of specific elements. Radium and Cobalt-60 were the notable ones. Dormant viruses became virulent in these energetic emanations. The horror filled him again. Medical practice was attempting the cure of cancer with these very radiations! There had to be some light spectra which destroyed the viral activity altogether. He search through the periodic table. Electrified argon and neon also brought intensified virulent activity from dormant viral cultures. He actually utilized argon lamps to grow virus infected tissue cultures with greater rapidity. But there had to be a spectral range which killed these terrible death-agents.

No light seemed to have any effect on their crystalline structures. This is why it was possible for him to view viral activity under intense light in the first place! No light spectra of any intensity was able to destroy these quasi-living crystals.

Then he thought of crystals. How could we destroy a crystal? What do chemicals do to germs...dissolve them, take them apart...shatter them?

He had done this very thing in 1917 with protozoa and large bacteria. He knew it was possible to shatter these kinds of pathogens by the application of a sudden electrical impulse. His early attempts with small radio transmitters and simpler microscopes proved somewhat effective. He used Telefunken output tubes to produce the impulses. Operated by a small generator, this simple device projected fifty radio frequency watts to his samples.

His original inspiration applied to larger pathogens. It therefore needed no excessive frequency, short wave being sufficient. It was certainly possible to interpolate the necessarily superhigh resonant pitch needed to shatter any microbe. But viruses? How high would this pitch need to be? If not attainable, could he use some much lower harmonic of this fundamental at greater power levels? Could he find the lethal pitch for every found pathogen?

Equipment was quickly assembled. He needed a generator of extremely short duration electro-impulses. Direct current electrical "spikes" of quick duration, when applied to a gas filled discharge tube, would project electric rays toward an infected sight. The tube could not be a simple high vacuum. That would release dangerously penetrating X-Rays. X-Rays would stimulate the BX strain into increased activity. No, the projection tube required a very light gas, one whose response was almost instantaneous. The gas he desired would be one whose mass would in no way interfered with the impulses.

Hydrogen was used in special high power thyratrons: quick acting high voltage switches used in diathermy machines and (later) in radar systems. Old X-Ray tubes often failed in their operation because they became filled with hydrogen and helium mixtures. Such X-Ray tubes were generally discarded. His new projector was one such old X-Ray tube. He tested its output, adjusting the excitation circuit so as not to release even soft X-Rays. The tube glowed, a good sign. This meant that there was sufficient gas for the release of electrical rays. Dr. Rife set the polarity so that the tube would pulsate in electropositive spikes of specific duration.

Power was ready. Pathogens cavorted boldly in view. Poised at the Prismatic Microscope, he fired the X-Ray tube. Turning the tuning dial near the specimen, he would know the lethal pitch by watching the pathogens. When these "exploded" he would mark the setting. If this method worked, then he could methodically correlate each lethal pitch with its pathogen. Soon, a catalogue of lethal pitches would be amassed. With this Dr. Rife could wage victorious wars against every disease in existence.

Dr. Rife swept through the diathermy range, which he calculated should vibrate these viruses to pieces. Empirical evidence always contradicts the theoretical. Quite below the calculated extreme frequencies, the BX virus suddenly dissolved. He switched off the transmitter and sat there quite amazed. The scene in the microscope was unreal. Not a fraction of a second at the lethal pitch and the specimen was reduced to a glomular mass. The viruses were stuck together in shattered fragments! He had successfully "devitalized" them.

Fine tuned lethal frequencies now filled his catalogue. With great precision Dr. Rife determined every lethal pitch as planned. Armaments of light against legions of shadow. Analysis of the electropositive impulse showed that its radiance was penetrating, intense, and unidirectional...more like invisible light rays of pure electric force. What then was this strange lightlike power? Experiment proved that virus cultures were absolutely incapacitated, congealed, and destroyed by the electropositive impulse. The power of an extreme form of light? Had such light ever been seen before?

This energy had been accidentally generated in 1872 by Thomson and Houston. Not waves, but rays. Electrical rays. A forgotten phenomenon. Unidirectional electric impulses of great power radiated electric rays, not waves. These rays penetrated all kinds of matter, whether stone and steel alike. The resultant sparks could be drawn from every insulated metal object in the large building in which the experiment was being performed. Not radiowaves, but electric rays.

Later in that century, Nikola Tesla accidentally observed the same electric ray production. He studied the phenomenon exclusively, developing impulse generators and electric ray projectors. When speaking of electric rays which evidenced a light-like nature he referred to this phenomenon. Not radiowaves, but electric rays. New light. Dr. Rife had rediscovered this phenomenon. Tesla spoke of his own "millimeter rays", mentioning their "bacteriocidal" value. This same phenomenon had vindicated Tesla's words. Therapeutic properties were demonstrated when precisely controlled.

FORTRESS

Whereas the destruction of virus cultures on a quartz slide was easily accomplished, the destruction of pathogen cultures in human hosts was not. Rays had to penetrate through skin, musculature, and bone; a considerable resistance through which to travel. Rays might lose their original accurate pitch in this transit, destroying the intended action altogether.

Fortuitous and strange, the pathogens were found to be some two thousand times weaker than body cells. This meant that pathogens could be destroyed by the radiant impulse method without harming the patient. How sublime. Pathologists had treated micro organisms as chemical systems for a century, working overtime in order to find each specific chemical dissolving agent. This method treated all germs as mechanical systems, dissolving them with vibrations.

He himself had been exposed to the instantaneous blast without harm. When adjusting the rates to annihilate ordinary viral infections, he noticed that he became drowsy and tired for a few hours. Determining the cause of this as the resultant toxin release after infective agents were coagulated, he recognized the need for a de-toxifying agent. Physiology had to be prepared for the curative impulse. Exposure would release large amounts of toxic pathogen fragments into the bloodstream all at once. The ray cure had to be metered in doses. Body tissues had to flooded with special fluid electrolytes to aid the enhanced and rapid elimination of these toxins.

To stimulate deepest shattering action, the patient had to be bathed in a "carrier field": an electrical body permeation in which the impulse light rays could penetrate into and through every body cavity. Superficial exposures would not completely cure the patient. This light ray energy had to permeate the body completely. Dr. Rife conceived of method whereby patients could be enveloped in a harmless body-permeating electric field of acoustic frequency, while the intense electro-impulses of short duration would be simultaneously projected. In this manner, efficacious electro-radiant impulses could shatter specific pathogens throughout the infected body with no harm to the patient.

Dr. Rife utilized two banks of oscillators with which to generate his primary and secondary impulse fields. Acoustic generators supplied the primary field of "immersion". A diathermy machine was coupled to a powerful transmitting amplifier to provide the shattering impulse. Two radiant energies were thus employed to destroy pathogens in vivo. Dr. Rife's catalogue of lethal rates always gives a pair of lethal frequencies per pathogen.

Dr. Rife discovered that virus cultures were not safe from the radiant impulses from the special ray tube. Fixed to the lethal pitch of a single pathogen, the rays were unerring in their message. Selectivity was the hallmark of the Rife curative method. Several pathogens could be assembled adjacent to one another. Choosing the lethal pitch for one of these, the others would remain unharmed. The target, however, was utterly destroyed.

Dr. Rife tested the lethal effective distance of his rays, determining the safe placement of patients from the radiant source. Pathogen cultures did not seem "safe" anywhere near the device at all. Arranging the tube at one end of his laboratory, Dr. Rife brought cultures out to increasing distances from the radiant tube. In a final amazing experiment, he took cultures away from the laboratory in sealed containers. It was found that radiant tube emanations operated effectively on viral cultures up to an eight mile distance! Metal cabinets did not protect viral cultures from the deadly ray effects either, being ray conductive. Even when locked in aluminum cabinets, the entuned light-like rays destroyed their pathogens wherever found.

This represented a major medical discovery of greatest value to all humanity. This principle actually made possible curative broadcasts. Entire populations could be electrically "vaccinated" from single monitored sites. The world potential of this system was staggering. Now the outbreak of epidemics could be controlled without the time-consuming need for individual inoculations. The radiant lethal message would eradicate specific pathogens in several simple broadcasts. The constant monitoring of socially prolific germ populations could be maintained by continual public health "broadcasts".

CONQUEST

He ran his entire staff through varied frequency exposures. Infections of all kinds each dissolved before The Ray. Dr. Rife was able to isolate the pathogens of infection and destroy them with the mere turn of a dial. The specificity of the raytube device was so precise that singular germ strains could be individually mass-targeted. Cured by the flick of a switch!

Firing the tube in the lab provided a continual source of inoculation. After a time, so little toxicity was present in staff members' bodies that the drowsy effects were never again encountered. They did not contract any illnesses. Not even colds.

After a time, Dr. Rife rarely used gloves when handling the viral specimens. Furthermore, neither he nor his technicians ever contracted any of the diseases which were handled. The Raytube "inoculated" them all against every disease. He reported these findings to the community, while himself remaining the designer and developer of the system.

Dr. Rife, a research pathologist, never used these devices in medical practice. Other physicians desired the units for their own purposes, recognizing the potential for curing human suffering. Dr. Lee De Forest supervised the design and assembly of many oscillator components for the Rife System. W.D. Coolidge himself (General Electric) willingly sent Dr. Rife hundreds of X-Ray tubes which were altered with a mixture of hydrogen and helium by Rife and his technicians. These improved tubes were tested so that they would project only the desired electro-impulse rays. These noteworthy references best recommended the Rife Raytube System to medical practitioners of the day.

Hearing of these wonders, numerous physicians began requesting that smaller, more portable units be designed. Soon, Rife Raytube devices were being assembled and given to physicians for limited use in their own practice. When properly operated, these devices returned successful reports, effecting complete eradications of infections and cures of various conditions.

There were never any adverse reports concerning the Rife Raytube Instruments. Neither could there be. Rated at such safe peak performance levels, no harm could possibly come from the portable devices.

The careful and reasonable monitoring of patient progress, the Rife frequency devices were bringing about a therapy revolution. Strep throat could be cured in an instantaneous exposure, seated in a physician's office. A specially designed gargling solution was given to remove the resultant toxicity from the site.

In 1934 Dr. Milbank Johnson, Chief Medical Director of Pacific Mutual Life Insurance Company, established a therapy center for cancer treatment in Scripps Castle, San Diego. A staff was brought together from specific institutions including Dr. G. Dock (Professor of Medicine, Tulane University), Dr. C. Fischer (Children's Hospital, N.Y.), Dr. W. Morrison (Chief Surgeon, Santa Fe Railway), Dr. R. Lounsberry, Dr. E. Copp, Dr. T. Burger, Dr. J. Heitger, Dr. O.C. Grunner (Archibald Cancer Research Committee, McGill University), Dr. E.C. Rosenow (Mayo Clinic). Dr. Rife functioned as a general consultant in matters of system therapy.

Using a Rife Raytube system, the team received cancer and tuberculosis patients. Fifteen cancer patients, each pronounced hopeless by medical experts, arrived at the clinic. Each evidenced progressive states of the disease. A few patients were ambulatory. Treatments with the Rife Raytube method were routinely applied. The dream was becoming real. Humanity was at last receiving its help.

Recognizing the critical condition of their patients, it was decided that exposure time would be raised to three minutes duration. It was discovered that exposures could not be repeated daily without necessary long rest periods. These critically ill patients could not withstand the extreme resultant toxicity released into the system as BX viruses were shattered. Emotional depression often resulted until the ray-dose was safely assessed. The team conferred hourly to assess the progress of each patient. Excessive exposure to the rays could result in severe lymphatic infections and blood poisoning. Therefore three minute treatments were repeated every third day, the rest periods necessary for blood detoxification.

Soon, the ray had done its work on the once-terminal victims. Constant blood and tissue samples revealed no BX viral presence in these now fortunate individuals. In sixty days' treatment time, and after examination by several physicians, each was released as cured.

Though under continual surveillance, no relapses occurred. The treatment was revolutionary. The results, thrilling and complete. Moreover, they were confirmed by a special medical research committee of the University of Southern California. Three more clinics were opened with Dr. Johnson as General Medical Supervisor. Other participating physicians included Dr. James

Couche, Dr. Arthur Yale, Dr. R. Haimer, Dr. R. Stafford with a mounting number of participating physicians. Clinics were operated between 1934 and 1938 having such a number of cures that it is difficult to list them all without simply reprinting the Rife files. Each of these cases were sent out and corroborated by other (non-participating) physicians.

In 1939 Dr. Rife was formally invited to address the Royal Society of Medicine, which had recently corroborated his findings. He was requested to bring all possible films, slides, and apparatus with great enthusiasm. Dr. R. Seidel reported these findings and formally announced the Rife Raytube System therapy for cancer in the Journal of the Franklin Institute. (Vol. 237 no.2 February 1944).

The formation of the Ray Beam Tube Corporation was announced, through which several models would become available to the medical world within a short time. Highly skilled hospital staff members and leading physicians were very receptive to the proliferation of this therapy. Here was a new means for controlling and eradicating any kind of disease by the press of a switch. This therapy would inadvertently challenge pharmacological methods, raising human standards to a new and lofty height. The dream seemed ready to materialize.

INQUISITION

Rife found both himself and his staff members under a strange series of attacks by unknown agencies. During this time, and under very mysterious circumstances, Dr. Johnson died in a hospital bed. Brought there for completely minor reasons, he was found in bed. The local chapter of the Medical Association proceeded to bring Dr. Rife to the San Diego Supreme Court, but lost their case (1939). Dr. Rife could not be charged with malpractice, being a research pathologist and designer of medical instruments.

This repugnant offense unmasked the heinous resentment behind which many powerful individuals had previously been camouflaged. The court action itself caught Dr. Rife quite unaware. A visionary, his entire life had been dedicated to humanity. Alleviating human suffering was his life theme. Here now was strong evidence that factions within the Medical Establishment were actually mobilizing against proven therapeutic methods. Cancer itself and other equivalent maladies were being cured. Why then the assault?

Growing opposition from deeper factions of the Medical Association brought pressure on Rife Treatment clinic staff members. Threats and other unprofessional pressure tactics forced members to leave the team in quick succession. In campaigns clearly waged to malign Rife and his findings, the Medical Association assailed remaining participants in the clinics until Dr. Rife stood alone.

Deeper than the verbal show of malignancy by other colleagues was the horrifying and insidious motivation, the implication behind the attack. Why would anyone wish to destroy so great a world-advancement? Who was betraying civilization in this critical instance? Of all betrayals and of all personnel, who in the Medical Profession would seek the eradication of such monumental discoveries? Dr. Rife's mind reeled under the weight of these thoughts. This was not mere resistance to a new idea in a time of ignorance. Pasteur experienced that indignity. No. This was a willful, calculated resistance in a supposed enlightened time.

Horribly shocked at the entire scenario, Dr. Rife literally became unhinged in court. Trembling and weeping, he could not come to terms with the sheer hatred and vehemence exhibited by his antagonists. "Why...why are you doing this?" he repeated. The prosecution could not have produced a better effect. Seeing this weakness as the very means by which to eradicate Rife and his discoveries, they continued to attack Dr. Rife openly. Calling him continually to the witness stand, they succeeded in destroying this frail hearted man of humble greatness. In short, the prosecution forced his total collapse.

Dr. Couche was compelled to desist operating Rife Therapy clinics under threat of malpractice. The Medical Association ruled that no society member who maintained use of the Rife Raytube System would be permitted to continue medical practice in the United States. Morris Fishbein, major AMA stockholder, treasurer, censor, editor, and controller extended his legal arm to inform each member of the Rife team of the impending legal process. All Raytube units would be recalled, impounded, and destroyed by Federal Court order, under penalty of fines and imprisonment.

LIGHT

All participants willingly returned their Rife units except Dr. Couche and Dr. Yale. These two surgeons later stated that for twenty-two years after this action, they continued to successfully treat and cure thousands of patients with the Rife Raytube devices which they secretly maintained. Dr. Yale published a large and concise chronological account of patients treated and cured in his practice throughout that twenty-two year period. Notwithstanding the fact that sixty percent of severe (cancer) cases brought him were medically inoperative, incurable, and hopeless, Dr. Yale confirmed that all of these persons were yet alive and living happy, full lives.

The Rife Microscopes challenged RCA and its lucrative electron microscopes. The Rife Raytube System would eradicate the accepted lucrative pharmacological methods everywhere. Such developments did not inspire challenged corporations. Dr. Rife developed a therapeutic means which works. This is all too evident by the rage of those who assailed him.

Systematic eradications of this priority level speak of social control on a vast and hideously deep-rooted scale. Implications necessarily involve corporate trusts and governmental agencies. The notion that disease proliferation is permitted for the continuance of pharmaceutical interests is too terrible to reasonably consider. Federal Officers came to impound the entire Rife laboratory all too late. Several faithful technicians had already purloined every piece of the priceless equipment, taking laboratory components and valuable documents across the Mexican border where they yet remain. John Crane maintains the priceless surplus.

Fishbein, the editor and chief censor of the AMA saw that Rife's name would be stricken from all previous publications, that no professional journal would dare publish anything by Rife, and that no mention would ever be made of Rife's achievements in formal proceedings. Inescapably linked with the pharmaceutical trusts, Fishbein's actions were all too conspicuous.

Social control has become a dominant theme since the second World War. Modifying and regulating social thought through both legal and financial steerage has brought natural discovery and true technological development to a standstill. World changing discoveries can be made but not proliferated. Cures for diseases can be proven, but not implemented.

Has the world now entered a new barbaric and vulgar time where medical wonders have become a regulated property? The historical evidence proves out these thought lines. Balancing profit against cost, it is clear that outright cures are far less profitable than exceedingly prolonged and profit-effective "treatments". Statistical analysis of social "disease incidence" mark the yearly expected gross earnings, a profit margin of untabulated measure.

Would the honor once laid upon the development of wondrous disease cures now be shunned, the cures themselves being suppressed at will by business managers? Would compassion for suffering humanity, concern for the elevation of human living standards on a worldwide scale no longer be a major medical theme?

World Society is driven by the unmodified flow of natural scientific discovery. At the fundamental level, such discoveries are truly socio-providential. While previous epochs simply endorsed and socialized each new natural discovery, newer attitudes have suffused the world from financial "sites of infection".

In the past, medical discoveries were never questioned or resisted. They were always looked upon as absolutes: if a medical cure for disease was found, it was taken as it truly is...a miraculous providence. Not even the most ruthless financier would dare interrupt the flow of medical discoveries in past times. This state of ethical acumen has not continually been honored.

When the records are actually examined, when the billions of research dollars have been computed and balanced against the true research effectiveness, we find a staggering disproportion. How is it that medical research of the nineteenth century, far less equipped and funded, produced definitive cures which have become medical "standards"; while contemporary medical research, best equipped and super-funded, has not produced a single cure of equal social importance in the last thirty years? Dr. Rife had the answer toward eradicating all virus potentials. Perhaps, because it was not a pharmacological one, his devices have been "legally restrained" from social proliferation.

A few moments' calculation reveals the effective ability of research to find a chemotherapeutic vaccine against each deadly virus. The calculated time exceeds several millennia. But Rife found the only reasonable technique for destroying any virus infection at will. The answer was not a pharmacological one. Eradication of his techniques at this early stage of development would be reasonable if one were heavily invested in chemotherapies. The systematic eradication of many such (recorded) cures is revealing.

Medical authorities have stated that "no means has been found by which viruses may be destroyed". Recent evaluations of "recaptured" Rife Raytube units contradict this statement. Dr. Rife treated germs as mechanical systems, not chemical systems. Vibration killed pathogens by the flick of a switch. A single such device could be easily tuned to destroy all deadly pathogens. His is the only device which can destroy viruses.

UCLA Medical Laboratories, Kalbfeld Lab, Palo Alto detection Laboratory, and San Diego testing Lab all had stated that the Raytube System is absolutely safe to use. The FDA went out of its way to publish and maintain Federally directed rulings on the Rife Raytube System, refusing to make further statements concerning its historically proven effectiveness in thousands of cured cancer cases.

A great gathering of esteemed colleagues of the medical and research professions came to honor and support Dr. Rife after the entire court affair. Friends who were too frightened to stand and fight at his side were now smiling, drinks elevated. But the man who was asked to stand and receive honor saw through the charades.

The seer saw the thick shadows which enveloped the professionals and other dinner guests. Armadas of pathogens were drumming their war drums again. Soon on the march, they would devastate humanity once again. It seemed that not one of the esteemed guests cared. The Rife Raytube Therapy was the only time in history that viruses could be selectively and dynamically destroyed. No chemotherapeutic agencies were ever required in the process. The mere closing of a switch could achieve these undreamed wonders.

Dr. Rife had developed and implemented what no contemporary medical research group has ever conceived. And, by the end of World War II, was prevented from ever doing so again on American ground. The cheers and accolades rang on, while standing ovations lasted for more than fifteen minutes. The now frail and ghostlike discoverer looked away. Far off and away. Searching through the shadows, searching in his own darkness...for new light.

Nikola Tesla

Sr. Antonio Meucci

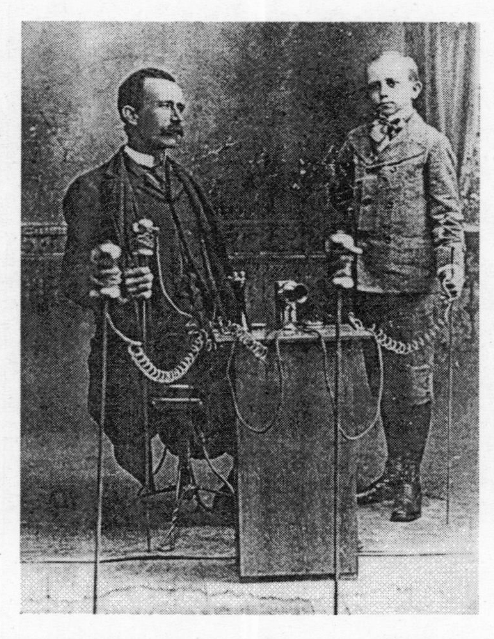

Nathan B. Stubblefield (with son Bernard)

Dr. Royal R. Rife with his microscope

Dr. T. Henry Moray

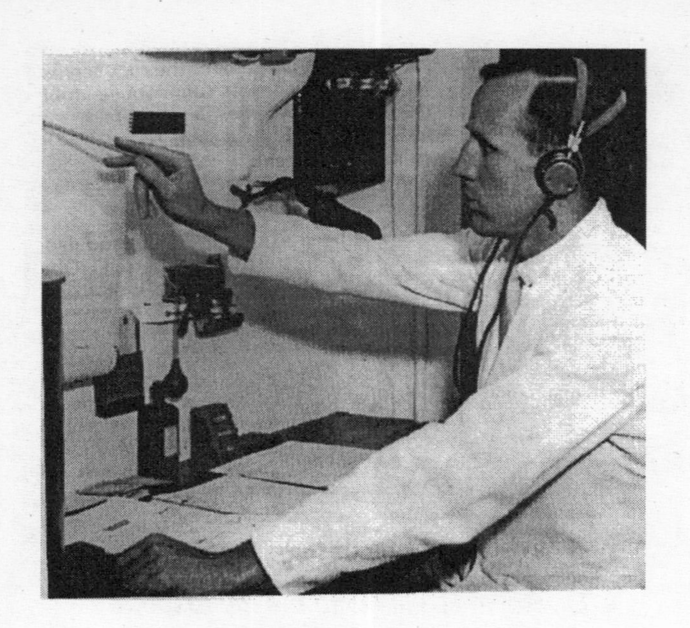

T. Townsend Brown

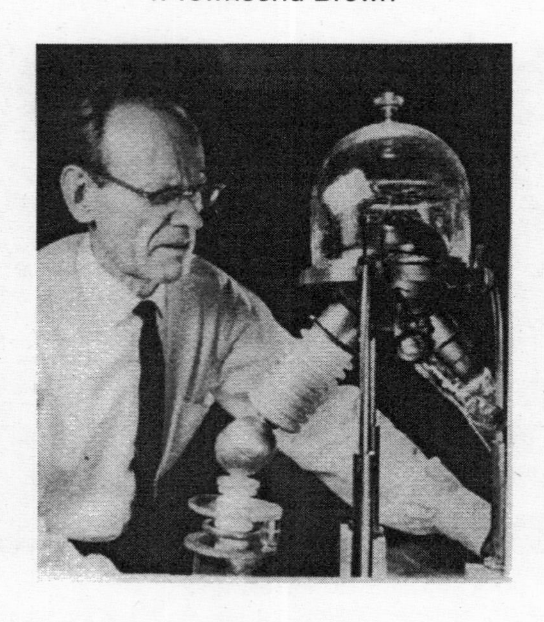

Philo T. Farnsworth with Fusor Apparatus

CHAPTER 6

ENDLESS LIGHT: DR. THOMAS HENRY MORAY

DREAM SEA

This wonderful biography requires considerable examination of archetypes, their suffusive power, and the world in which they materialize. To comprehend the power and import of what Dr. Thomas Henry Moray truly discovered, we must pass through the waters of time most arcane, until we emerge fully prepared to appreciate the wonder of his find. Why this is so will become most apparent as we progress through the narrative. His story begins, strangely enough, in the sea of dreams and archetypes.

Histories, dreams, sagas. Told by ancient philosophers and bards, the great epics contained dim recollections from an archaic world. Singing the glories of lost kingdoms and lost civilizations, poets transported their listeners into another age. Hearing rhythmic elaborations of verse, ancient audiences glimpsed enchanted visions of towering mountains, fabled citadels, and magnificent cities. The bards brought them into the very portals of The Golden Age. Utopia. After hearing of them, the mysterious lost worlds seemed easy to reach.

Poets, bards, and historians provoked their listeners with symbols, images, and themes. Lives were always filled with a special sense of clarity, purpose, and direction after the bard sang. The telling of these long sagas provoked great inward excitement, stimulating aetherial hope and dream quest. In strong evocations of rhyme and verse, of mythos and pathos, individuals were infused with new and impassioned desire. The sagas remain the engineworks of civilization.

Shared archetypes activate human hearts toward specific desire. In the mouth of bards, each archaic saga initiated great and historically memorable human works. Lovers sought pure love, adventurers sought ultimate adventure, and travellers journeyed forth for unknown horizons. Seekers all. Seekers after the ideal world and its wonders.

The long entwining themes were set with little jewels. Unfamiliar wonders suffused these worlds. Exotic and mysterious artifacts. Familiar accoutrements and accessories of fortunate Utopian dwellers. Invariably, the marvelous inhabitants of lost worlds employed strange and magickal technologies. Hearing these legends, one was always inspired to recapture or reproduce the magickal artifacts and accessories described by the bards.

Enthralled with themes of love and valor, the beauty of silvery princesses and strength of bronzed heroes, young listeners sought to emulate the heroes. Adventures, excitement, thematic focus and purpose. Life was explained in the song. These wonders, so very evocative, are yet difficult to comprehend unless they were in some way real. It was known that many poets were once themselves noblemen of high degree. Educated and knowledgeable. Until they themselves were changed by the song of former bards, they knew lives of luxury and riches. The sagas drew these men away from the ease of villas and marbled estates with their irresistible power. More aged bards and troubadours had passed to them their precious treasuries of verse and epic, until the heart of legend and myth found new strength. Transferred among truth seekers, glorious mythos survived the centuries. Mythic credibility finds an essential heart of truth, being drawn of histories and arcane priestly records from distant lands.

On sapphire blue-splashed sea coasts, the naturally radiant and great ascending mountains of violet and green founded the worlds in which Utopian lovers dwelled. They were the royal originators of civilization, immortals whose goodness and wisdom was boundlessly expressed in mighty civilizations of archaic splendor. Cloud-piercing towers of silver, dazzling beacons of pure white light, and fabled harbors where golden pillars heralded immortal seafarers home signalled the approach toward wonder.

Mountain terraces, ringed by strange lightning mirrors, pierced the air with deadly star blue rays in defense of the citadel. Peace and respect reigned supreme. Large emerald green crystals worked in the heavens, bringing blue and sweet perfumed winds. Distant isles were traversed beneath the waves in dolphin-like vessels. Golden framed mirrors of Mercury, gifts of unknown age and origin, conveyed travellers among disconnected silver terraces set like jewels in the sea mountain world.

Wide plazas of white stone, pyramidal temples of crystal, floral avenues, gardens with soft green pools and waterfalls. All were suffused with the topaz yellow radiance of lanterns which never extinguished. None were ever ill in these Utopian lands, except for the illness of love. All lived for centuries. Special elixirs and violet lamps blessed young and old alike, bringing health, wholeness, and joy to their recipients. Large shields of onyx, set in unknown metals, permitted one to see across the waves to distant amethystine islands. Copper dialettes to show the time were everywhere, controlling the very hour. Mighty artisans directed twisted pipes of tin, magickal winds shrieking and lifting great stones through the air.

And the wonderful people who inhabited these lands were always persons of nobility, enlarged minds, virtuous and wise, living full lives of joy. Their persons were glorious, radiant, possessed of rare abilities. They seemed always adorned by desirable magickal accessories; rings, buckles, bracelets, necklaces, jewelled brooches, belts, capes, helmets, shields, swords...all the tools of magickal technology.

Their homes were wonderful and rich, flooded with fine fare and luxurious furnishings having magickal aspect. Their nights were flooded with the wonder of stars, planets, and radiance; the radiance of rare lanterns and hearths which never consumed fuel or ceased pouring forth their effluves. Utopia was never more clearly portrayed.

By rhythms of small drums and soft strains of harps, poet-sages sang and spoke heartfelt fables. Tears ran down his bearded face and onto a colorful robe when he sang of lost love. Musicians played softest music when the bard became emotional incapable of speaking further. Anger and authority quivered in his voice when heroes vanquished evil sorcerers. And the songs went on. Young children of noble families laid gifts before the honored bard. Misty white grapes, bread wreathes with raisins and honey, crystal cold mountain water in terra cotta pitchers, and golden goblets of purple wine.

The telling of the tale often took a few days. During this time, an entire civilization was raised and destroyed. The tale told, the aged bard was escorted by a great host of devotees where, just as mysteriously as he arrived, he vanished off into time.

The themes of these timeless tales are powerful elements. We each yet resound with them. The imagery is the power. Of these tales there is no end. But the images, the themes, the quests, the magickal artifices. Something in these evokes the very deepest of desires. We each wish for that eternally radiant world.

The power of the tale is the archetype which projects them forth. During contact with the tale, listeners are absolutely engaged with the worlds described. Group transcendent experience and metadimensional contact. Phenomena with which Pythagoras was well familiar unified societies in the telling of sagas. The tales told by bards are eternal themes. The images, persistent reminders of historical realities. They emerge from visionary worlds, having timeless import. Being messengers of the eternal sea, the bard functioned as prophet for his people. The very mention of the mysterious worlds, their artifacts, and accessories captivates the mind of society with lost connections from archaic worlds.

Why do these specific images have such a deep and powerfully evocative effect on us? Each image and metaphor which performs this marvelous effect on us is, in reality, a visitor from a forgotten land. The images haunt and hurt us with a strange desire. They are archetypes, symbols, runes, communicating to us directly from an uncharted sea. The symbols are eidetic discharges from the mysterious sea of dreams, the consciousness in which we fuse with our world. The themes, images, and symbols persist for centuries, flooding the world mind. Archetypes travel from person to person, a message from eternity. Archetypes annihilate time and distance. All those who merge with the archetypes momentarily find themselves in the same timeless locations, viewing the very same scenes.

The archetypes project forth from the unfathomable deep, the dream pool of humanity. Found in fables, they are relics of lost archaic civilizations. These images have great power. Archetypes are quasi-material realities, demanding our active attentions and creative cooperation. The great armada of images proceeding forth through the epic poets have powerfully evocative effects on society. It is they, in truth, which are responsible for slow forward movements of technology.

Visitors from the very deepest wells of consciousness, symbols and archetypes have stimulated our beings with their radiant power of vision and desire, impulse and motivation. They generate technology. Humanity has for centuries felt their mysterious urge forward. The chief desire of certain individuals has been to find the wonderful tools of magick described in legends...to locate the natural loci where dreams and Nature "fuse" in material solidity. Sufficiently motivated travellers and wise archaic masters of natural science, claimed the successful materialization of specific archetypes. Forgotten lands and fabled treasures were located. Large silver mirrors were pulled from vine covered lagoons. Gemlike lanterns radiated cold green light.

There are those who insist that dreams, images, and symbols are metaphysical ephemerals lacking material reality or ability to materialize. Humanity is subject to all the archetypes and symbolic pressures which surge up from the timeless sea. Dream waves ebb and flow in the mind of humanity. The dream potential is the true and prolific power which suffuses and drives the world. Historic dreams actually found their natural material expression. Great and notable natural discoveries were preceded by haunting dream images, by which the discoverer was driven. After thorough search throughout natural locales, the dreamed thing was remarkably found in its material form. Natural "correspondence" between dream and nature has been the common historical experience. Certainly endless dreams of seacoasts materialized from the dreams of Columbus.

Those with ability to articulate dream visions produce notable developments in art and science. Symbols, images, and visions of dreams project mysterious visionary artifacts, of which artists and scientists attempt reproduction. The accuracy of their success is wholly determined by the accuracy of their response to the dream image. Artists and scientists alike seek articulate reproduction of their dream symbols in each, their various media. The gracious gift of technology is a deliberate result of both emerging dream treasuries, and the artisans who reproduce them.

Society is helplessly moved in the great tide of dream images. Each epoch is determined and defined by the dream images which provoke the "epochal theme" or "zeitgeist". The archetypes surge in waves, relentlessly expressing themselves and materializing as arts and technologies. Archetypes, though metaphysical, find material correspondence. Romantic artisans know this. When the outward flow of inner experience merges with the inward flow of outer experience, then dreams locate their material correspondence with rare precision.

Dreams materialize when we look for them in natural settings. Sensitive and attentive to this mystical crystallization of thought in matter, the wondrous and continual surging of consciousness reveals strangely haunting "similarities" and "synchronicities"...between things seen in dreams, and our day to day world experience.

ETERNAL LANTERNS

Every symbol, artifact, and accessory of the Utopian world emanates light. Transported earthly visitors first remark at this suffusive light. Their host joyously explain that "the light comes from everywhere". This magickal radiance is the light of consciousness itself. Mythologies and histories alike are replete with reports of the "endless light".

The radiant beauty of mythical "eternal lanterns" stimulates and provokes timeless scientific desire. Their endless effulgence remains their single most fascinating aspect. Can such a light source exist in the material world? The mystery of that eternal radiance could neither be compared nor equalled on earth. Firecraft did not reproduce the quality of Utopian ideal radiance. Such eternal light projects joy and wonder. Neither their radiance nor the vivifying thrill ever fade with time.

The image and symbol of the eternal lantern permeates mythology. However separated in distance or time, mythologies the world over each speak of eternal lanterns. The symbol of the eternal lamp is a haunting object which we most desire. A wordless message of hope and life beam forth, most directly representing divinity.

Believing that eternal lanterns once existed in the fabled world, ancient qualitative science sought them with a deep faith that they could be found. Ancient conquerors made extensive search for eternal lanterns and other such "fabulous" treasures. To locate, recapture, of unearth these lamps provided a formidable conqueror's quest. Likewise, numerous radiant sources were discovered and described by credible naturalists.

As centuries passed, certain scientific artisans claimed that they had successfully reproduced eternal lanterns. According to some bards of the Middle Ages, there were lands where dreams had literally materialized. Kingdoms and rare wonders lay to the east, possessors of magickal technology. Following this trail of legends and marvels came Marco Polo. Among the material proofs which he brought home to Venice, there were also new legends and reports. Marco Polo told of palaces and kings, kingdoms and artifacts, exotic natural wonders and anomalies. Caravans of archetypes and symbols.

In the city of the great Khan, there were hundreds of fabled rubies, thousands of gold tablets, and millions of standing soldiers. Marco stated very plainly that, in the outlying provinces of the Khan's empire, radiant stones and magickal accessories were commonly employed for a great variety of purposes. He had seen some of these marvels with his own eyes. One particular legend which he was fond of retelling is rarely heard or mentioned today. It centered about the fabulous Prester John, mythical King of the East.

Prester John sat in a magickal throne room, a great flooding radiance shed by special rare gems. The unearthly light of his throne, an undying light. He employed the radiance of these rare gems to render his throne room sacred, enlightened, vivifying, and quiescent. His knights were continually flooded with strength and love for their cause because of these radiant stones.

Through the agency of special viewing stones, Prester John gained instant knowledge of distant events. Empowered to project peace and benevolence to distant warring lands, he watched and interceded over whole regions through these magickal means. He was able to project help to those lands through rays which came from his magickal stones. Famines reversed, plagues eradicated, joy restored, Prester John was the protector of nations who did not know him. Prester John ("Pastor John"), the mystically advanced Christian King, is a notable story of Mongol origin. The eternal lantern, one of innumerable archetypes, persists in mythologies the world over.

Marco claimed he had seen black rocks used in Cappadocia to produce light and heat. In parts of the same region, he claimed to have seen "black oils" taken from bubbling earth pits for the same purpose. Scholars rejected everything he had to say when they heard these two reports. Long after his passing, when coal and petroleum were later discovered by Europeans, all the words of Marco Polo were heeded without question. The wonder of eternal radiance, Prester John, and magickal technology continued to occupy human curiosities throughout the following centuries.

All lands and peoples have the eternal lantern in their dream treasuries. Nordic mythology ascribed "eternal lanterns" to the gnomes, who both inherited and manufactured them. The gnomes used their mystical lanterns to light gem-studded subterranean palaces. The lanterns themselves had names, archaically crafted by famed gnome masters. Made of radiant stones, they continuously emanated soft colors and an atmosphere of great delight. The magickal lanterns themselves were fabricated from rare glowing elements and gems.

In fables, mystical eternal lanterns are made of humanly inaccessible minerals and elements. Legends continually remind us that magickal elements and gems have archaic world-origins, remnants of the lost world. Uncorrupted by the touch and taint of mortals, they frame the evidence of a first creation. Their properties, pure and sacredly honored, emerge from the dream world.

Fables teach that all beings naturally seek these materials. Certain beings, gnomes the most frequent species, covet these lanterns with a rare viciousness. The "radiant stones" reveal the first world Nature and all its wonder, the "lost elements" of which the old world was made.

In the fables, the magickal elements are said to yet exist in the deepest recesses of the earth and in special secret mountains. Spiritual prowess is required to both recognize and retrieve the minerals. The "hidden folk" always see what mortals cannot, plucking magick gems and mining magick metals from their archaic repositories with ease. Gnomes, faeries, elves, and angels jealously retain the secret of radiant stones and radiant lanterns. When humans manage to obtain them, there are consequent complications.

The magick elements and radiant stones are always wonderful to all who behold them. Their radiance is divine. Mysterious beings reverence the appearance of the radiant stones. Humanity especially cherishes and desires them. Elves cynically remind us why we have lost both the first world and the wonder elements of which it was made. The wondrous gems and metals invariably come from "forgotten archaic ages". They are "first created matter", "sacred gems", and "starry metals". They are the material of the old world.

Fables report that these wonderful elements come from the times just after the beginning of creation. Having been buried in the angelic rebellions, some remain in the deep recesses of earth. Others, having been thrown among the stars, reside in the stardust, awaiting the time when they may return to bless humanity. They hold the key to human conscious progress, requiring only humility before they may be discovered.

Mystery minerals and radiant gems are often found where natural catastrophism is at work. Radiant stones are loosened from archaic imprisonment by strange events which the "hidden folk" worriedly pursue. They jealously guard their treasures from "bumbling" humanity. Wonder elements are found in the dearth of volcanic explosions, flung up from mysterious metaphysical depths. Some fall to earth from space, glowing and pulsating. Whether thrown out of earth or space treasuries, they are usually found by adventurous humans whose lives become transformed. What these persons do with their treasure usually determines their fate, a moral lesson concerning the abuse of power.

There are peaceful ways in which the radiant stones are located in some tales. There are those who see magickal glowing pools of water by night. Venturing in, they manage to find the rare "wish-granting" glowing gems. Some only appear during certain astrological seasons, under specific "heavenly signs". They are seen only by sensitives who, with greatest care, find them radiating their light when touched by the crescent moon. Mysterious visitors often add a "pinch" of magick dust into the mixtures of old despondent alchemists. When this alchemical projection has performed its work, the molten metals become joyfully radiant.

"Lost elements" grant their bearers strange abilities and miraculous powers. The abilities they give often require a small exertion of mind, with a resultant magnification of intent. Wearing the radiant gems, one can become invincible or invisible. One can pass through walls. One can fly, lift great weights, bring lightning or storms, and perform unheard superhuman works. These mysterious materials link dreams directly with the world. They are windows through which archetypes flow directly into our world. Wish amplifiers. The very artifacts of Prester John. Magickal technology requires the radiant stones. They are the lost talismans, the lost instrumentalities which guide and extend human consciousness. Radiant stones are the reagents of every lost technology.

Those who find them become heroes or villains. The legends tell both of the location and loss of mystery elements. They may be found, granted, or stolen. Each means of obtaining them has its consequences. Magick rings are given to humble recipients by angelic visitors. Mystical necklaces are given in return for humanly kind gestures to children. They are fabulous rewards from gnomes, faeries, and sprites. Little children and humble old folk often accidentally find magickal materials by the deliberate design of "hidden ones".

Radiant crystalline jewels are often found by humans who enter unfamiliar glowing caverns and grottoes through accidental circumstance. Wanderers, lost in a storm, find caverns of wonderful radiant treasuries in places they can never relocate.

There are those whose lust for power drives them to acquire the lost elements through evil sorceries. There are mysterious alchemists who stumble on the strange minerals, taking them by night from their "sacred resting place". Working their excessive "treatments and chemical labors" until the magick is released, they often become the terrors of the countryside.

Evil alchemists, betraying the sacred calling, challenge rule and dominion. Working feats of mind magick and enslaving the populace, they portray the dangers associated with the misuse of magickal technology. A hero is usually summoned by a rival good wizard to destroy such evil alchemists. The fabled confrontations always pits magick against magick, the hero having been given an equally great talisman by which to succeed. Love is the key to winning the battle. Love stimulates the radiant gems. Greedy and ruthless hunters steal them from forgotten temples, plucking red glowing gems from the eyes of idols with rude steel knives.

When the "wonder elements" are lost, they are lost through misuse and pride. They are often taken from the wicked by mysterious and protective guardians of humanity. Jealous and hapless fools accidentally drop their radiant stolen prize into wells, seas, burning pools, and crevasses of ice. The magickal elements and radiant stones are often lost just after certain worthy persons have been blessed by their magick. These materials seem impossible for most humans to keep.

The loss of the wonder elements is always attended by great sadness and regret. Those who lose them are shunned, their villages fading into a disappointing and lackluster future. In the closing verses of these stories, one is always taught to expect their re-emergence. Hopeful that the rediscovery of lost elements will raise society into a clarified future, there are always promised signs by which we know of their appearance on earth. World conscious revolution always requires the radiant rocks.

These mysterious "lost elements" and "radiant stones" continually emerge from the deepest memories of humanity. Their re-emergence in the social symbolic lexicon is irrepressible. The greatest themes of modern science fiction are all archaic in origin. Their symbols, however updated and modified, have arcane roots. The archetypal image of both the eternal lantern and the magickal elements reappeared with the "Green Lantern" comic book series of the 1930's, where the eternal emerald lantern from another world conferred its power to one who would justly bear the magick ring. Charging the ring by the eternal radiance of the emerald lantern, the ring bearing knight swears to protect the world from evil. He then learns the secret psychotronic power of the ring.

While the images of Prester John's fabulous court room illuminated the minds of those who thrilled to their hearing, a new and unexpected materialization was historically chronicled.

So dramatic an episode in human history, it yet demands a complete bibliomantic examination. There was an instance in time, a true psycho-social event, in which archetype met human desire in material form.

RADIANT ROCK

During the late Middle Ages, a great variety of remarkable "radiant rocks" were suddenly and unexpectedly discovered. These discoveries emerged from the mountainous regions of Central Europe in continual waves. It was the countryfolk who found them. Approaching in a timid, childlike wonder they saw the impossible. A powerful and unusual radiance was found emerging through the very rocks of the mountainsides. Light seemed to be streaming out of the hillsides! The brilliant green and blue mineral light was seen among familiar mountain rocks after sundown. As reported, the light was brilliant, far above the brightest light of the full moon.

Most of the first fortunate individuals who made these discoveries were not scholars or craftsmen. Typically, they were shepherds, mountain villagers, pilgrims, and wanderers; the innocents whom angelic hosts historically visit with messages of love. The glowing rocks seemed to be windows on some underground domain. Possibly from unknown cavern worlds beneath! Nights were spent watching the rock radiance in absolute wonderment. The radiance was considered an apparition, a divine event. The light was beautiful, glorious, and unfamiliar. It brought with it the atmosphere of another world.

Here was a new revelation, a new kind of light which did not come from fire. The rocks gave forth a bright light, bright enough to illuminate the faces of those who stared into them in disbelief. Though shining for hours with no other energy source, this cold rock-light would not die! Those who discovered these strange glowing rocks believed themselves to have been favored by miraculous visitation, answers to fervent requests.

Stories began emerging from different parts of the world concerning these very same radiant rocks. Some of these were given names. Some were associated with the saints. Others were simply named for those who found them, or the place from which they were dug. A few brave souls were not afraid to attempt sampling the find. Unearthed, pieces were brought indoors. Though separated from their parent rock mass, they continued illuminating cabin interiors with their rare and cold radiance. Undying fire. Cold, pure endless light!

Most of the rocks came from familiar regions, yet their radiance was never before seen. Innumerable opportunities existed for the observation of the brilliant and spontaneous radiance. But, why had no one reported such wonders? Had no one ever seen these lights before? The shepherds who frequented those very places had crossed and recrossed their grazing paths for countless centuries. Yet, they had seen nothing unusual. Certainly nothing this noteworthy was ever observed by anyone familiar with the very terrain in which the manifestations had taken place. This equally profound and perplexing mystery now formed additional evidence concerning the radiant rocks, one which surrounded them with an additional and inescapable aura.

It was apparent that these were divinely inspired events, utterly new creations. This was the accompanying awareness which boosted their fame into public consciousness. Clergy and monastics became enthralled...humbled by the notion that science and theology were not, as most had assumed, separate experiences. Archetype, visionary desire, and natural reality had been brought into material fusion.

The radiant rocks foretold a coming dawn, an Age of Light perhaps! For the people of this time period, the phenomenon was material evidence of the ancient faith. The glory of these special radiant crystals brought about a curious form of devotion. Taken as miraculous signs, each were displayed before persons of both low and high estate. Cathedrals displayed them for the common folk. The rocks were seen as sacred artifacts, befitting the treatment rendered toward relics. Private viewings, accompanied by all the pageantry of liturgical service, were held before the courts of kings.

Here was a new breath of promise, a silent comforting word. In this atmosphere of prayerful silence, the miraculous stones radiated their eerie green or blue light to the wonderment of all who beheld in silent awe. Certain varieties were exceedingly brilliant, some reporting a radiance far above the brilliance of full moonlight. In a few other instances the report of radiant brilliance approaching sunlight was confirmed by credible authorities. Those who worshipfully gazed at these rocks in the silent cathedrals awaited metaphoric answers. It was difficult to imagine that former scholars had separated innervision and external Nature, calling the one "fantasy" and the other "reality".

In the clear light of these wonder rocks there was some powerful sense of a lost innocent world and its wonderful radiant vitality. A new and striking atmosphere seemed to capture all the scholars who approached the brilliant rocks. Gazing into their wondrously bright and seemingly endless radiance, one had the sense that the Divine Presence was mystically shining through "corrupt matter". It was light from death, the complete antithesis of the fire paradigm. Here, the production of light did not require the death of living thing. The light did not emerge through the consumption of fuels, the soul of the green forest. This matter was not "corrupt". This was a rare kind of matter, unknown except for the legends and fables which accurately described them in every detail.

With each discovery of a new radiant mineral, more proof was added to the ancient belief that Creative Light had not indeed abandoned the world of tears. Where death seemed the ruling power, it was the light of these rare gems which heralded sweet proof that Divine Love was yet reaching for humanity. Those who both studied and reported their experiences with these radiant rocks certainly knew the comparative illuminating power of candles and bonfires. When they likened the brilliance of these strange rocks to full moonlight or of sunlight, they were neither being scientifically primitive or emotionally excessive.

It is astounding to recount the historical density of such finds during the early Renaissance. Arcane journals and manuscripts contain anecdotes of these discoveries, although found with difficulty in widely scattered fragments. They are the remains of notable past events in the indelible historic record.

Scholarly minds reeled under the dream-impact of possible new technologies. The wonders which could emerge from the proper implementation of these rocks would take some time to fully develop. Would these rocks respond to the mind? Would they influence actions without contact? Could they be used to move heavy objects? Could they make one invisible or invincible? It is very significant that those who sought the development of radiant technology used the very archetypes of legend to guide their efforts.

There were those who now ran to the mountains in search of radiant rock. Many new varieties were found, but few gave the rare radiance produced by the original finds. Large crystals of fluorite and barite were torn from mountain scarps. Having been exposed to intense sunlight, each was then methodically taken into dark chambers for examination. In the clutches of secular scholars who sought purely mechanistic explanations, the comparatively weak glow of these rocks also remained incomprehensible. They did not equal those which had previously been found. Words and logic failed to explain what Nature had revealed.

Late Renaissance science was searching everywhere for more examples of this wondrous rock luminescence. As continual examples of these fluorescent phenomena were found, science could do nothing more than simply collate the evidence. Soon, a large collection of minerals and experiences had been patiently compiled. Yet, none of the forthcoming finds quite equalled those early and spectacular events which heralded the discovery of the famed radiant rocks of Europe.

Never was such a singularly religious treatment of the phenomenon ever seen again; and it is indeed curious that the radiant intensity of later retrieved rocks, mostly barites and fluorites, did not match those reported during the first spectacular wave of discovery. The original stones surfaced in the courts of various nobles throughout the Renaissance, the coveted possessions of rulers. Their appearance and disappearance followed the rise and fall of their power. It has been impossible to trace the whereabouts of these rare gems whose great radiant intensity was "frightening". Discoveries of natural radiance such as these never made their appearance in Europe.

In the absence of newer reports, these remain inexplicable. Spontaneous sources of light made their appearance throughout the following centuries, but the truly great radiant light sources seemed to have dimmed once more. In this, the scholar is faced with a mystery. What is the reason for their appearance and disappearance down through the centuries?

NEW RADIANCE

When the element "phosphoro" was discovered in 1669, the frightened alchemist Brandt fell down in silent prayer. He is often depicted in this poise at moment of discovery. Fascinated preoccupation with radiant rocks and other related phenomena always re-emerged with each new century. They came with great regularity during the Eighteenth Century, mostly associated with bioluminescent phenomena. Luminous insects, fungi, coral, fish, mushrooms, and so forth. The light they gave was neither eternal, nor radiant. These were again heralded...but not with the great sense of awe or religious reverence rendered to those first wondrous manifestations.

There were those who grew accustomed with the "disappointing" nature of quests and natural finds. This sad tendency became the "expected outcome" of belief in any kind of visionary artifact. Disappointment and hopelessness was associated with the quest for dream archetypes in material nature. Nevertheless, several significant discoveries continued to overthrow this negative world view, vindicating those who expect the natural world to surprise them.

A mysterious radiant stone was discovered in Connecticut during the latter 1600's by a Mr. Steele. Living in East Haddam at the time, Mr. Steele discovered a truly marvelous and precious stone which he claimed visibly radiated incredible volumes of light. He confided secretly with his landlord, a Mr. Knowlton, that he would soon be able to procure it in secret. He referred to the stone as "the carbuncle", relating that a huge sum of money would be theirs to share if only the secrecy could be maintained until time of disclosure. Mr. Steele seemed to infer that he had found a large deposit of the white material, and this increased the earnest expectations of his landlord.

By night, Mr. Steele brought the "white rounded carbuncle" back to the boarding house under thick covers. Despite the attempt at cloaking the stone, it glowed with an intensely penetrating radiance. In the dark, the light grew to an incredible and anomalous proportion, far exceeding that of sunshine. This material was secreted into the cellar of the house, one "not having any windows". There, Mr. Steele "worked on the material by night", performing chemical operations on the substance.

Despite the thick stone walls, the light of this stone "shone right through" to the outer meadows. So great was its penetrating strength that the entire house appeared illuminated by fire, being seen at very great distances by curious others. In addition to this mystery, large and continual booming sounds were heard surrounding the stone and the house. Mr. Steele stated that these sounds emanated from the stone. He labored on the stone every night until it was impossible to hide the secret any longer from neighbors. Mr. Knowlton, the landlord, thought it bewitched by Indian sorcery and angrily warned Mr. Steele to cease his evil acts.

Wrapping the carbuncle in sheet lead and taking on a disguised appearance, he fled from the town by sailing ship. Because of the stone's remarkable luminescence and equally powerful thunder, it was impossible to hide the stone. It has been presumed that the sailors, superstitious and frightened of the accursed carbuncle, simply threw him overboard with the object in his grasp. England, Steele's intended destination, was never reached by him. The stone, of course, was lost.

The large mysterious stone was original dug out of a very specific hillside, known by the residents of the area. From descriptions given by the now-late Mr. Steele, a local cleric found the very spot. The cleric is the same gentleman, a man of impeccable character and irreproachable honesty, who wrote the journal from which the tale was derived. He spoke of the existence of the mystery carbuncle as a fact, in greatest confidence.

The cleric said that lightning was frequently drawn into the hillside at that very point, loud booming noises constantly emanating from the place regardless of the weather. The place was also known to the native Americans, who treated the district as a sacred spot. Puritans believed the place bewitched, and avoided it with great terror. No mention was ever made of the strange substance again.

While sounding completely fanciful, such stones were actually observed in New Guinea by numerous traders who managed to penetrate the high mountains of Mount Wilhemina. These adventurers reported that native villages employed large "balls of stone" to brilliantly light the night darkness. The giant glowing stones were exceedingly bright, resembling "suspended moons". These filled the jungle region with their radiance, giving a surreal quality to the place. The source of the mysterious light found, those of the expedition were completely astounded. Poised high on very large pedestals, the huge white balls of stone glowed with a brilliance equal to that shed by electric lanterns. Their light did not fade with time.

Another such account came through Ion Idriess, a famed Australian writer. Aborigine elders, while recounting island history to him, reported the existence of "the booyas". These were large balls of stone which glow with an eerie magickal light. Three of these stone "scepters" were known in the area. Poised on tall pedestals of bamboo, the light shed by the booyas was so bright that it enveloped its spectators. Held up toward the sky, the stone flashed with a brilliant cold green light, and was thus "charged". Villages thus illuminated by greenish white brilliance were seen far off at sea.

The diary of a conquistador (Barco Centenera, 1601) told of a similar, if not exact, stone ball lantern. The setting was Paraguay, in the city of Gran Moxo. There he reported discovering a huge stone pedestal, some twenty four feet in height. This pillar was surmounted by a huge ball of stone which shone with such brilliance that it illuminated both the lake and the inhabited area.

The English Colonel P.H. Fawcett reported hearing of cities in the same South American jungles whose people employed a similarly strange means for illuminating their night times. These were the very same kinds of cold green balls of stone, poised on very tall stone pedestals. Colonel Fawcett, of utmost integrity, sacrificed his life while seeking the ruins of these lost cities. His qualified opinion was that these places were "contemporary remnants" which retained the forgotten knowledge of...much older civilizations.

There are those who claim to have retained some portions of these legendary materials. Nicholas Roerich, symbolic master artist, travelled to Mongolia in 1925 after learning from monks of a famed stone which "fell from Orion". Securing the stone on behalf of the monastery in which it was housed, he claimed to have discovered its amazing mind-expanding qualities. Both Nicholas and Helena Roerich accompanied a fragment of the stone to a neighboring monastery where it was enshrined.

These sacred stones, elements of the first world Nature, were said by him to radiate a sharply defined consciousness. The monks attested to the ability of this stone to "maintain peace and elevate consciousness to all the outlying districts". In these mountainous lands, timeless traditions preserve what centuries of European history would have erased. Could this have been the very stone which Marco Polo had reported? Was this one of the "magickal stones" employed by Prester John? The couple returned home, encrypting the truth of this remarkable find in a series of mystical books ("On Eastern Crossroads", "Legend of The Stone", Abode of Light").

ELECTRIC ROCK

Numerous explanations for the phenomena of luminescence and phosphorescence flood the technical literature of the day. Each seek mechanistic explanation for the remarkable radiance of certain materials and apparatus. Despite these academic speculations, the phenomena of luminosity and phosphorescence represent some very essence of our deepest dreams and mythic desires. Strange light sources form the heart of all artifices of legendary magick. These are fascinating archetypes which surpass our merely intellectual fixations. Why their mere mention fills us with an awesome reverence has much deeper source.

Each new technological epoch is always accompanied by the emergence of new and remarkable light sources. From the latter part of the nineteenth century until the middle 1930's there was an abundant emergence of such devices. While many of these "lanterns" required the forced generation of energies, there were a significant number of devices which did not. In the following account, we will see that the fables of "lost magickal elements" and "radiant rocks" are grounded in truth.

Dr. Thomas Henry Moray, an electrical engineer, began research on aerial static generators in 1910. He succeeded in deriving usable electrical energy from the earth's electrostatic field. Many others had achieved similar results in the century preceding Dr. Moray. Patents of "aerial batteries" fill the archives (Vion, Ward, Dewey, Palenscar, Pennock, Plausen). Their remarkable efficiency required only the establishment of elevated stations in appropriate places, each differing in the actual mode of extracting the atmospheric energies. Some of these aerial battery systems successfully provided the utility requirements of small factories and telegraphic exchanges.

Dr. Moray was fascinated with the concept of drawing electrical energy directly from the environment. His initial and primitive tests brought a modest amount of electrical energy from his aerial battery design, producing clicking tones in a telephone receiver. With this device, he gradually developed enough atmospherically derived energy to light a small neon "arc lamp". Dr. Moray was encouraged, but not satisfied with these results. There had to be a way to get much more energy from the environment.

Static-field systems are necessarily tall aerial structures, requiring a great deal of space. Some designers used large balloons to hoist their static collectors. Others simply utilized fixed structures: large mountain-poised screens, point-studded poles. Benjamin Ward used an astounding "directional chute" which "funneled" electrostatic winds. Aerial batteries relied on the surface area of structures to absorb electrostatic charges. The larger the system, the more the available electrical power. Moray wanted to miniaturize these large systems. If miniaturization was to be the design goal, there would be a necessary and revolutionary change in the approach.

While investigating the output of his device, he discovered a feature of the natural static energy which had somehow been overlooked by other aerial battery designers. The electrostatic power had a flimmering, pulsating quality to it. He learned of this "static pulsation" while listening through headphones which were connected to telephone wires. The static came in a single, potent surge. This first "wave" subsided, with numerous "back surges" following. Soon thereafter, the process repeated itself. The static surges came "like ocean waves". Indeed, with the volume of "white noise" which they produced, they sounded like ocean waves!

These peculiar waves did not arrive with "clock precision". Just like ocean waves, they arrived in schedules of their own. Dr. Moray was convinced that these were world-permeating waves. He came to believe that they represented the natural "cadence of the universe". This intriguing characteristic suggested that small amounts of pulsating electrostatic charge might be used to induce large oscillations in a large "tank" of charge. The resultant oscillating power would be applied to industrial use. But experiments in these avenues were not very promising.

Dr. Moray believed that the earth's natural electrical energies were derived from the mineral content of the ground. He therefore began examining minerals with a rare devotion. Everywhere he went, mineral hunting was the first impulse. These minerals became quite a collection. Each was examined in his small laboratory to discover any possible new electrical properties which might reveal the truth of his ideas.

There was practical truth in his first suppositions. The early days of Radio utilized mineral crystals to detect signals. Tesla was perhaps first in announcing that selenium crystals could detect the special rays with which he was principally occupied. Thereafter several different personalities claimed to have "discovered" the crystal detection method.

Essentially solid state in nature, the method uses mineral crystals to "detect" radio signals. Fine wires ("catwhiskers") touched mineral surfaces at specific "sensitive points" and were tuned with a small coil. One could receive radio signals without batteries by employing this detector. When connected with an aerial, a ground rod, variable coil tuner and headphones, the resulting "crystal set" provided a means for receiving strong radio signals. As children, many of us had these little crystal sets. They are still popular and may be purchased for a few dollars in science shops today.

Since the crystal was the key to better radio reception, experimenters were searching the mineral kingdom for new and more sensitive detector crystals. Early radio magazines taught the eager hobbyist how to mount special radio-sensitive mineral crystals. There were several favorites of which radio hobbyists were very fond. A brilliantly reflective metallic crystal having a silvery blue cast, Galena was the mineral of choice. Pyrite, otherwise known as "fool's gold", is a close second to Galena. As a radio detector, Pyrite gives a good signal strength in the headphones. Some preferred Molybdenite, a mineral which is especially sensitive to infrared energy as well as radiosignals. Radio amateurs were always trying new mineral crystals to see which ones amplified radio stations with greater strength and clarity. Each had their favorite crystals. Little did the radio enthusiasts know why there was mystery in this quest for "radiant crystals"!

Advancing the science of crystal radio detection, certain researchers discovered that contact-combinations of minerals gave stronger signal amplifications. When carborundum and silicon crystal nuggets were pressed together with little springs in a metal tube, the signal strength was enormous. Small battery voltages turned these mineral detectors into amplifiers, early transistor-like detectors. Bi-mineral and bi-metallic hybrids were tried with good results. Mineral-metal, mineral-mineral, even multiple minerals in contact with two different metals produced enormous magnifications of signal strength.

Some detectors incorporated carnotite, a radioactive mineral. This greatly increased conductivity with a resultant enormous signal magnification. Coating aerial points and catwhiskers with autonite, another mild radioactive mineral, produced amplified signals. There were small companies which manufactured synthetic minerals ("Radiocite" and "Russonite"). These claimed superiority in the signal-boosting ability. There were scores of other sensitive minerals which gradually appeared in the journals, some natural and rare, and some synthetic compositions.

In light of these wonders, a full scale assault on the mineral kingdom was launched by both private and government research labs. It was discovered again that the mineral world is not simple. The mineral world is mysterious and locale-specific. Each ground site where minerals are found evidences unique mineral combinations and admixtures. The special qualities of identical mineral species can vary completely among geographic points.

Galena contains critical element traces which differ completely among mining sites. It was found that Galena, taken from certain special sites in Kansas, gave an excessively clear signal output with high volume. This natural Germanium rich variation placed it far above other galena samples in radiosensitivity. In addition, these crystals visibly differ from Galena taken from other places, having a rare blue green hue.

GROUND ENERGY

Crystal sets were the ruling radio technology in the early 1900's. Listening to radio stations through a crystal set is still a treat! The sounds are startling, clear...and loud. An old neighbor told how his great-uncle so tuned the crystal set that "everyone in the house could hear the music". Most crystal radio sets are "unpowered" radios. The entire study field of unpowered radios has been receiving enough attention lately for the publication of several books on the subject.

The anomalous strength of crystal radio signals has much to do with design, conductivity, crystal quality, aerial surface, and ground contact. In many cases, the aerial may be completely eradicated when the ground connection is "right". In fact, the aerial can be eradicated, but not the ground. It is the ground connection which is all important.

Amateurs are rediscovering that specific groundpoints emanate radio signals in sufficient quantities to power loudspeakers! Recent reports from three independent researchers reveal that crystal radio reception through ground connection alone is shockingly powerful. The developed radio power in these crystal radio receivers was so strong that volume controls were installed to limit the output sound!

According to each report, crystal received radio signals actually seemed to "grow in strength" with time. The anomalous growth is one which numerous researchers have noted. The pattern follows the "vegetative growth" which Reichenbach observed in his "Odic" energy and which Stubblefield saw in his "electrical earth waves". Vegetative growth patterns appear as a gradually increasing signal, reaching frightening volumes at the maximum. In one such instance, the weak signal strength required headphones. Within several days however, the headphones had to be replaced with a small loudspeaker. This loudspeaker was then replaced by a much larger diameter horn. Finally, the horn had to be disconnected periodically because neighbors complained of the "outrageous booming sound". Vegetative signal growth. Not an electrical characteristic.

One may view the crystal radio receiver as a tuner of crystal Od, the radio signals and other electrical values merely appearing as epiphenomena (Meinke). Recall that Reichenbach hoped to use Od in a new non-electrical technology, and that later pioneers developed Radionic instruments to these ends. The engineering convention of the early Twentieth Century had not yet realized the active appearance of these more fundamental energies in their circuits as did their predecessors in the telegraph industry.

The idea of obtaining and using "ground energy" is covered in secrecy. What would happen to fossil fuel companies were it even suspected that vast electrical energy could be simply pulled from the ground at specific points? These energies began "making their appearance" during the years of telegraphy. Well placed telegraphic ground plates were able to operate with energy simply taken from the ground. Several early telegraph lines historically continued signalling among stations, though their batteries had been "dry and dead" for several years! I spoke to an engineer who saw this kind of system operation when yet a teenager. Seeing this strange system in full working order so impressed him that, developing that rare taste, he forever sought such anomalies as a lifelong passion. Numerous articles from the last century retell exact details concerning these phenomena.

It is possible to demonstrate its principle with ground rods and galvanometers. Yes, there is great energy in the earth, vast natural energy which is accessible only in specific points. But the true and fundamental identity of that energy has been questioned. Most qualified investigators observe that ground energy does not "begin" as electricity. Electricity from the ground only appears after several natural stages of transformation. Vegetative growth. This is evidenced in old telegraph lines where measured currents do not provide adequate wattage for the activities which are thereafter observed in the components. This was especially true for the forgotten chemical telegraph systems, where scarcely any electrical current managed the successful exchange of strong signals.

The forgotten science of selecting "special ground sites" is re-emerging among VLF radio researchers. No two ground sites are ever the same. It is possible to probe around in a garden with simple meters and metal rods to prove this claim. Touching carbon and iron rods to the ground registers as currents ONLY when specific points are touched. It is fascinating to find extremely active sensitivity spots immediately adjacent to points which produce absolutely no response in meters. The effects measurably increase despite rod separations.

In no manner can these be referred to as "electrolytic" or "battery actions", since the requirement for best energy extraction by this method is dry ground. Rainwater destroys these effects. Moreover, it is only when the right ground contacts are made that one will watch the meter "pin". There the meter will remain until the rods are removed. Such energetic discharges can continue for months!

Removing the rods, however, produces a more astounding phenomenon. The meter, dropping to "zero", does not rise again when the rods are replaced in their very same groundpoints. One can lift one rod out of its well, watch the meter drop, and then instantly replace the rod with no resultant energy rise. Ground energy withdraws in a manner suggestive of "biological irritation". Each of these phenomena may be demonstrated to personal satisfaction with very simple apparatus.

THE SWEDISH STONE

Dr. Moray travelled to Sweden as a missionary of the Mormon Church in 1911, visiting relatives for a summer. He frequently hiked through the lovely green meadows and blue mountain ridges to examine and collect more minerals. Here, in the historical land of gnomes, he found an unexpected treasure. But it was this part of his biography which separated Dr. Moray from fictions and fables.

It was during one such mineral expedition that Dr. Moray found a soft, silvery white mineral which greatly attracted his attention. Despite his great difficulty in obtaining the necessary parts for even a simple laboratory examination, he found that silver catwhiskers produced electrical rectification. When the silver contacts touched the mineral, the stone would pass battery currents in one direction. Believing that this material might be useful as a new industrial radio product, he stored a good quantity of the stone for his voyage back to America.

Dr. Moray never elaborated on the "discovery" portion of his story. He mentioned only that he obtained the material from two separate Swedish sources. The first samples were crystalline, being found in a hillside outcropping of rock. The second, a smooth white powder, was scraped from a railroad car in Abisco. Both materials were identical in composition. No doubt, he wished to secretly preserve the location of this mineral lode for future use, since he never told of the exact source location.

Endowed with extremely peculiar electrical properties, the stone provoked great surprise. Implementing the mineral as a crystal radio detector, he discovered several unexpected phenomena in quick succession. Dr. Moray discovered that radio signals were so amplified by this mineral that headphones were destroyed by the current. He then scaled up the output to accommodate a very large loudspeaker. Tuning in any station produced excessively high volumes of sound without external power! The fact that these crystal radio receivers required no extra power when achieving these activity levels suggested new experiments. His assessment of the mineral was mystical. His perplexing comment had no prior equal in the electrical world.

"What I have found is a mineral radio detector, having self-amplifying abilities".

Now, bright blue-white sparks were observed playing along the thin wire connections around the mineral at specific station settings. Mysterious! Incredible! The stone provoked a world of theoretical "problems". Any device with "self-amplifying abilities" is necessarily drawing its power from somewhere. But, from what place of origin was this "Swedish Stone" drawing its self-amplifying abilities? According to convention, there were no such sources to be found.

Unable to yet find an answer, he modified his theory on earth electricity. He came to believe that natural earth electricity was entirely developed through special minerals like the one he found. Minerals, he reasoned, were able to modify more fundamental energies which emerge from the ground. During the process, electrostatic charge is developed. This is why the earth maintains its charge. It possibly explained why the earth static charge also "flimmered" in pulsations. Whatever causative energy was manufacturing the charge in minerals was obviously a pulsating one.

There was no precedent for the peculiar behavior of this mineral. No existing electrical explanations for the activities whatsoever. Despite his inability to comprehend why the stone performed these marvels, Dr. Moray continued his empirical experiments. In the process, he developed several modifications of his original theory. Based on strong intuitions, he formed a doctoral thesis in 1914, postulating the existence of a "sea of energy" from which all future power would be derived.

CRUCIBLE OF THE STARS

These events all took place in the obscure privacy of Dr. Moray's early life. In 1917 he was married. He managed to obtain several successive engineering jobs, all the while pursuing his dream of deriving energy from the earth. Between 1914 and 1921 he was unable to pursue this experimental work for any sustained time, the happy duties of work and family life absorbing all his attentions. His engineering employment record is prestigious, considering the time frame. He was employed as Designer and Engineer by the Utah Power and Light Company, Engineer for the Phoenix Construction Company, Assistant Chief Engineer for the Arastard Construction Company, and Division Chief Engineer for the Mountain States Telephone and Telegraph Company.

Having labored on his mineral through a sudden inspiration, he now advanced the operation of the device, a cylinder of eight inch diameter and six inches height. Utilizing an aerial and a good ground rod, Dr. Moray successively powered both a 100 watt incandescent lamp and a 655 watt heater. It was found that deeper ground rods produced visibly brighter lights. The world did hear from Dr. Moray in 1925. There, in Salt lake City, he began performing wonders with his new "energy receiver".

Several local witnesses observed these experimental proceedings. It was seen that power increased with increasing ground rod depth. He now approached several authorities in hopes that the invention be given proper treatment in an established research laboratory. Then, he anticipated that the industrial groups who handled power generation would be the most reasonable people to contact for the deployment of his revolutionary technology.

In that October, Dr. Moray brought witnesses from the Salt lake City General Electric Company in order to formally disclose his discovery. The demonstrations proceeded as before. Dr. Moray showed that the removal of either the aerial or the ground caused power to fade, proving the external source of the energy. Several qualified witnesses arrived from Brigham Young University to observe the device in action. Themselves engineers, he allowed them to examined and dismantle the device completely so that there would be no accusation of fraud. They even tore the boards open to see if there were any concealed transformer coils or batteries which would account for the tremendous power production. No fraud could ever be found in this severe examinations. The one component which Dr. Moray would never show the others was the size of a pocket watch. In it was his precious "Swedish Stone".

In a strange way, in a most remarkable way, Dr. Moray had found an eternal lamp! This caught all the authorities off guard. Careful to watch over and protect every rigidified dogma and institution, the guarded hierarchies of self-centered authority was shaken from their foundations. Academic authorities declared that Dr. Moray's device was simply receiving energy from powerlines or local radio stations. They demanded that he take the device to various locales in order to pinpoint both the true source of the energy and establish operation criteria. No industrial development of the device would be considered, they stated, until he complied with the stated requirements.

Taken across the countryside in hopes of realizing the failure of the device, these academicians were thunderstruck when the device continued operating. It was impossible to find a place where the output showed any slight drop in power. The device worked during snowstorms, rain storms, and in deep mineshafts. It was sealed in a metal vault, dragged many miles away from power lines, and taken to mountainous terrain which did not sustain any radio reception.

In each location the device suffered not one degree of diminished output. Appliances usually tested simultaneously in these desert locales included thirty five lamps each rated at one hundred fifty watts, a one thousand watt hand iron, a heater, and a fan. Moray himself wished to test whether the device would drive a plane or submarine. The device was taken up in a plane and submerged in a water-tight container at the bottom of a lake. In each case the output remained the same.

Once, while demonstrating the device before a group of electrical engineers, a sudden sustained dark blue spark of some eight inches length was observed. When the engineers saw this, they were truly amazed, as the spark discharged from one of the output leads into midair. The incoming energy had actually momentarily exceeded the sustaining ability of the receiver. Obviously, this energy was coming in "from the outside". It was later found that this blue "spark energy" was a current which could penetrate several solid plates of glass.

Some analysts identified this display with "high frequency electrostatic" energy, but this was an impossible conclusion, since the glass-conducted current could perform all the same energetic wonders as if passing through metal wire. The question was then to address the exact nature of the receiver's current output. Was this a different kind of electrostatic energy? Were the output currents of a different "species" altogether?

Now, Dr. Moray guarded his detector with special care. The new component was no longer the size of a small pocket watch. The modified detector looked like a small crucible with its cover welded shut. This, Moray removed and placed in his pocket whenever the tests were completed. He later revealed the nature of the device, stating that it contained a piece of the soft "Swedish Stone" and "a few extra additives". A few close associates were permitted to examine this component. There was nothing more than what his drawings showed in its metallic shell. The whole secret lay in the nature of the mineral, the lost and miraculous mineral. The radiant stone. What remarkable truth did Dr. Moray learn? How had he managed to transform his device into this veritable powerhouse?

SPACE RAYS

Years before this demonstration, the absence of reasonable academic explanations forced Dr. Moray to plunge into study. Preparation for his thesis required a substantial personal library, which he had gradually acquired. In the process, he secured several rare fifty year old volumes on radioactivity. These included the theories of Nikola Tesla and Dr. Gustav Le Bon, both of whom had each published extensively on the subject.

Tesla's theory of radioactivity has never been adequately appreciated, despite the fact that he was first in demonstrating the existence of cosmic rays. When announced, after a period of intense investigation, Tesla was heckled by the American academic community. But, long before Henri Becquerel and Marie Curie investigated radioactivity, Nikola Tesla first called attention to the notion that matter was spontaneously converting into energy. This process, Tesla stated, was an eternal one. A steady external shower of stimulating "cosmic" rays was penetrating all matter. These "cosmic rays" bombarded and disintegrated all matter. The process was measurably increased during the daylight hours because, Tesla stated, the bombarding rays come from the sun. These external rays were possessed of incredible electrical potential.

Tesla cited particle potentials exceeding "one hundred million volts". He said that he had measured these potentials with "special" detectors. These might have been selenium detectors in vacuum bulbs, as used in his radio receiver patents. The electrified particles were constantly bombarding all materials, causing radioactivity to be observed. Tesla stated that all matter was in the eternal disintegration process. He claimed that the more dense metal elements were more easily recognized as "radioactive" because the dense materials were "better targets". According to this viewpoint, radioactivity was the manifestation of externally sourced agencies. Tesla insisted that the true source of radioactivity was outside, not within, matter. Ray bombardments "from cosmic space" were his explanation.

Tesla defined true cosmic rays as an entrant light-like effluve having incredible penetrating power. These were in no way similar to the conventional cosmic rays detected by Gockel (1910), Hess (1912), Kohlhorster (1913) or Robert Millikan (1925). Tesla viewed his discovery of these light-like effluves as holding the only promise for energy application. According to Tesla, the energy of these effluves greatly exceeded those of cosmic ray "particles".

When Moray read these ideas, he seemed to find a piece of the puzzle which so eluded explanation. Another researcher, a contemporary of Tesla, succeeded in advancing the "external bombardment" theory of radioactivity with new experimental proofs. Dr. Gustav Le Bon, a Belgian physicist, examined and compared ultraviolet rays and radioactive energies with great fascination. Concluding from experiments that energetic bombardments were directly responsible for radioactivity, he was able to perform manipulations of the same. He succeeded in diminishing the radioactive output of certain materials by simple physical treatments. Heating measurably slowed the radioactive decay of radium chloride, a thing considered implausible by physicists.

In each case, Le Bon raised the radium temperature until it glowed red hot. The same retardation of emanations were observed. He found it possible to isolate the agent which was actually radioactive in the radium lattice, a glowing gaseous "emanation" which could be condensed in liquid air. Radium was thereafter itself de-natured. Being exposed to the external influence of bombarding rays, the radium again became active. The apparent reactivation of radium after heating required twenty days before reaching its maximum value.

Dr. Le Bon was utterly dumbfounded when, forcing theory into fact, other colleagues announced the "immutability of radioactive decay". He also perceived where their erroneous logic would ultimately lead when they cited "internal instability" as the source of radioactivity. Separating themselves once more from the external world of energy, they would lose more than they imagined themselves gaining.

Le Bon disagreed when physicists began isolating the heavy metals as "the only radioactive elements. He had already distinctly demonstrated for them that "all matter was to a degree radioactive". He was first to write books on the conversion of ordinary matter into rays, an activity he claimed was constant. He showed that this flux from ordinary matter could be measured. Le Bon stated that the reason why all matter was spontaneously emanating rays was not because they were contaminated with heavy radioactive elements. Ordinary matter was disintegrating into rays because it was being bombarded by external rays of a peculiar variety.

The external source continually bombarded matter, producing a continual energetic release, "an effulgence" of energy. Because the energetic disintegrations of matter occurred under focussed sunlight, he first cited a special photoelectric effect. But, judging that nuclei were actually disintegrating in this process, he saw the need for a new and "extended" photoelectric effect.

Matter disintegration and conversion into energy were both described in several treatises written by Dr. Le Bon thereafter. He described matter conversion processes which may only be termed "photonuclear" in effect. He suggested that photonuclear reactions take place in all solar irradiated matter. Light itself could convert ordinary matter into pure energy. Exposure to focussed sunlight could demonstrably convert certain light metals into "energetic emanations".

He traced the actual portion of the solar spectrum which first manifested this matter-converting ability, isolating it in the deep ultraviolet bands. In these experimental arrangements, full focussed sunlight fell on metal plates with their remarkable electrostatic radioactivity the result. But, the natural process of radioactivity in all materials proceeded without help from experimenters. What portion of the solar spectrum existed which could continually bombard matter and produce the observed radioactivity of all metals?

Le Bon stated that there were invisible and highly permeating solar spectra whose power could pierce buildings. They existed beyond the light rays, beyond the deep ultraviolet. Through the use of special sensitive emulsions, he demonstrated the unique identity of the mysterious rays. Dr. Le Bon showed that he could take photographs of outdoor scenes right through laboratory walls. He distinguished them from ordinary infrared rays. "Dark light" he called it, identifying it with Reichenbach's Od luminescence. "Dark light" was a part of the solar spectrum which could sufficiently penetrate all matter and disintegrate it.

PHOTONUCLEAR REACTORS

Radioactivity was a very slow process. Elements were not being consumed every day in a rapid, uncontrolled process of dissolution. The tremendous amounts of released energy did not measurably diminish the mass of any source metal. Why was the radioactive process so very slow and moderate? What natural condition moderated the otherwise annihilating matter-dissolving process?

In Dr. Le Bon's thesis, the photonuclear reaction involved a special "coupling action" which naturally existed between element and ray. Each element responded to a specific light energy series. Whenever the proper ray struck near the specific element, there was a disintegration "reaction". It was not necessary for the ray to strike the element dead-center. The mere proximity of the ray to an atom of the element was sufficient to stimulate atomic disintegration.

Disintegrated atoms could produce very distinct products of particles, forces, and rays. These products and the nature of the radioactive disintegration was determined by the element used and its specific range of resonant rays. One could design a reaction by appropriately arranging elements and rays. Not every ray which struck near an element could provoke the photonuclear reaction. Specific rays and specific elements were necessarily brought together before the reaction could begin. A "fortunate providence". Furthermore, that this bombardment was a true reaction and not just a simple collision became clear by experiment.

According to Le Bon, the intensity of the bombarding rays was not important. When deep ultraviolet rays couple with the proper "resonant" element, showers of identical rays were liberated. The photonuclear reaction was a high-yield electron reaction. Careful theoretical survey of the photonuclear reaction reveals its activity to be a high yield electron reaction. Proper entrant photons from space observably stimulate electron cascades in specific materials. Each photon was capable of stimulating the emission of countless others in a chain reaction which completely swept through the material.

In this mounting cascade, prolific volumes of atoms were disintegrated. Only the presence of elemental "impurities" blocked the continual disintegration of certain elements. The photonuclear process was a chain reaction of far greater significance than those which rely on the release of slow neutrons. Since a single such ray could stimulate the dissolution of a great number of atoms, energetic emissions were constantly being radiated from the most ordinary of materials. Only the spurious and scattered nature of solar light prevented the complete annihilation of all terrestrial elements. Nevertheless, the incidents in which proper rays and elements were "resonantly" disintegrating were sufficiently high to produce measurable radioactivity in all matter.

Dr. Le Bon stated that there were solar rays having far greater potential than the deep ultraviolet. The stimulation of radioactive emissions took place in "successive stages". The complete disintegration of matter occurred when very specific gamma rays resonantly coupled with the proper elements. Such ultra photonuclear reactions released unimaginable amounts of energy. These photonuclear reactions were "complete". There were no intermediate particles formed by these disintegrations.

Thus, only specific gamma rays could completely disintegrate a specific resonant element. It was by these rays that matter was completely converted into pure energy with no intermediate particles. Thus, in the absence of these gamma rays, the reactions were "incomplete"...far less than the "peak" possible conversion into pure energy. These incomplete reactions produced the various particles and rays which physicists were studying as "radioactivity".

Continuing his thesis, Dr. Le Bon discussed the cosmic condition, calculating the amount of energy potentially released in his photonuclear process. It was only the rarity of specific gamma rays and the equal rarity of their resonant elements which prevented the world from dissolving in a flash. Thankfully, the earth surface elements which would dangerously explode by this process into pure energy had long been providentially been dissolved. Stars were eternal reminders of this potential.

Le Bon showed that rare gamma ray bombardments had indeed penetrating all matter, creating the steady conversion of matter into energy in the stars. What Dr. Le Bon had succeeded in demonstrating was never really appreciated by those who later became enthralled and entrapped by thoughts of nuclear fission.

He could arrange the focussed action of such stimulating rays with resonant elements, incalculable volumes of energy being derived under control. The output would last for an eternity. An eternal lantern! Dr. Le Bon was the very first theorist to cite "intra-atomic" energy as the future world energy source. He also was first to design and operate special reactors for the conversion process. Light metals being the "fuel" for his photonuclear reactor, he stated that matter would not be radically diminished even when the disintegration continued for a century or more.

The photonuclear process was one in which all sorts of strange intermediate particles could emerge. Some of these might not be of the "known" particles. Dr. Le Bon also spoke of these as "aetheric" conversions. There were two extremes in the photonuclear process. One, a range of partial conversions produced numerous particles and rays. The other involved the "complete conversion" of matter into energy.

Le Bon showed how specific resultant energies could be determined by "designing" the photonuclear process. One can theoretically tailor these reactions to produce heat, light, motive force, projective force, attractive force...whatsoever kind of energy is desired. His "intra-atomic energy" was the result of tailor-made reactions in which physicists could determine the energetic outcome with precision. One could produce pure electrostatic flux with no other dangerous emissions by reactively coupling the proper rays and elements together.

The absence of dangerously penetrating rays prevented the complete dissolution of terrestrial elements into energy. All naturally observed radioactivities were incomplete and "haphazard". Solar energy normally contained insufficient concentrations of both deep ultraviolet rays and more transcendent rays to release uncontrollable amounts of energy. Nevertheless, Dr. Le Bon declared that the photonuclear process could be harnessed.

In astonishingly simple experiments, he repeatedly demonstrated that the proper resonant coupling of rays and simple elements did release sufficient charged particle volumes to surpass those of the so-called natural radioactive elements. He performed this feat with both magnesium and tin in highly focussed sunlight. The resultant pure electrostatic release exceeded the radiant output of radium itself!

When colleagues protested that he had simply evoked the "photoelectric effect", he proved them wrong by demonstrating the two effects side by side. The photoelectric effect, erroneously attributed to Heinrich Hertz, was actually discovered by Nikola Tesla. It was found that ultraviolet rays could stimulate the emission of electrons from light metals. The output from this effect was enormously magnified when specific rays were matched with specific elements. When this resonant coupling was arranged, the emanations were anomalous and prolific.

Le Bon charted the resonance of rays and elements, beginning in the deep ultraviolet. Each experiment showed that pure electrostatic energies could be extracted from the partial disintegration of light metals when very specific deep ultraviolet rays were filtered. He also showed the tiniest introduction of gamma rays, could stimulate complete and rapid energetic conversions in the appropriate resonant elements. Where did the gamma rays come from? Stars radiated gamma rays earthward. It was the crucible of the stars.

SEA OF ENERGY

In these texts Moray found pieces of what he was looking for. The answer to his energy source possibly lay in rays, released from the sun and the stars. It was intuitive guidance perhaps, but this is all he had to work with. The Le Bon photonuclear theory was the closest anyone had come toward providing Moray with clues toward explaining the performance of his energy receiving mineral. He came to believe that the crystalline lattice of the "Swedish Stone" was intercepting certain of these rays. The necessary research now would involve determining their exact nature. He would also need to discover why his strange silvery white mineral was able at all to intercept these rays. Additionally, why was the ground connection always necessary?

If the crucibles of the stars were converting their matter into radiant energy, then the energy received would be an eternal source. For the moment, he looked up in thanks. The stars were supplying all the energy the world ever needed. Manifestly, here was Providence again at work. How he had managed to find the mystery substance was in itself a chance which no one could have foreseen. How he would use its power would now determine both his and the world's destiny. Eternal lanterns!

According to Le Bon, universal matter is turning into energy in an unceasing photonuclear process. The process occurs in the stars, driving their vast expulsions of light and other energies. In doing so, they too were special radiant sources, flooding space with all kinds of rays. The sun was such a source. Being so very near to the earth, its permeating influence could not be ignored. The sun expelled prodigious volumes of rays throughout its space. Earth received a great gale of these rays. Some of them were visible. Most were not. Natural radioactivity was the result.

All materials were theoretically being bombarded by these permeating rays. This was especially evident when certain materials were exposed to focussed sunlight. If the more invisible rays of the sun were the most likely candidates for the Swedish Stone's activity, then there should be energy maxima and minima throughout the day.

As Moray read of Le Bon's elegant tabletop experiments, he could not help but marvel over the apparent conclusive quality of his statements. Certainly, they were pale in comparison with those which he himself was obtaining. But it was the Swedish Stone which made the difference. What Dr. Le Bon did not have was "the stone". Something about the structure of this stone permitted a powerful electrostatic emission possibly under the ordinary influence of solar rays. Its phenomenal output greatly exceeded that which was experimentally produced by any single element of natural origin.

The mineral possibly held the secret to intercepting special solar rays. The mineral held the secret to releasing the staggering electrical output which he was learning to harness. Adding his own experimental verifications to Le Bon's theory, Dr. Moray studied photonuclear process in materials other than elemental ones. While le Bon studied pure metals (tin, magnesium, lithium, caesium, and potassium), Dr. Moray focussed his attention on the crystallography of minerals.

If it were possible to discover which rays from space activated his mineral, then perhaps it would also be possible to improve the operation of the detector. Isolating the specific rays which he believed were responsible for the "Swedish Stone" phenomenon would give more credence to the Le Bon explanation. In addition, if it were possible to match synthetic microcrystalline structures with their "proper" ray energies, then he could develop even more powerful electrostatic emitters.

The study was completely revolutionary. Dr. Moray advanced the Le Bon theory to the point where it became a plausible thesis. He was convinced now that a "radiant sea of energy" suffused the earth. Moray repeatedly stated that this "sea of energy" continually permeated the earth in energetic gusts. The rays he proposed were responsible were "from beyond the gamma ray bands". Recognizing that these naturally prolific energies and their strange dynamics required a special interceptor, Dr. Moray stated that:

"The most widespread and mightiest of the natural forces has remained so long unrecognized...because man lacked the reagents necessary for the proof of its existence".

Dr. Moray had found one of those reagents in the form of a mineral crystal. An eternal lantern! Nikola Tesla always spoke of the means by which the energy of space rays could become available to humanity. Moray named his device the "COSRAY" receiver, believing that mineral intercepted cosmic rays were causing the material to disintegrate. The disintegration process in his mineral detector was not complete, electrostatic charges being the photoreactive products. The dream of endless power was in his possession. But there would be necessary new research in order to greatly intensify the output of the receiver.

RADIOACTIVE IMPULSES

Though his earliest experiments produced several kilowatts of "electrical energy", it would require much longer development for his mineral to be worked into a completely potent energy source for humanity. Dr. Moray observed in his early experiments that the extremely powerful "electrostatic" energy came in powerful intermittent pulses. In circuits outfitted with a few radio tuning components, he found it possible to "sharpen" the effect. The operation of the device was never stable in those days.

Large bluish sparks often sprang out of the connector wires, this effect having been observed in public several times during early trials. The escaping electrostatic energy was wasted because the receiver could not sustain the incoming power levels. This represented a potential of energy which could be harnessed under the proper conditions. But how to sustain the tremendous power? After having studied the theoretical writings of Dr. Le Bon Dr. Moray believed that he had found the most probable explanation for the intermittent electrostatic bursts.

If solar energy was an incoming gale, then that gale might not be completely homogeneous. What appeared to be a steady stream might, on closer inspection, be a completely disruptive flow. Judging from the observed solar surface, it was more likely that solar rays were being expelled in intermittent violent explosions. Both the activity of the mineral and its unpredictable pulsing quality could be explained if one assumed that very special space rays were arriving on the earth's surface in solar "gusts" and "explosions". In his own terms, they arrived "like huge ocean waves". "Radioactive waves" he called them.

So. The blue sparks appeared because the intermittent gamma ray pulses arrived in unpredictable bursts. Now as he watched the radiant receiver producing its intermittent bursts of electrostatic energy he understood what was occurring. The individual impulses were potent. Each contained enough energy to run hundreds of appliances for the fraction of time during which they were received. Taken over a longer time period their intermittent nature made them generally ineffective for common utility. The solution to this major obstacle was a means by which the initial surge could be stored and thereafter "spread out" over a great time period. New kinds of energy components were therefore devised by Dr. Moray for this very purpose.

Dr. Moray developed several novel circuits in which these components were combined and staged. Several "COSRAY" receiver models were developed in the process. The intermittent impulses were applied to special capacitors through equally special electrical "plasma switches". Received burst energy was "leaked" into these components. The system then spilled out their contents into each successive output stage until the accumulation was too enormous to contain. Back surges were blocked by appropriate components. The volume of electrostatic energy poured forth like a sustained lightning bolt.

Dr. Moray developed a strange plasma "tube" in which the mineral was poised. There were several features of this "tube" which may be best comprehended when studying the diagrams. The mineral was poised against the inside surface of a metal crucible, pressed there by several other small beads of another substance. A straight metal catwhisker touched this conglomerate of beads, the lead from this was drawn outside the crucible. The whole assembly was filled with an inert gas and hermetically sealed. The crucible was referred to as the "tube". It was a hermetically sealed solid-state composition having a metallic envelope.

This component, with all of its parent samples were soon placed in a large safe after each experiment. Experimental developments in these regards were completely empirical and thoroughly painstaking. But, very gradually, he had developed a number of models whose output energy was truly impressive. His continual work with the design was accompanied by continual private exhibitions. Between the years 1925 and 1929, he performed the tests before hundreds of qualified personnel.

In its most powerful embodiment, Dr. Moray described his "bucket brigade" of multiple staged components. A few minutes were always required in order to "tune in" the receiver. Earlier models required an initial "spark start" which was supplied by a small handheld frictive generator. Once tuning was established, the energy would appear immediately, lights instantly reaching full candlepower. Dr. Moray successfully developed 7500 watts through radiant energy conversions. The most powerful embodiments each supplied 50 kilowatts. This output could run the needs of a small factory throughout the day.

On several occasions there were obvious environmental "interferences" which temporarily stalled the activation phase. But once these were overcome, the device operated with a familiar constancy which defied all physical rule. Sharp hammer blows or physical impacts commonly interfered with the detection process. This was due to the frailty of silver wire contacts touching the "Swedish Stone" inside his special metal tube.

The smaller device served as a solar observatory, a completely unexpected instrument. Dr. Moray could make direct observations of solar pulsations. The device followed solar patterns, producing its most powerful outputs during the day. It also showed a very minor corresponding change during the night. Considering that the outputs were so enormous, Dr. Moray was not disappointed. He was simply fascinated.

The precious mineral became "more precious" when he discovered its rarity. Identifying the material with local varieties of the same, he found to his very great amazement that only the Swedish Stone produced the energetic output. This material was completely phenomenal in nature.

Demonstrations with his vastly improved detector model still required an aerial "absorber". The pulsating electrostatic output more powerfully manifested with taller and larger surface area aerials. No activity occurred without the aerial and ground connections. Activity also vanished when the aerial was "shorted", touched, or approached in any way. The earliest photographs show a large outdoor aerial to which the receiver was necessarily attached. The unit was made less attached to this physically stationary poise when a small copper aerial was strung across the room. The need for even this aerial was also eventually replaced by an internally connected copper plate.

Comprehending this requirement demands study of earlier writings. The cosmic rays which Tesla reported were "ultra material particles" which were capable of passing through glass. Tesla observed how projective cosmic rays of this kind were literally drawn into metals. This metallic focussing effect allowed a capacitative surface, effectively extending the space connectivity of the Moray detector tube. The detector "tube" was really a sealed metal crucible, the metallic products fused in place and filled with argon. It was reasoned that space rays were focussed by the detector's metallic envelope, just as certain X-Rays can be focussed by appropriate metal forms. The copper plate conductively exposed more of the mineral to surrounding space, making it a more efficient focal point for the specific stimulating space rays.

The one feature which Dr. Moray was never able to dispense with was the ground connection. Ground connectivity alone made this device operate, a characteristic which was studied with great intensity. His almost daily displays had notable witnesses from the region. The strange "electrical machine" became a topic of great interest among the scientific and corporate communities surrounding Salt Lake City.

Being the size of a tabletop radio receiver, the knob-covered mahogany box generated some seventy five hundred watts of electrical energy. Photographs show that the output from this device was conducted to external appliances by two heavy cables. The total output was strong enough to brilliantly light fifteen two-hundred watt lamps. Additional power from this device operated a small flat-iron, a heater, and a fan.

He continually claimed that the detector was receiving radioactive signals from the sun and stars. Moray's device utilized these radioactive signals to a degree which does not seem possible when contemporary principles are applied. Despite these theoretical boggles, history repeatedly teaches that empirical discovery rules and often contradicts what existing science considers "possible".

Mysteries surrounded the output currents, especially when they were applied to electrical household appliances. Currents from the Moray receiver was able to raise the candlepower of ordinary household lamps far above their normal output rating. When operating such incandescent lamps, the filaments themselves never lit. But the gaseous spaces above the filaments became a brilliant ghost-white. Moreover, the temperature of the lamps remained externally cold to the touch. Several persons at various times reported that physical movement near the device, whether of the whole body or of hands, could throw the tuning off. This would cause the lights and other attached appliances to "go off". His new receiver used several special "tubes" in parallel. These "tubes" were hermetically sealed metal containers. A dear friend, Gabriel Mes, machined parts for Dr. Moray. Through Mr. Mes, several European craftsmen in England and Germany fabricated special parts for the tubes. Alfred Burrell, a local jeweler and watchmaker, soldered contacts and fine silver wires. No one else was ever permitted to see or handle the interior of these strange metal tubes. Dr. Moray referred to them as "boosters" in his patent application of 1931 (application 550611).

The boosters were delicate. Large vibrations would dislodge the internal contacts, causing the operation to fail. In another large family gathering, Dr. Moray demonstrated his receiver using a long copper wire as the "absorber". Someone asked what would happen if they touched the wire. Dr. Moray said that the lights would go out. When then asked whether this would cause shock, they were informed that it wound not.

Currents from the receiver produced strange optical phenomena. Photography became very difficult when using the light from lamps lit by the receiver's output. Large dark spots blacked out most of the photographic detail immediately over the lamps, while the surrounding dark spot boundary gave a sharp clarity. This clarity was a visible effect, persons noting the stark details which these lamps conferred throughout their surroundings. In addition, these dark spots had defined diameters, extending beyond the lamps perhaps only half a foot.

The dark spot phenomenon was anomalous because the actual light radiated by the lamps photographically extended beyond the dark spot. Rooms photographed by these lamps were awash in foggy grey regions which seemed to hover in mid-air near people. While the fogging effects were captured on film, no such phenomena could be visibly detected. Fogging could be explained if invisible ground-sourced arcs were discharging from the device into the open air surrounding the receiver. Careful examination reveals that this is so.

The darkspot which surrounds the receiver is a brushlike discharge of a very special energy. It extends in a confined perimeter surrounding the receiver. The ground connection seems implicated in this process, identical radiant black discharges having been observed in grounded radionic tuners. Dr. Moray understood that his energetic "waves" were incredibly small impulses of enormous power. Was the earth capable of receiving, sustaining, and vibrating back such gamma ray impulses? Was the earth the absorbent medium, his mineral being the converter? New theoretical analysis was demanded by these penetrating questions.

Motors, rebuilt to accommodate the energy supplied by the receiver, ran at extremely high speeds. They, like the lamps, also ran "cold". Dr. Moray reported that when they ran in the dark, they were surrounded by a violet corona. The paradox in these matters came when the current was applied to ordinary resistive heating units. When properly resisted, the currents produced heat. Dr. Moray showed this effect in ordinary electrical floor heaters which became red hot. In his numerous public and private demonstrations, Dr. Moray showed that the energy receiver could light conventional lamps, rewired motors, and power heaters.

In the largest COSRAY receiver, Dr. Moray successfully converted cosmic rays into fifty kilowatts of electrical energy.

Part of his lost art which made this fifty kilowatt radiant receiver possible was a special "stage-by-stage" amplification. It was by this staged reception and amplification (quoted as his "bucket brigade" amplifier) that all the received energy was handled, being absorbed among several simultaneous channels and "spread out" into one energetic flow.

Dr. Moray described his multiple staged process as one by which "a small spark was expanded into a bonfire". The radiant energy from space was received through ultrapure Germanium, whose cosmic ray response came through "seven window frequencies". The Germanium had to be ultrapure, since contaminants (arsenides) would absorb and block released electrons of the photonuclear reaction. When measured in the laboratory, each tube measured the unheard capacity of one Farad!

Dr. Moray, a remarkably gifted electrical engineer, necessarily redesigned power transformers to efficiently handle the excessive voltage from his receiver. A peculiar "step down" process simultaneously involved both an impulse decrease and a voltage decrease. This, achieved through successive transformers stages, successfully brought down the rapidly impulsing high voltage into a low frequency high-current state. The developed currents were definitely not simple electron currents, since they failed to operate conventional motors without rewiring, and operated heavy amperage appliances without heating.

No professional investigator was ever able to comprehend the operation of the Moray device in electrical terms alone. While the attributes of currents derived by the device seemed electrical, their overall effects were decidedly of a different nature. Physicists continually pressed Moray to repeat his explanation of his principles, hoping to get more information on its secrets. The device, so obviously successful, was a true engineering anomaly. In fact, presentation of the device was a "moment of truth" for many. How very fortunate these men were both to have met Dr. Moray and to have personally witnessed the operation of his receiver!

There were those colleagues who were each genuinely thrilled by his discovery. Some postponed judgement on their assessment of the device, declaring that the actual operation might depend on some new "battery" effect. Chemical actions in the detector tube, they said, might be causing a temporary strong emission of energy. Only a timed test of the detector would prove this out. If there was such a chemical explanation for the tube performance, then time would show a gradual waning of energies received. The engineers were very anxious to see this test performed. It would necessarily be conducted in a "sealed" enclosure. They quickly added that such a possibility, while detracting from the lofty concepts expressed by Dr. Le Bon, would be noteworthy enough if found true.

This devalued critique revealed a curious ignorance in so-called professionals, whose reputations were more highly prized than the "moment of truth" which had arrived. Moray insisted that the device relied not on any battery action, but on receptive sensitivity to the suffusive "sea of radiant energy". There were other protocols and agendas within each of these tests of course. There were those whose fears were being continually confirmed with each successful test. These individuals reported back to their superiors, confirming the potential threat of the new technology.

Moray patiently obliged all of the scientific seekers during these new public tests. The improved device was repeatedly taken out beyond the inductive limits of powerlines, into remote desert areas. Grounds were driven deep, the aerial was erected, and the device was tuned. Once power appeared, the system was sealed and left alone. Four, five, seven days...time mattered not. The radiant energy device continued operating. No one ever mentioned the excessive heat which should have come from the sealed trunk, were the current an ordinary electrical variety. No. Instead, the device ran "cold".

Certain academicians, fearful of what his discovery meant for existing theory, took him to task on both his methods and theoretical assertions. If it were possible, these professorial committees might have convinced even Moray that his device "didn't work because...it shouldn't work". Academicians now wished to study Moray's diagrams and materials "on their own". He freely gave them all the pertinent drawings and diagrams concerning the device, but never parted with his original Swedish Stone material. This was secretly and safely locked away from imminent theft.

These professionals, eager to "get their hands on the gadget" cited Moray's "paranoia" as proof that the device was a fraud. Governmental monopoly being the theme of that time period in America, the extent of a highly coordinated "daisy chain" would now reveal the true extent of its boundaries. The REA was constantly intercepting Dr. Moray's grant proposals through professional individuals who were connected to nationwide REA activities. These frustrating occupations stalled the development and proliferation of Radiant Energy technology for two decades, while justly deriving no commitment from Moray.

Wishing now to simply publish his findings in the professional journals, he found that "obtaining permission" to do so would "necessarily" come from University authorities. Opposition now came from University bureaucrats who "refused to handle" the information. The very individual who previously congratulated Moray for "having achieved the impossible", now wrote damaging letters to the very agencies from which Moray hoped to receive funding. Declaring that Moray had "not sufficiently proven the validity of his claims", these letters seemed to be appearing in every energy-related government office long before Moray's formal proposals arrived.

In an aggravating display of smug arrogance, academes began to play the "word game" with Moray. It apparently was expedient to prove that the device did not really work! Physicists examined his reports and plans, returning indefinite conclusions. These non-committal verdicts so enraged the patient Moray that he decided to take his material directly to the government institution which would grant him exclusive manufacturing rights to his designs. If no one else was interested in the discovery of the century, then he certainly would undertake the manufacture and distribution of COSRAY receivers if need be!

The next step was to obtain a patent. In 1931 he decided to assail the patent office with numerous applications. Basing his patents on several different claims, he had hoped that at least one application would be accepted. True to the formula, each such application was returned without explanation. Each was stamped with the official "REJECTED" seal. No title, treatment, or adjustment which he made ever seemed satisfactory to the Patent Registry. After this obvious stall action was repeated far too many times to recount, he reached for his last ounce of patience. Trying to take opportunity from the words of those who once sought to diminish his discovery, he inwardly cringed and wrote what he considered his "very last application". In this, he cited the operation of the device the result of a "new battery action". He thought that, should the patent court officers not officially recognize the validity of his past descriptions, they would at least grant patent licensing on this descriptive basis. His firm resolve was to stop applying, a costly process, until the obvious shady identities behind the patent rejection process clarified themselves and came forward.

Interceptions at a high level were preventing the proliferation of his revolutionary technology. In fact, coupled with academic cooperation, no journal would ever publish his data. In effect, no one would ever hear the very existence of the Moray device. If possible, no information would ever escape from the inventor's own immediate neighborhood. Isolating and imprisoning the inventor to small town perimeters became the new regulatory device.

The Patent Officers rejected his last claim, churning out the response which he also expected. Clearly, there was more behind this merry-go-round behavior than a simple misunderstanding of technical descriptions. He was being stalled for very deliberate reasons. There was no hope of making the Patent Court accept his findings. His was a "no win" situation. Utilities, engineering groups, university personnel, publishers...far too many agencies had already visited the Patent Officers with connections and claims of their own in the matter of Dr. Moray.

Thereafter, he permitted numerous repeat examinations by those who obviously believed him to be a fraud, but eventually lost patience with the skeptics and critics. Here was proof of an enormous phenomenon before their very eyes and all their professional expertise could manage was doubt! Dr. Moray could not waste time with them, moving his research into new avenues and applications. There were many associated phenomena he would discover while experimenting with the Swedish Stone. But social pressures would now seek to divide his time, and far worse.

SOCIAL CRUCIBLES

This phenomenal activity eventually attracted the wrong kind of attention. Of course the radiant energy received by his wonderful mineral was absolutely free, and everywhere plentiful. Its industrial proliferation would trigger a revolution in the power and light utilities around the world. This possibility was not the favored topic of discussion among the local board room members. During the Depression, the REA (Rural Electrification Association) was anxious to "addict" as many isolated rural families as possible to the utilities. Forcing such people to accept the electrical utilities may have been "security" for those whose patronage was buying out the government.

Depression was long and hard for all working class Americans. It was difficult to imagine that the upperclass was moving into newer investments and higher ideals, while millions of children were starving. Therefore and unfortunately, socialism was the attractive dream of many disgruntled Americans. In those days a seductive utopian dream of economic equality, socialism had none of the negative connotations which are associated with Stalin and the Cold War. The Soviet Union was simply and naively viewed as a land where workers were all equal, moving corporately toward a common ideal.

To starving, jobless Americans who watched the rich driving through town in expensive cars, socialism was the war cry of the times. But this, of course, was the "party line" with which socialists drew outsiders in. Daniel and John Magdiel, close friends of Dr. Moray, had become members of the Communist Party. Having gained reputation as American Communists, they independently decided to "help" Dr. Moray. Perceiving that their friend was being stalled by the industrialists for good reasons, they decided to take his "cause" to the Russian Government directly.

In 1926, Daniel Magdiel travelled to Russia with this in mind. The naive plan was to interest a nation whose policies on new technology were "unregulated". The story of Dr. Moray now took an unwilling turn in a direction of political intrigues. Among the many academes and industrial officials who visited Moray, one individual was distinctly not an American citizen. Colonel Yakavlev, an official of the Communist Party, was a representative of high esteem in Russia. He kept his fanatical devotion hidden deep within a smiling exterior, while shopping for new technology.

The American Government was so busy eradicating and regulating new technological developments on behalf of its old family patrons, that it did not recognize a vulnerability to foreign privateers. The naive academes and industrialists who attended Moray's exhibitions did not restrict Col. Yakavlev, seeing in him no potential future threat. Military interests would gradually be attracted, but only after this foreign presence expressed interest in Moray. Superficial liberty is seductive. Obvious when Moray was imprisoned in his own neighborhood by resistive industrialists, the determination to block the

Radiant Receiver was a foregone conclusion.

The indifference to revolutionary ideas has repeatedly proven deadly. And the regulation of revolutionary ideas is deadly. The ideal of a new and better world condition completely escapes the urgent pursuit of governments. Governments frequently act only on behalf of their controlling patronage, an old and traditional reflex. In this instance, the successful military penetration of the national interior by a "vacationing" Soviet Colonel was inadvertently made possible because of economic depression.

Had American investors seized the Moray system and implemented it, their fortunes would have been unlimited. What now occurred "looked very bad" for the inventor in the eyes of later investigators. In 1929, Col. Yakavlev invited him to New York. There, the two men were to meet with "superiors" at the AMTORG Trading Mission. Moray found himself, however, in the offices of General Electric. Meetings occurred after work hours in secretive fashion. Moray did not like this one bit. He left New York abruptly and returned home.

Daniel Magdiel contracted with Dr. Moray to construct a large and completely equipped research laboratory in Salt Lake. Moray was so pressed for capital to develop his technology that he accepted the grant from the young Soviet Union through his friend Daniel. Moray performed research with absolutely no political consequences in mind. He simply needed the money. Daniel Magdiel later moved to Mexico in 1952 for obvious political reasons. By 1933, foreign nations were not the only ones interested in Dr. Moray. Government "regulatory commissions" on electrical utility were the very first to plague Moray.

Contact between the Rural Electrification Association and Dr. Moray began in 1939, with the approach of several officials. Mentioning that Daniel Magdiel and the REA Administration were "going to have a meeting", this individual brought several engineers along. Suddenly, interested parties began arriving at Moray's laboratory. The REA called in a "scientific expert" who, in a very short span of time, attempted to both destroy the receiver and eradicate Moray.

While entering his darkened laboratory, Moray was actually confronted by gunman. One of this company of assassins was the "scientific expert" sent by the REA. A scuffle ensuing, guns were fired among both assailants and Dr. Moray, their intended victim. Moray drew his revolver and fired back with unerring accuracy. They never again harassed Moray, although a "liquidation" threat was made by a government agent in the hearing of young John Moray. Agents were continually harassing Moray to "come back to the work or face the consequences". Moray had anticipated death threats, having replaced all the car windows with bullet-proof glass. He continually carried his own revolver.

These more visibly honest expressions of vehemence stand as warnings. Those who will continue imagining that government bureaucracies are legitimately interested in democratic ventures are gravely mistaken idealists. Doubt, anger, and murder. These three proceed from the heart of the fearful, propelling the self-destructive "conspiracy" of human nature away from the ideals. Away from wonder. Away from the lost and glorious world.

SOCIAL TOUCH

Retrieved patent applications of Dr. Moray have been closely examined. The examiners rejected claims for the radiant energy receiver despite the working model and the statements of credible witnesses. Apparently there are times when demonstrations and working motors are not "sufficient proof" in the Patent office! The technical basis of rejection was a mere semantic tool used to keep the device out of the industrial complex. Examiners claimed that current could not be developed in his device because it "ran cold". How curious that several crystal radio receivers were given license that year. These sets "ran cold"! Either the examiners are not sufficiently educated or quite aware of the annual agenda...as it alternates from year to year.

Though Moray's own patent was never granted, it is most curious that another virtually identical application appears in 1926. Patent 2.032.545 by H.B. McElrath antedates that of Moray, and is virtually identical to it! Complete with a working model, this five-stage amplifier utilizes special radioactive minerals to operate phonographs, public address systems, as well as radio-television receivers. Output volume is strong. The device can amplify radio and phonograph signals without the application of external energy, and requires no tube replacements.

There are several other examples of these patents which have been retrieved, the McElrath system not being the only one to emerge during that time period. Several inventors (Blackmore, Hubbard, Winkelman, Ainsworth, Burke, Farnsworth, Hart) developed remarkable energy amplifiers and electrical sources which implemented small amounts of unrefined radioactive materials. What is more remarkable is the cavalier manner in which the McElrath patent and others, though licensed, never reached the consumer market. None of these patented devices were ever mass-produced. After studying sufficient numbers of such patents, one realizes that corporate connected regulators and examiners established an active search for all emerging energy devices.

All patents having to do with free energy are purposefully blocked from reaching social scales of proliferation, as is evidenced by the great number of patents which never reached the market place. All we have are the official documents, proof that the designs both really existed and that they were successfully operated as described. Certain principle authorities of the University were given a folio of drawings and notes by Dr. Moray in confidence. The drawings surfaced years later, when Bell Laboratories patented their "transistor". The very same designs. Transistor development was the direct outgrowth of Moray's research, being directly derived from his own early models. Improbable? The attorney in charge of Bell Laboratory transistor patents was the very same person who handled the Moray patents.

Dr. Moray maintained his own research laboratory throughout these years, working as a consultant for the radio industry. Various radio companies employed his expertise in the design and manufacture of superior vacuum tube receivers. The golden age of radio saw notable developments in circuit design by Moray. Working for E.H. Scott Radio Labs, he pioneered the development of their famed "Philharmonic", "Imperial", and "LSB" receivers. These chrome-plated grand consoles produced an uncharacteristic "velvety warm" AM and Shortwave reception which sounded like FM.

The notably unusual Moray radio circuits made possible the reception of small stations as far away as Antarctica! On one occasion witnesses clearly heard Admiral Byrd broadcasting from "Little America". Company brochures and journals show Moray and E.H. Scott in Scott's industrial radio laboratory. Later specializing in the design of efficient vacuum tubes and vacuum tube circuitry, it was not difficult for Dr. Moray to find numerous such consulting positions. When Moray went to work for Hammerlund Industries, he developed their "Super-Pro" series. The clarity achieved in these designs was reminiscent of sounds heard through crystal radio receivers.

Employing principles learned through his work with the Swedish Stone, he designed true cold cathode tubes. In these, radioactive materials were used in place of thermionic cathode emitters. Applications of these tubes for continuous high-output operation were employed in military designs. Dr. Philo T. Farnsworth independently developed numerous such radioactive cathode tubes during the same time frame. His caesium coated "multipactor" design is a pure photonuclear reactor whose anomalous outputs baffled radio engineers of the day.

Dr. Moray's talents were admired and sought by numerous radio companies. He helped the Fisher Radio Company by designing the famous Fisher Model 50 stereo amplifier. All of these designs built by Dr. Moray had characteristic coil structures in which signals were very efficiently built up to maximum volumes through successive stages, a development learned through working with his radiant receiver.

Among fellow engineers and technicians he managed to find several warm-hearted persons who perceived genius in this gentle man. Gaining their confidence, he invariably discussed his notions of the "sea of energy", always arousing intense excitement. No one doubted his words or his claims. No one qualified his statements. He was respected by all with whom he came in contact. Friends were completely convinced of his claims. Their friend and colleague had a world-revolutionary discovery, a discovery which would change society completely.

Whereas professorial concerns balanced empirical fact against dogma and patronage, Dr. Moray found opened minds in the nation's industrial workshops. This was his first best clue about truly influencing the scientific community. Thereafter, his approach maintained this personal touch. Numerous lectures were given concerning radiant energy and the possibilities of employing radiant energy principles. No doubt Dr. Philo Farnsworth, a young man living in Salt Lake City at the time, read of these early talks. Both men independently pursued similar inspirations.

FUTURE TECHNOLOGY

Besides the principle radiant device, his power receiver, Dr. Moray demonstrated several different wonders over a period of many years for countless witnesses. Visions of the future, several thousand persons witnessed these remarkable technologies, whether during birthdays, barbecues, prayer meetings, or formal and informal lectures. Hosts of neighbors, relatives, friends, and guests each beheld the full range of his liberal hospitality in these regards.

Dr. Moray never stopped discovering the remarkable new properties of the Swedish Stone. He generalized the principles learned from Le Bon's original thesis, developing revolutionary applications of photonuclear reactions in other materials and chemical compositions. Three distinctly different applications of the general principle were produced over the years following his original discoveries.

The first of these was an accidental effect, found during his experiments with the radiant receiver. This strange discovery manifested while attempting to "tune" the Stone with an early grounded radio receiver. He found to his very great amazement that he was tuning, not radio stations, but local neighborhoods! Headphones attached to the device produced a world of local sounds where no microphones were present. The sounds he heard were human conversations and common workday sounds. Tuning into these eerie vocalizations, he eventually travelled to the very spots and identified the very voices and sounds heard from so very far away. The device was no microphone.

Sealed in a belljar, its connections were solidly drawn under the bell to outside headphones and ground. Tuning mechanisms were all housed below the thick glass tank. It has been reported that this function would only work when the Moray "tube" was pointed at the ground. Several photographs show this mysterious "secret listening device" or "sound pickup device".

Dr. Moray displayed the listening device for family, students, engineers, and friends alike. Each was able, with unerring precision, to locate the neighborhood spots into which the device had penetrated. It was imagined that this device received sonic vibrations in the radiant envelope just at ground level. Tuning with the device permitted a strange "lateral ground sweep" of this envelope.

In one demonstration, two headphones were connected to the device. Once tuned, Dr. Moray handed the headphones to two different persons. Three others were asked to go out in front of the house and carry on a conversation. Dr. Moray said that they would be tuned to the three individuals. Upon listening at the headphones, the conversation was distinctly heard. Each listener could clearly distinguish who was speaking. In addition, the sound of rain falling on the pavement was also clearly heard. During this demonstration, one of those listening decided to "tune in" for himself. One turn on the sweep knob, and the amazed younger listener began to hear other conversations and sounds. He later wrote that he distinctly heard the whistle of a train, the voice of a station master calling out "all aboard", and other simultaneous conversations. He identified the sounds coming from the local railroad station...more than five miles distance from the tuning site.

During this entire demonstration no one carried an external transmitter. The doubters were all put to silence when the young man mentioned what he heard from the railroad station. No one at the railroad station was carrying any kind of transmitter. Sweeps could be made of the entire surrounding area by turning the dial. How this is possible challenged the very heart of electrical science. Questions concerning the basic notions of earth energy were asked.

How were these distant living sounds being derived from a ground wire? Did the sounds associated with human activity somehow have a "biological" effect on the radiant energy environment? Government agencies were most interested in these devices. Years after his unfortunate experience with the REA, Dr. Moray took the listening device with him to Fort Monmouth (New Jersey) Radio Signal labs in 1950, where he developed the system in a top secret government research project. He rarely mentioned the device again.

A second most amazing area of discovery which Dr. Moray engaged concerned therapeutic ray-devices. After working with his radiant energy receiver Dr. Moray began noticing radium-like burns on his hands. In an effort to cure his own problems, he delved into the mysterious surrounding biological healing. Radio frequency stimulation of the body was often used to speed healing process. Broken bones evidenced a rapid repair time when exposed to certain radio impulses. Cuts, bruises, and some burns were also healed by the use of specific radio impulses.

Dr. Moray then studied the use of various radiant therapies, with particular regard for radium and cobalt therapies. He determined that, while each of these methods had their specific effect and use, a more penetrating radiation would outperform all of these systems. He began experimenting with systems of his own, developing several remarkable tubes for the projection of rays. During the research he had found that certain ray energies could enhance tissue repairs without harming the body. He now applied the theory of Le Bon, intent on producing novel and unknown rays through photonuclear reactions.

He arranged specific elements and radioactive composites in low pressure gaseous tubes. The idea was to stimulate a near aetheric disintegration of matter, releasing deeply penetrating radiance which was far less energetic than gamma rays. Near light like emissions was his goal. Tesla generated these in high vacuum tubes with carborundum buttons. The light from Tesla's tubes provoked physiological stimulations of a healing variety. After a thorough series of experiments, he wrote several articles on the subject. In his short treatise on "Alpha, Beta, and Gamma Ray Therapy" he wrote:

"Because the fundamental radioactive process does not originate in the electronic structure... on the surface of the atom...but in the center of the atom, deep therapy is possible over a long time period...".

Dr. Moray developed and used his special "ray applicators" with the deepest conviction that their penetrating powers would render therapy without damage, himself having been the living proof. The theory behind his devices was profound. Learning the radiance emitted by tissues during self-repair, he could apply the same radiance artificially to stimulate repair. Light-like penetrations could stimulate deep tissue healing with specific precision if made gentle enough.

"These rays will penetrate one half inch thickness of lead...and yet they will not injure healthy tissues because of the internal "phantom" characteristics...and the nature of the active material used...".

These mystifying statements indicate the revolutionary nature of the Moray therapeutic devices, large bell-shaped tubes and blown glass containers (patent 2 460.707). Careful study of this design reveals four distinct ray tube forms and several possible variations of each. His therapeutic ray tubes utilized a variety of gases and radioactive materials in judicious proportions and combinations.

With these devices and others like them, Dr. Moray now freely investigated the strange world of gamma and "aether ray" energies. Taught in science classes as deadly, Moray found out that gamma rays could neutralize the radioactivity of natural ores and perform other wonders. Remarkably, these patents were granted. There are those who yet declare that Dr. Moray revealed the secret of his radiant detector in this patent.

According to reputation, exposure to radiant output from these Moray ray tubes was non-hazardous and "thrilling". Some of these devices employed windows of quartz or of ruby glass to project the rays. Output from these large belljar tubes is body-permeating, made to invigorate the entire being. Their stimulating and refreshing influence reportedly produced an invigorating response similar to intravenous injections of vitamins (Bearden).

The YAROM tube ("Moray" backwards), is a blown glass device which is not unlike a cylindrical Zworykin iconoscope tube (Lehr). Electrically activated by 250 Kilovolt impulses, electronic currents are directed toward a strangely configured multi-staged target of unknown compositions. When operating, the device releases a soft, healing pink light. This permeating light appears within tube, easily travelling through the blown quartz walls to the outside air. Hands can block the light, but longer exposures prove their permeating effects. Moray claimed that these rays originated deep within atomic nuclei.

Exposing various materials to the output of this comparatively small raytube, Dr. Moray found it possible to stimulate the growth of crystals and metals. The divided gold content of mining soils were actually made to "grow" by exposing them to specific gamma rays. This led to other more dramatic research objectives, his third revolutionary development.

ENERGY INTO MATTER

By 1961, Dr. Moray was describing a means by which received radiant energy could be "directed anywhere", an obvious reference to a new development. This would have coupled his radiant energy receiver with the ray tube applicators.

His principle forte being metallurgy, Dr. Moray applied his knowledge to various related fields of study; crystallography, metallurgy, and radiant energy. Comprehending these principles very thoroughly now, he was able to "design" crystalline and metallic compositions whose response to radiant energies, whether natural or manmade, would produce specific radiant products, whether special rays or particles.

In addition to the use of crystalline materials in absorbing radiant energy, Dr. Moray explored the possibilities of converting rays directly into matter. His experiments in these regards received notice when, in 1965, he addressed the 68th national Mining Conference in Denver on the subject of transmutation. Originally begun in 1945 as a method for raising the yields of soils taken from gold mines, Dr. Moray rented and employed a linear accelerator at high personal cost. The accelerator, rented from VARIAN associates, obtained very specific energetic electrons for his process.

Exposing various materials to the output of this particle accelerator, Dr. Moray found it possible to stimulate the growth of crystals and metals in these tailings by special treatments. This process closely followed his work on tissue curative rays, an obviously analogous theoretical development. John Moray's statement went as follows:

"...the process we have described is in reality a crystal growing bath, activated by irradiation. As the gold atoms come into being by transmutation, they become 'gregarious', resulting in the crystal formation. From this observation there is every reason to suspect that low-grade ores and mine tailings provide, not only seed for crystal growth, but also a nuclear environment which is well advanced, or uniquely favorable for the formation of the precious metals by...transmutation...".

In a rare and amusing episode, Dr. Moray found that the scant precious metal content of mining soils were actually made to "grow" by exposure to specific electron energies. Gold, silver, and platinum microcrystals were found throughout these "tailings", but in widely scattered "seeds". These rays experimentally demonstrated ability to cause the "organic" growth of tiny gold crystals, scattered throughout these soils. Processes developed by Dr. Moray included mixing the tailings in numerous chemical baths.

His numerous references to the "reagent" and the "environment" indicate that these "catalysts" were the most important feature of his process. The lumpy claylike slurry was poured into large disc-shaped molds of varying depth. These were exposed to electron bombardment on a conveyer belt system made entirely of wood and resins. There were significant "interferences" whenever plastic or metal (especially copper) was employed in the ray chamber.

Early experiments required the "aging" of slurry mixtures, an essential feature of the process. Later statements indicated a new success. With special improvements in catalytic solutions, Moray found it possible to mix the slurries and expose them directly to the rays without "aging" them. He mentioned that the cost of these solutions was not more than fifty dollars per one hundred gallons.

Gold obtained from this 8 million electron volt LINAC system was efficiently produced. But Moray wished to boost this efficiency beyond all expectations. He therefore began to study transmutative reactions within the slurry with greater attention. When it was found that particle bombardments were often not consistently flowing through the samples, Moray designed a special "resonant chamber". This "undulated" with the incoming blasts, producing incredibly high yields. He measured their ability to transmute in terms of cost. Total yield versus total cost.

In tailings which initially assayed at 0.18 ounces Au per ton, the resulting yields were often as high as 100 ounces of gold and 225 ounces of silver! Moray estimated from his results that increased gold percentages ranged between 107 and 329 percents! Once exposed to these electron blasts, the mildly radioactive buttons necessarily had to "cool down". These buttons were then treated with gamma rays. This "treatment" rendered them "neutral", Moray having discovered how to de-nature radioactive matter. These de-natured ore samples were sent to assay offices and analyzed. Their yields confirmed by chemists, Moray proceeded with the manufacture of gold for a time. He then turned his attention on the possibility of raising the levels of weak uranium deposits by his method. The results of these experiments were classified.

It is said that he later designed a small system of his own for the production of gamma rays in prolific quantities. Designed and operated along parameters which embodied his rare knowledge of rays and metals, these were implemented in the gold-growing process with greater success. The use of gamma ray fluxes greatly outdid the performance of cumbersome LINAC devices which were more costly to operate and maintain. Dr. Moray employed radiant bombardment in a special element transmuting process of his own design.

He produced coppers and leads having astounding refractory qualities. Impossible to melt below two thousand degrees Fahrenheit, the lead was the wonder of every metallurgist who received his samples. His coppers were so strong and heat resistant that he employed them as bearings in his high speed motors. An undisclosed alloy, made by the Moray process, could withstand twelve thousand degrees Fahrenheit without melting. W developed extraordinary metals and alloys.

MYSTERY MINERAL

Dr. Moray continued researching phenomena which the Swedish Stone produced, but recognized that he would eventually "run short", having used so much of it in his past experiments. Multiple staging in his last few designs required much of the material. Realizing that his progress toward industrial aims would rely on the artificial synthesis of the mineral, he therefore subjected the Swedish Stone to a complete microanalytical profile.

Dr. Moray considered that, perhaps only specific chemical parts of the mineral were the real "active" components. In addition, perhaps also there were also components which actually limited or "blocked" the photoreactivity. A synthesized compound would eliminate the blocks and maximize the activators. Far more power would then be received and converted to usable energy. Nominal sized COSRAY receivers could theoretically produce gigawatts of electrical energy.

For a moment he considered that, perhaps nature had achieved what could not be humanly achieved. In this case, he knew where to obtain the mystery mineral again in large quantities. Himself an excellent metallurgist, he eventually succeeded in synthesizing his rare material in a laboratory furnace. Moray mentions that it was only after synthesizing the material that he realized the rarity of his original find.

We do not know if Dr. Moray improved the material beyond the Swedish Stone composition. What we yet know of this synthesized material is that its primary ingredient is ultrapure Germanium. Dr. Moray constantly complained to chemical supply houses that their Germanium was "not pure enough". Germanium is derived from euxenite, argyrodite, and germanite. Of the three, euxenite contains radioactive elements and several rare earths. Euxenite decomposes into [Y Er Ce Ti Nb Fe U O], Argyrodite into [Ag S Ge], and germanite into [Cu Ge Ga].

Dr. Moray determined a small radioactive content in the Swedish Stone (Lehr). The various "doping" materials in his synthetic mixture include zinc sulphide, iron sulfide, bismuth, and three other secret elements in "combination". These latter chemicals were never known by outsiders, but have been revealed as a combination of thorium, caesium, and radium sulphide.

Careful analysis of these materials, with sensitive attention to their combined functions, was achieved several years ago by an extraordinary electrical engineer and visionary (W. Lehr). His conclusion is that the Moray component is a photobiased diode which is sensitive to a specific resonant series of signals. Beginning with the X-band, and increasing through to the gamma ray series, the Moray device is a high frequency band-passing gate. A nonrectifying diode. The system responds to specific incoming signals as well as those which are "transduced" through the radioactive materials which are included in the mixture.

It has been hypothesized that the Swedish Stone was rare variety of the mineral Spodumene. Spodumene grows in gigantic crystals, some of these measuring in excess of forty feet in certain locales. Brilliant pink or white, its crystalline surface is smooth. When fractured or crushed, it becomes a smooth silvery white material. Large Spodumene deposits are found in Sweden. Spodumene decomposes into [Li Na Al Si O].

Whether his Spodumene sample contained Germanium, we will never know. Perhaps his knowledge of the Le Bon photoreactions permitted the design of a distinctly new radiant-receptive mixture. It is not generally known what, in fact, is being done with the material and the detector. Some have suggested that Dr. Moray may have destroyed the essential parts of the device. Those who worked with Dr. Moray attest to the absolute validity of his claims. Each witnessed the operation of the radiant energy receiver. Each comprehended somewhat of its essential secrets. Each attempted in some small manner to duplicate these findings. Each subsequently developed certain designs which demonstrated varying degrees of success in releasing anomalous outputs of electrical energy.

True to the archetype of discovery, this is not the only time such a mineral has appeared during this latter part of our century. In another rare documented instance of haphazard discovery, Arthur L. Adams, a retired electronics engineer, claimed to have discovered a smooth silvery gray "electroradiant" mineral in Wales during the 1950's. He found the mineral with a device of his own design. When fine wire contacts touch the mineral surface, high voltages are produced. They are strong enough to constantly sustain a sizeable current in an external load.

This mineral (Adamsite) produced prodigious amounts of electrical power in proper circuit configurations. When sliced into thin layers and stacked among metallic contacts, the power output is greatly magnified. When dipped into water, the output increases. When the stone is then removed, the water retains an ability to produce electrical power for hours.

British authorities managed to seize the material and all the inventor's research papers, claiming to be studying the material for "future social distribution". We are not likely to see this mineral from the British Government (or any other government) until a social change is demanded.

Other energy-receptive devices which use electroradiant minerals have appeared this century. Devices built by several different inventors supplied enough power to light their own homes for years after disconnection from the power utilities (Amman, Molinet). The device by Amman (1930's) used special "chemicals and minerals" in conjunction with electrical capacitors. With this device it was possible to operate an entire household worth of electrical appliances. A subsequent development proved that the device could power an electric car indefinitely.

More recently, an amazing example of this technology has been produced by Gene Molinet in 1982. His development was the result of an astounding observation made while repairing an airplane radio. Removing a crystal component, Mr. Molinet received a heavy shock. He then undertook a complete study of crystals and magnetic fields. His device was observed by an electrical engineer who reports that it somehow utilized Galena and magnets in specific spatial relationships, requiring a firm ground connection for its successful operation.

ENDLESS LIGHTS

Numerous voices throughout the years have expressed desire to reproduce the effects obtained by Dr. Moray. All are after the missing "Stone"! This is certainly one of the most tantalizing mysteries in the annals of lost science. Many opinions have been tenaciously held by researchers concerning the means which Dr. Moray employed in his COSRAY receiver. We also find a great many theoretical propositions and equally as many technical approaches in this quest.

There are indeed several major problems which face those who wish to probe the Moray mystery. There are also several related discoveries of importance help in clarifying our perspectives in these regards. In addition, there have been a significant group of researchers whose work, in combination, reveal the most probable explanations for Dr. Moray's remarkable achievement in radiant energy reception.

The logic trail shifts with the findings and moves with the conclusions. But the rewards for society are great, provided one can match an empirical demonstration with the theory. Above all things, we must not imitate those whose fixation on textbook-approved conventions lead into blind alleys. Building and toppling our house of cards is not frustrating for those who thrill to this quest.

It is imperative that, when academic conventions fail to supply adequate conclusions, one must actively exercise human metacognition over statistics. The prerogative to choose alternative possibilities. To look outside "the facts", and look around "the barriers" of dogma. Therefore this short section will serve to air some of the many ideas historically offered in explanation of the COSRAY detector. The goal, we remember, is reproducing his results. The reward is benefitting humanity with an energy revolution.

In the absence of hard empirical data, the search for the Moray detector is very much a process of eliminations! When considering the available radiant energy of the natural environment, one is faced with remarkable contradictions. These contradictions, however, may not invalidate the obvious demonstrations which Dr. Moray engaged. Researchers have studied a shortwave radio phenomenon known as "bursters" and "drifters". The "bursters" come as short but transcendently powerful electrical signals.

Radio "bursters" remain in one frequency, pouring all of their power into the receivers which entune them. They seem to "stand in place" in a columnar manner, growing in strength with increased time. When once entuned, the small input seems to attract all the available incoming energies until the receiver can no longer handle the power. Bursters destroy receivers and perform other strange electrical feats. "Radio-drifters" are related to bursters, differing only in their frequency "drifting" nature. One investigator has judged the actual energy content of radio-drifters as exceeding a megawatt (D. Winter).

Evidence of staggering incoming power, the energy of the drifter is difficult to explain with conventional theory. Recall that Dr. Moray first became aware of the power potential in space energies while employed by the Telegraph and Telephone Company. His extensive preoccupation with the oceanlike surges, heard through the long lines in headphones, prompted all of his successful research. There are those who therefore believe that Dr. Moray was tapping the energy of the potent auroral electrojets which constantly surge in the ionosphere above us. They therefore do not cite cosmic rays or any such radiant sources in explaining what Dr. Moray "realistically achieved".

These researchers believe that his discovery of the Stone provided a material, a semi-conductor, capable of very high frequency avalanche conduction. The very high voltages, instantaneously released in substantial capacitances such as telephone lines, were capable of flowing through an external circuit to power several appliances. In essence, they believe that the Stone permitted the construction of a high frequency diode having solid state negative resistance (Lehr). It is a common observation that certain impulse energies cannot "pass" through silicon diodes, while effortlessly finding conductive passage through Germanium diodes.

His employment of the early aerial and ground elements provided the capacitance through which the electrostatic energy was absorbed. It was probable that Dr. Moray further enhanced the threshold conductivity of Germanium with special radioactive additives so that it would respond with both speed and increasing saturation to the mounting electrostatic energies. Primarily developed in the large capacitance of his ground connection, these energies passed unnoticed by most experimenters. What was needed to tap this tremendous energy reservoir was a "low threshold switch". Therefore, the examination of the Moray device proceeds as a study of conduction bands and quantum potential energies in crystals.

Some writers proposed that Moray had developed a "cosmic ray diode". In this model, the Moray receiver is treated as a transducer in which cosmic rays drive electron currents. The special material is the ray sensitive material in which this conversion process supposedly occurs.

If we use the available potential energy of a single cosmic ray, we find that it could raise only one ten-thousandth of a watt's worth of electrical energy. If cosmic rays were intercepted by the pellet, producing extensive photonuclear cascades, then why was the ground connection needed at all? According to the calculated values, a detector the volume of that used by Moray could never intercept enough cosmic rays to achieve the demonstrated outputs. The stone pellet itself did not offer sufficient interceptive cross section to be the whole generative center of Moray's device. Clearly the stone pellet was part of a much larger "organized activity" involving the ground.

There is an alternative model which focuses attention on the necessity for ground connection in the device. While aerials could be eliminated from his apparatus, it was not so with the ground connection. One recalls that Dr. Moray was never able to do without the ground connection. Also the successful operation of the device required a "tuning" procedure. What was he "tuning"? Both good ground connection and sensitive tunings were indispensable for obtaining the enormous energies demonstrated in his COSRAY detector.

The volume of cosmic rays intercepted by a volume of ground is vast. Ground entrant cosmic rays would stimulate the activation of vast free electrical volumes. Any section of ground would effectively become an available "interception plate" of vast size. These free charges would "leak up" into any radio ground connection, producing significant and conspicuous "static" power when tuned to certain frequencies. "Tuning into" these radio bands, the ground "interception plate" would pour its vast electrical surpluses into any detector, producing lightning like discharges.

This model sounds plausible, but why would the Swedish Stone be the necessary item then? Why would not any radio receiver locate these frequency bands? The infrequent observation of "bursters" lends this objection some credence. Could bursters simply be a rare "radio detectable" solar emission, one in which electrical impulse are surging at lower frequency? The pulse which Moray was able to constantly receive might then have been occurring in the microwave band (Lehr).

This model might lead in the right direction provided we shifted our attentions to an alternate kind of energetic spectrum. If ground entrant cosmic rays produced a special and distinct subatomic "vibration" in the ground, one which could not be entuned with ordinary radio receivers, then the mildly radioactive Swedish Stone acts as a special kind of receiver...both tuning and receiving ground-wide "radioactive impulses".

Such "radioactive waves" would appear as a "white noise" in a grounded radioactive detector. Specially tuned settings would release the energy into the receiver. We might liken the process to rain falling on the surface of a lake. As the rain comes in sheets and gusts, the lake surface becomes "rimpled" with the disturbance. Tuning into this violent surface of energy would be equivalent to using a straw in order to obtain a coherent flow. Getting a gush of water from such an incoherent vibrance is nearly impossible. A wider "pipe" would not enable a greater extraction of energy. Moray did not rely on ever larger cross-section ground connections. This would be the requirement if we were utilizing radioactive white noise. If such a draft of energy were coming through the fine wire contacts on the pellet, the delicate connections would burn away before any energy could ever be extracted.

When considering "ground secondary radiations", one would have to account for the "reflective surface" within the ground capable of absorbing the cosmic rays and "vibrating". What would be the exact nature of the "absorbent" ground medium? Would it be free electrons? Is there a possibility that the actual medium of absorption was something more exotic than electrons...some subatomic particle sea of which we are yet unaware?

The entire notion of cosmic ray absorption requires the ability of a fluidic medium to absorb the projectile energy of entrant cosmic rays. If the fluid were "free electrons", then the cosmic rays would pierce it like bullets fired into water. The resultant energetic violence would produce no coherent pattern, only a "frothy electrostatic disturbance". A slim possibility might exist for entuning the major "harmonics" of such a frothy white noise, obtaining an electrical output. But the incoherence of all resultant "white noise" energy is still the reasonable objection to this otherwise good model.

The "ground plate" theory might more reasonably apply in this instance. Where individual entrant cosmic rays might effect only white noise disturbances in either an electron sea or exotic subatomic sea, sudden gusts would suddenly "deform" a large ground region of the absorbent medium. The resultant whole regional deformation and recovery would represent a coherent energetic impulse of great power. These gusts could be entuned, the short and unexpected impulses being maximized in the proper circuitry. This is exactly what Dr. Moray described.

The sudden gust may produce a powerful electrostatic potential which appears above the normal background of white noise as "oceanic waves". Furthermore, if the sudden burst energy is electrostatic in nature (received into a terrestrial electrical "tank"), it is certainly received as a singular impulse with a very rapid decay among progressively lower harmonics. In this latter case, the gust impulse would enter the white noise sea, being lost in the incoherence.

In addition, we do know that conventional cosmic rays evidence abrupt showers when measured across a broad ground region. Rare interstellar cosmic rays, mostly nuclear fragments from stars, manifest in "showers". But these do not coincide with the periodicity required for the Moray effect. Solar flare activity would account for a far greater energetic flux, manifesting the "bursts" and "impulse waves" which Dr. Moray observed. We must also remember that, both Tesla and Le Bon considered natural radioactivity as the sign of "true cosmic rays". What conventionalists measure and call cosmic rays do not explain the constancy of radioactivity, and cannot be that of manifestation to which they referred.

If solar cosmic ray showers were not a steadily observed event, then Moray's "cosmic rays" are not conventional. If that is so, then we must determine the nature of both entrant showers and absorbent media. If the earth "energy absorber" proves to be another energy stratum, like the aether of early Victorian physics, then we must discuss how such a fluidic energy can be conducted into the metal wires of Moray's device.

And what if the Moray energy receiver is not be responding to "conventional" cosmic rays at all? To what then did Moray refer when speaking of "cosmic" and "radioactive waves"? There are those researchers who believe that Moray "cosmic rays" or Moray "gamma rays" may be completely different entities than those which academes identify. Dr. Moray believed that it was "neutron bombardment" which were responsible for the powerful electrostatic emissions coming forth from his mineral. One may then consider that Moray agreed with Tesla when speaking of cosmic rays.

To be specific, the Teslian definition of cosmic rays had nothing to do with those which are conventionally described or studied by astrophysicists, being "light-like effluves". Tesla said they were not easily detected. If proper detection of these effluves requires radioactive materials, then the most noteworthy work toward this achieving this goal was performed in special galena radio detection circuits by Daniel Winter.

According to the theories of Tesla and Le Bon, radioactive materials are the dense targets of external energetic streams. These streams were said by Moray to come in sudden "gusts". Mr. Winter proved this principle. Galena crystals were touched with carnotite-tipped catwhiskers. Remarkable electrostatic "spikes" appeared when sensitive spots were touched, the magnitude of which was almost impossible to measure. The device was receiving an anomalous vast energy which came in sudden bursts.

Coupling two such detectors across a space of several feet, it was possible to show a phase-heterodyned signal which existed as a tension between the two receivers. In other words, the lagging response between each carnotite-galena detector showed that "radioactive" waves were travelling across space. Moreover, these waves were of small impulse lengths, seeing that they could be heterodyned across such a short detection space. The waves were those which "caused" the radioactivity of the carnotite samples. Energy bursts appeared when the waves surged. That they were travelling was revealed through the phase-lag between each receiver.

Such phase lag in radioactive materials can be easily see through careful darkroom examination of radium dial clock faces. Maintaining the dial in a constant dark condition for days permits a most amazing observation. What is seen supports the theory of Le Bon. Radioactive luminescence very obviously "flimmers" from edge to edge in endless processions, as if induced from the outside.

One uses this phenomenon as a visual detector of these external radio-inductive waves, observing sudden soft gusts of light which often spread through and across the luminescent matter. This wavelike luminous manifestation was the phenomenon which Victorians marvelled over in a device known as the "sphinthariscope". The flimmering waves reveal a specific band of external energetic induction, the true cause of radioactivity according to Le Bon. Small radioactive additions to appropriate conductive materials might produce the requisite detector of Teslian cosmic rays.

There are those who reduce Moray's detector to an "atomic battery" (P.

Brown). Victor Hart, a researcher who once worked with Dr. Moray, developed a special tube of his own. Witnessed by several credible witnesses, this tube resembles a Farnsworth Multipactor (Brown, Lindemann, Redfern). Cathodes are housed at opposed ends of the tube, the anode being a screen at the tube midpoint. The vessel is filled with helium and argon gases at low pressure. The target anode is a brass screen coated with a carnotite mixture (probably with caesium).

Activated by a four thousand volt discharge, the tube springs to life. Electrical meters which are attached to the device "go off scale" with the increased output, the tube emitting thunderous sounds and a blinding white light. In addition, there is a radiant flux which drives distantly placed Geiger counters offscale. The device is clearly an energy amplifier, taking electrical power at one volume and magnifying it. The agency of this transformation is, very obviously, the natural radioactive material used in the target.

Perhaps the ground surges were not "radioactive" in nature at all. Both the use of an ordinary ground connector and the simple tuning mechanism indicates a radionic activity. In fact, this is the most satisfying model to date. It organizes all the divergent aspects of the Moray receiver quite effortlessly. It does require a new perspective however.

The Swedish Stone might have served as a conductor and magnifier of Od. Such Od radiance could never stimulate electrical charges for Baron von Reichenbach. Perhaps Dr. Moray had found one of those "mineral gates" where this did occur. It may have been in the nature of the mineral itself to effect conversions of Od into electrostatic currents. In citing this possibility we encourage the re-reading of the Reichenbach biography and consider his science of Radionics.

Remember, though Moray was able to replace the aerial with a copper plate, Dr. Moray was never able to eliminate the ground system entirely. The absolute need for the tuning device indicates a necessary radionic tuning component. With the detector in firm ground connection, the mineral became Od radiant. Od currents grow in time, saturating and magnifying in ground lines of any small cross section. Od discharges across space, producing black radiant auras and smokey white photographic traces when arcing into nearby people. The radionic model effectively unifies all the aspects of the Moray receiver, and keeps with the essential simplicity of his descriptions. Obtaining that "magickal mineral" remains the last key to the process!

ETERNAL RADIANCE

There were and are those who would assault the story of Dr. Moray, relegating it to the myths of lost dreams and their dreamers. Yes, discoveries are dreams which never disappear entirely from the social psyche. They are messengers from the world soul, the sea of dreams, the land where eternal light is forever radiant. The mere existence of the idea is the proof that such a world exists. The mere appearance of an idea invades the inertial world where dead minds see no light. Similar discoveries with those of Dr. Moray remain as myths in common places, plaguing those who seek to resist and regulate them. Lost wonder elements of the first world do not yield to the proud. They continue to manifest among the humble.

Resisting the flow of discoveries maintains dynastic fortunes. Addicting society to any particular utility secures that utility for centuries if need be. Those with desire for absolute economic control produce a self-defined "status quo".

Protecting the threatening information became an obvious priority among certain power groups. While resisted by a continually accumulating industrial resistance, the social implementation of "lost science" is a statistical certainty. The imposed forgetfulness, brought about by those whose financial concerns outweighs their social concerns, does not eradicate Discovery. Discovered things, we know, have a mysterious way of haunting the social consciousness.

The archaic disappointment of fire which split the human psyche, produced qualitative and quantitative sciences. The archaic disappointments rule quantitative science in the physics of thermodynamics. The archaic curse was turned to advantage by modern moguls, who employ the all consuming need for fuel into a means for gaining profit. The world system operates on the notion that wonder cannot be real, that no light can exist for eternity, and that light dies in the absence of fuel. Dr. Moray was one among many who discovered a means for breaking the fire-fuel chain.

Throughout the entire time in which Dr. Moray taught, lectured, and conducted his many varieties of research we see the inspired man at work. Dr. Moray made demonstrations of his original discovery before thousands of guests and assembled witnesses with regularity. Each was permitted to see, touch, and handle the radiant energy receiver. Furthermore, Dr. Moray had the scruples of a deeply religious man, whose ethical and highly moral character stand as proof enough of his integrity and essential character.

Dr. Moray's lifelong preoccupation with mineralogy, crystallography, and metallurgy produced equally astounding related developments in three different areas of study: distant communications, medical technology, and metallurgy. The search for an endless lamp, for eternal radiance, is a dream image which yet beckons scientific dreamers. A rare few, like the wonderful Dr. Moray, have discovered and seen its power.

CHAPTER 7

ELECTRIC FLYING MACHINES: THOMAS TOWNSEND BROWN

CAVERNOUS SPACE

For the enthralled onlookers who reported the mysterious and luminary "aeroships" during the 1890's, cavernous space seemed to be opening new secrets and potentials for humanity. The whole nation watched the night skies for signs of strange crafts, ships "from an unknown world". Aeroship sightings swept the country long before the press could reach and contaminate the more susceptible with the furor of panic and mass hysteria. It was the only such mass event in recent time in which unidentified flying objects were sighted, not by media-precipitation, but through direct and continual experience.

The townsfolk and farmland residents of the yet agrarian American Society were bewildered with the source of these sightings. Here was experiential contact, but contact with whom...or what? The first aeroships were ghostlike in appearance. Though fixed in their outward cylindrical form, they often appeared semi-transparent and vague in detail. Their silence was another feature which positively enthralled those who accidentally beheld their serene aerial passage.

Gossamer fabrications, their solid geometric shapes gradually acquired other mystifying attributes. Like a vision, which forms from mist and slowly clarifies to sharpness with time, the aeroships "became" identifiable as some bizarre craft for transportation. Colored lights, flashing lights, searchlight beacons, turbines, sounds...the sounds came after a sizable population saw the objects, and...vaguely "human personages".

Those who looked into the stars were the fortunate recipients of a new and fast coming dawn, where dream symbols were actively weaving the future. A new revelation was suddenly permeating the American mind. Books and gazettes were flooded with tales of aerial abductions. Townspeople shared what aerial visions they nightly saw. Local newspapers were astir with the reports. All thoughts turned away from the earth and focussed on the stars, looking for signs of the strange crafts and their whereabouts. The "mysterious visitors" who made their nightly, silent aerial courses across Midwestern wheatfields seemed vaguely linked with a lost time and a forgotten world. There was something dreamlike in their nature. Dreamlike, yet solid.

Were they the embodiments of some inventor's mad schemes, or were they phantasms of the collective symbolic world? Sighted over California, New Mexico, Texas, Nebraska, Iowa, Omaha, Kansas, Missouri, Wisconsin,

Michigan, Illinois, Ohio, Delaware, New York, the early sightings of aeroships signalled a new movement in the sea of dreams. Soon, human art would join that movement, producing physical crafts which mimicked the first "aerial ghost-ships".

Their movements seemingly had no boundaries or limits. German immigrants had seen these "demonic engines" in their homeland, from 1860 until the 1880's. Why had they seemingly pursued them across the Atlantic? Who were they and why were they demanding attention? What did these voyagers signify? Travelling over the houses of those who would see them, the ships could be described with greater accuracy. All of them were "cigar shaped" measuring some one hundred feet long or more. Better details were seen than those in which the aeroships "soared overhead at six hundred feet". There were mystery ships which came close to the ground, multiple witnesses of high credibility simultaneously seeing the ships land.

Whereas early sightings (1890-1892) were dreamlike and attractively benign, most persons were increasingly frightened by their appearance during the "mid-season" (1893-1896). The strange designs somehow seemed "hostile", though no hostile activities were ever associated with them. People were gradually sensing an insidious "invasion" of their world. Fearing that hordes of nameless, faceless armies would descend and do harm to thousands, ranchers took note and armed themselves.

All too numerous first aeroship sightings remained in the files of the paranormal, involving mysterious personages of truly unknown origins, languages, and abilities. Fears seemed confirmed when some aeroship shadowy "visitors" were seen during night flaps. Gradually clarifying from shadow to light, these mystery beings were observed by a great number of people. Standing amid intensely brilliant "search lights", strange figures were seen examining their craft. Certain of these strange figures spoke bizarre languages, hybrids of familiar dialects. In one case, the design seemed "oriental" in design. The aerial visitors seemed human, but their clothing was totally otherworldly and, somehow, futuristic. They certainly "looked different". Their languages were certainly no identifiable tongue. They came close enough to engage a contact.

Running toward the figures often resulted in their "immediate" withdrawal and ascent. They seemed able to de-materialize and appear overhead in seconds! Intent on remaining elusive, ordinary people were convinced that something supernatural was happening. The "mystery visitors" maintained a curious and dreamy separation from the humanity which they were stimulating. "They" seemed frightened of meeting and engaging people, as if power would be lost through the contact.

Late season aeroship encounters (1895-1899) changed dramatically. Some farmers and mechanics tried running near the ships, describing them as "canoe shaped crafts". They were often flooded within with a "greenish or bluish" light. Under the large housing, there were multiple portholes from which downward looking faces peered excitedly. In several cases there were turbine-like wheels, whose slow turning effected rapidly ascending retreats.

In one case, the mystery night visitors hoisted cattle away, strung by the neck with what appeared to be a wire rope. The red aeroship flew off toward the distant hills. Several of the "later mystery aeroships" were actually engaged in friendly conversation, dirigible hovering in plain sight. Aeroships now became "aerialists", the mystery seemingly solved.

For most, it became obvious that "inventors" were behind the entire phenomenon from start to finish. German inventors! Dirigibles began appearing everywhere. The names Autzerlitz, Eddelman, Tillman, Dolbear, Nixon, and Schoetler seemed to answer the question which frightened German Americans had asked. But these individuals had also seen the early ghostships, an anomaly which could not find a reasonable answer.

Nevertheless, most people were completely assured that the entire history of aeroships was an elaborate confusion of observations...secret societies, hoaxes, publicity stunts, and the like. Certainly a few of these last sightings were indeed the result of secret earthly aerial "clubs". Designers and financiers together undertook the early construction of dirigibles. There were several reports of such an enterprise. The device was huge, used hydrogen gas for lift, and sported several advanced osmium-filament searchlights for night time travel.

The inventor, a Mr. Wilson by name, came out to meet with intrigued townspeople. Sharing with them in a friendly conversation the secrets of his developments, he explained that his point of origin was a "little peaceful town in Iowa". Yes, he was an American, born in Goshen, New York. An electrical system employing "highly condensed" electricity provided propulsive force for the craft. Mr. Wilson added that he had undertaken the construction of five other flying machines such as the one which he flew.

Before leaving, he asked the sheriff, to give his regards to the local itinerant judge whom he knew by name. Asking only buckets of water "for his engine", he entered the craft. Lifting out of view to the many cheers of those who watched, he passed into history never again heard. Dirigibles and other such flying crafts were already becoming a Patent registry revolution; Patent 565805 to Charles Abbott Smith (1896), and Patent 580941 to Henry Heintz (1897), being two typical examples.

Researchers who have investigated the all too numerous mystery airship sightings observe that modes of aerial travel very swiftly became an international obsession among all too numerous youthful engineers. Thereafter, the world beheld a new era of experimental daring, as aerialists played their soaring games before the skyward looking eyes of wonderstruck admirers. Lovely designs appeared, first on drawing boards, and then in the skies.

Cylindrical balloons were wrapped in netting or canvas, and firmly fixed to a "well aerated" gondola, slung underneath. Some of these designs were truly compact and efficient. Engines, propellers, and rudders were all controlled by levers and wheels. The problems of aerial maneuverability were solved by a brilliant little man, a physio-type perfect for the aerial arena. Alberto Santos-Dumont, the aerialist playboy, incorporated his own private dirigible design...for engaging young belles along the shores of the Seine.

Descending from the clouds with his butler-assistant, he brought champagne and succulent delicacies for an occasional "chance meeting". Permission duly granted by governess attendants, butler was exchanged for belle, as the marvelous Monsieur Dumont flew away with his jewel. Never was the fairytale more complete. The socially accepted aerialist was never refused. To refuse Santos-Dumont was to refuse an honor of the very "highest" sort. Wealthy, eligible, poised, and proper, the silk scarfed bandit of the Parisian skies made his daily appearance over and about the lovely Champs Elysees.

Soaring aloft with his more adventuresome feminine admirers, he toured the Parisian skies. No one of these swooning mademoiselles could thereafter claim never to have been literally "swept off her feet" by a man. After a specified time, he easily settled his craft down again with the great skill and panache of an artistic lover. The damsels safely returned to their enthralled and permissive governesses, belle was sadly exchanged for butler. Hands were lightly kissed, a flower exchanged perhaps.

His timing was always impeccably precise. The "wrist-watch", which his friend Cartier first designed for his exclusive aerial use, had already become the rage of Paris. Aerial crafts, strange glass-covered instruments, flying goggles, wristwatches, drooping moustache, and special flying suits...the short little serious-faced man cut a comic, but somehow dramatic figure. Imbued with a sense of the visionary future, women flocked to him. In truth, he remains an historic figure of bizarre aerial gallantry.

Alberto Santos-Dumont justifiably received the most public acclaim in the early days of aerial transportation, a master of the art. His performances greatly endeared aerial transportation to the public as a science, art, and sport. In one exhibition, he successfully maneuvered around and through the Eiffel Tower. Photographs of the event are startling. The art of dirigible flying was perfected in him, the strange little flying man for whom dreamers owe a strong gratitude. Vive Santos-Dumont!

A never-ending armada of aerialists, hoping in part to mimic Dumont, covered the aeroship mystery for most bewildered people of the day with their grand public displays. Forgotten were the phantom-like apparitions of vague form, mysteriously floating like visions across the worldwide skies. Despite the historical closed chapter on aeroships, a single mysterious note of the most exquisitely haunting variety followed the development and deployment of dirigibles.

The story fucuses upon an elderly German gentleman, Dellschau by name. An early and forgotten researcher of aerial phenomena, he maintained records of all the aeroship sightings after 1850. The poor man clung to his precious notebooks until his passing at the remarkable age of 92. These books were later noticed at an aviation exhibition by an inquiring researcher (Navarro). The books are covered with drawings of dirigibles and other clippings, all from the middle 1800's. Among the numerous and rare newspaper clippings were bizarre designs for airships. Far too massive for realistic flight, they may have been attempts to sublimate the apparitions.

There are indications that Mr. Dellschau was a member of a secret society which, on further study of the arcane German dialect in which he wrote, had every aspect of a Jules Verne novel. According to the researcher who examined the notebooks, a group of sixty researchers and developers formed the core of this early Aero Club. The translation infers that aerial ships were tested and flown by the secret group in Germany during the 1850's, and afterward in California.

This anomalous report would explain all the previous sightings, both in Germany and in America, were it not for more important details. On close examination, there were significant inconsistencies with the claims and the designs themselves. The designs each seemed more like rockets, their actual balloon sections being far too small to realistically lift the indicated weight. There are those who would believe Dellschau's descriptions of "NB gas", the "weight nullifying gas", belong to a yet unknown lifting agent. Possibly obtained in the distillation of rare minerals, or in some electrical process, these bizarre explanations would be plausible for many who are aware of similar past discoveries.

Nevertheless, there is another explanation which, having a more macabre fascination, seems to be closest to the reality of both aeroship sightings and Dellschau himself. A reclusive visionary, he wrote in the manner of a mystic possessed by a great and awesome secret. The more extraordinary explanation for both European and American sightings seems to be found in recognizing that the sightings "followed" Dellschau himself wherever he travelled. May it never be said that dreams and visions, suffusing sufficiently empowered human beings, cannot spatially materialize.

ROCKET

The "mystery aeroship" sightings yet remain as true materializations of dream and reality, myth and engineering, archetype and design. Space-projected dream fragments have a curious way of moving through the stimulating revolutions which they materialize. With the expression of the aeroships now in material form, all thoughts of apparitional aeroships were dispatched to the world of dreams and dreamers.

Designers and builders undertook mighty works toward these more material ends, fabricating the grandest, most elaborate renditions of dirigibles. They were one latest wonder in a Century which produced so very many wonders. But those who watched the skies for passing dirigibles made of wood, canvas, glass, tin, and gas were suddenly taken aback. For there, there above the clouds where dirigibles puttered along, new aerial manifestation began appearing.

Dreamy in appearance when first sighted, earth-bound watchers were almost afraid to report them for fear of public ridicule. The apparitions which thousands began seeing and reporting were called "ghost rockets". These cloudlike apparitions were cylindrical with tapered ends. They sprouted prodigious quantities of smoke, while travelling straight across the sky at velocities which seemed fantastic. Like the first aeroship apparitions, these ghost-rockets were absolutely ill-defined and silent.

These crafts, if dirigibles, seemed totally advanced to those who beheld them. Wingless, rudderless, and silent; these devices defied all inventive reason. The ghost-rockets were seen in every nation. Their gradual "acquisition of details" is now a matter of the indelible historic record. Portholes, fins, wings, humans, each appeared in graded successions. In the same developmental manner as was experienced with mystery aeroships, human stimulations determined to build what they sensed. The dream sea, surging, suffused the world-mind with a new quest.

It was no wonder when the idea of space flight seized the imagination of all whose parents previously beheld the silent armadas of mystery aeroships. Edgar Rice Burroughs lived through the days when mystery aeroships were making their inexplicable journeys through both the night sky and the mind of society. A true visionary of his day, he thrilled readers with his Mars Adventure series.

John Carter, his central theme hero, was the earthman who was mystically "translated" to Mars after accidentally walking through a certain "forgotten cavern" of the Arizona deserts. The interplanetary gateway, an artifact of archaic magick, was indeed the most gloriously advanced means for travel among the distant planets. The beauty of this mythic dream portrays the archetypes well, as magickal doorways into other worlds consistently flood the symbolic lexicon of fables and legends the world over.

The Mars Series exposed young readers to the possibilities of interplanetary travel and contact with other civilizations. The extremely sublime dreamforms portrayed and represent by Edgar Rice Burroughs required another thirty years for their realization. Legendary experiments dealing with interdimensional travel continued to haunt American scientific society throughout the remainder of their Twentieth Century, mostly among privateers and natural philosophers. Among the works of several independent researchers it is said that these wonders were approached and actually achieved. In the inability to immediately realize the "gateway" symbol in material form, a mythic theme more capable of bridging gaps from existing technology toward possible new ones was forged. John Carter's mystery caverns and their magickal technology was forgotten. The modified dream quest, the image and frame of desire in the early Twentieth Century, became Rocketry.

Space was opened, a portal pouring forth its dream floods. The great rush of activities focussed all technological attention on rockets and their potential. Rockets into space! Even the heroic tales shifted their focus for the new theme. Buck Rogers and Flash Gordon appeared, embracing their young readers with a fresh new dream whose power derived from more mechanically accessible sources.

Rockets were not being developed by academicians. Too many physical laws taught them to be "impractical and futile". American academicians had difficulty with accepting the rocket as a viable propulsive mode for travel. But these "laws and restrictions" did not stop young enthusiasts bent on making history. Rockets were being made and tested by numerous rocket clubs in Europe. Experience taught that rockets (whether strapped on to sleds, trains, cars, boats, planes, or human rocketeers) were too unstable and dangerous to be taken seriously. Rockets were indeed unpredictable.

Films of the early rocket era reveal the often frightening scenarios of explosions, flying wheels, spinning sleds, and burning coveralls. Solid fuel rockets were too uncontrollable. Once ignited, there was "no turning back". One rocket train experiment was heavy enough not to fly away, but its acceleration was so extreme that the passengers simply blacked out after ten seconds' travel time. Some way had to be found by which rocket thrust could be "throttled".

Here, in America, science writers were busy upbraiding the designs of one Robert Goddard, a high school physics teacher who had been developing liquid chemical rocket engines of superlative power and performance. Goddard's liquid fuel rockets demonstrated the critical control feature so obviously missing in solid rockets. This was achieved through valves, which could be applied or shutdown as thrust was desired. Numerous articles appeared in Scientific American, refuting the very ability of rockets to operate in vacuum...in space. The writers of such outrageously non-scientific articles each offered their "reasons" why Goddard's scheme would fail. Such pen and ink assaults "proved" that rockets would not work in vacuum. It was said that rocket engines would self-extinguish in vacuum.

Among the carnival of ill-informed academic statements we find that singular "proof" which taught that rockets could not long travel through vacuum...not having "anything against which to push". No doubt, this onslaught came just as Goddard was about to receive a sizable research grant! With monies of his own, Dr. Goddard developed guidance systems, fuel-pumps, nozzle-coolant systems, directional stabilizers, and every fundamental component which appears in modern liquid chemical rockets. Government agencies were now thoroughly convinced that rocketry was an impractical scheme.

But the dream travelled among honest dreamers. It settled on a European rocket club which enjoyed their Sunday afternoon lectures. Theirs was a celebration of rockets, space dreams, beer, song, and pretty girls. This club gained prominence in accomplishments which sounded across their land. Their fame was very unfortunately discovered by their own now-fascist government. Despite an overwhelmingly enthusiastic endorsement from one Charles Lindbergh, the U.S. Government failed to deliver Goddard's grant. Interest in his complete patent collection went elsewhere: to Nazi Germany, to be exact.

Back in Zanesville, Ohio, a young dreamer was looking up into the night sky. The very thought of space travel and of visiting other near worlds thrilled the mind of young Thomas Townsend Brown. Tom studied existing rocket engines and rocket engine performance. These revealed great new possibilities for getting into space. In his mind and hands, a far better dream would be weave itself. It would be one which would challenge every fundamental doctrine of science.

He simply wanted to build a rocket engine. A new kind of engine. A small, compact engine which could use very little chemical fuel, and deliver gravity defying thrust. In order to begin this quest, he first took to the library to see what was known about rocket engines. The physics and chemistry texts which he consulted were not encouraging. Dead laws, walls, boundaries, restrictions, and limits were encountered at every turn of a page! They were the same writings used to turn Dr. Goddard's funding requests down.

Tom did not believe that Nature was not ironclad, certainly never limited by "restrictions". Books were not the face of Nature, books were descriptions of small pieces of Nature. It sure was funny how whenever Nature showed a new thing, the books were rewritten and taught again as ironclad truth! Despite his every search through the physics books, it seemed that conventional avenues of pursuit were walled in by laws which said "no" to his rocketeer dream.

The avid young mind was never satisfied with these academic "limits, bounds, and laws". This disappointing wall of resistance from which the young fifteen year old could not turn caused him to move into a new line of thought. Putting the heavy and disappointing texts away, his mind clearly embraced the numerous possibilities inspired by the very thought of space travel. There just had to be some better way to launch out into space! And he would find it.

ELECTRIC ARC

Tom Brown's mind raced. If chemicals could not supply enough thrust, then new fuels and systems could be developed. Potent propulsion systems might be discovered by combining numerous ideas together. There must be a way. Nothing would stop him. Zarkov's spaceship was surrounded by a ring of rockets, spouting electrical ignitions and mysterious applications. Maybe electricity held a secret, yet untapped.

Why did rockets work? Rockets worked because they arranged for the controlled explosion of their fuels. The explosion was shaped and directed by a temperature resistant "reaction chamber" in a single direction. The action of these escaping gases produced the reaction of the rocket mass. Newton was right in this case. The key to a rocket's thrust was the mass of the flame per second and its speed. The mass of a flame was a nothingness, so where did the thrust come from? It came from the velocity of the flame.

A small mass per second was multiplied by the high rate of explosive escape. This product gave the reactive momentum. Chemical explosions gave thrusts which were dependent on their "burn" temperature. The flame speeds could be measured against the speed of sound by several factors. Chemists of the day called this the "fugitive pressure", that is, the explosive pressure.
Tom stopped reading and thought. Could there be another means for reaching higher thrusts with a smaller unit? What could make a flame get even hotter? The hotter the flame, the higher the thrust. The higher the thrust, the smaller and more compact an engine could be. What flame gave the very highest gas velocities? What was "hotter" than the very hottest chemical flame?

The local drugstore had a neon sign in the window. This was always a fascination to Tom. Growing up, he spent time looking into its buzzing glass tubes to watch the red feathery gas which filled the sign with light. Now when he looked into the tube, he suddenly realized something of great importance in his study of rockets. Was the glowing neon a gas whose "velocity" was faster than a chemical rocket? Was the answer to his quest always right in front of his face?

An electrical rocket, of course! Electricity, lightning! These were things whose velocities were close to that of light itself! The highest velocities could be achieved through electricity. Now, here was something to really dig into. How fast would a gas move in an electrical field? It would have to be much faster than any chemical explosion could ever yield. Now he had a direction. Rather than having the texts guide his vision, his new vision would guide the use of texts.

Every book which mentioned electrical discharges gave unbelievable velocities for the glowing gases. Sir William Crookes described these molecular "mean free paths", the free space through which ions could accelerate in the applied electrical fields. Their velocities were enormous, far more than chemical explosions could produce. It was known that a very small spark could produce tremendous pressures (Riess). Such velocities should explode ordinary neon tubes, he thought. Why did this not happen?

Neon signs were low pressure gas discharge tubes. The constant electrical current which passed through them "pinched" the gas into a tightly constricted glowing thread, pulling it away from the tube walls. Tubes operated with constant electrical currents never exploded. But neon tubes were known to crack when the electrical current was applied in a sudden impulse. This released a tremendously explosive thrust...and at such low pressures! This meant that the velocities had to be incredibly rapid, since there was practically no gas inside the tube.

Tom studied on. There were cases when lightning exploded massive objects in which a small amount of air was trapped at normal pressures. Such phenomena taught that electrical discharge in impulses could be coupled with gases at normal pressures to produce tremendous thrust. Furthermore, there were tradesmen who employed this principle daily, when welding metals together. The local welder coupled high current impulses with various gases to weld metals together. He was told that handheld welding apparatus often gave quite a "kick" in certain applications. In addition, there were times when very massive metal objects were thrown clear of the brilliant arc, propelled at high speed by intense arc pressure.

Well then, he had all the information he needed. Several problems would present themselves, but that was the "fun" of engineering. Intense thrust was being developed in the welder's arc, and intense heat. Any rocket reaction chamber which employed electrical arcs would have to be made from new materials. The problem was not impossible to solve. Ceramics might be the choice over metals however strange that seemed at the time. This was new territory, and Tom was designing something new.

According to the mathematical tables which researchers had provided in texts, the velocities of gas molecules in electrical arcs increased with voltage. Higher voltage meant higher velocity components. The "thickness" and "brightness" of the arc depended on gas density and current. All three factors would produce an enormous thrust when properly arranged. These thrusts would do better than compare with those produced by chemical rockets. In a given volume of system space, an electric arc propulsion unit would deliver several more times the thrust of any chemical propulsion unit. This was a staggering thought. If this was true, why had no professional designer ever tried to build an electrical rocket?

Such an electro-engine could be small, compact, and efficient, exceeding the effectiveness of any chemical rocket. Such an engine could travel to the stars. This scheme could be produced with commonly available items. Local shops could supply the gases and arc electrodes. Such a powerful rocket would not be difficult to construct in a shop. The arc-flame of the small electrically powered spacecraft would be white...and small. It would offer aspects of control not dreamed by even Goddard.

That night he dreamed of space travel. Each story seemed closer to becoming real. He would build such an engine. He would both test it and fly it. He would produce an engine that would change the way the world thought of its upper space boundaries forever. Forever.

ELECTRICAL SPRINGS

His next thoughts were to devise the controllers for his magickal compact rocket engine. Electricity could be impulsed or continuously applied. Thrust levels could be controlled by "valving" both the electrical currents and gas volumes. Volumes of electricity could be raised or reduced by rheostats. Valves could control the gas "fuel" flow. Also, if normal pressure gases produced fabulous thrusts, and high pressure gases continued to show thrust increase, what would liquid gases provide?

Liquid gases, preferably of a heavy molecular mass, would provide the very highest thrust levels. The scheme seemed perfect. It was based on sound thinking. Were there any "weak links" in his thinking? He examined every possible flaw, finding none. His tiny little engine might lift a ship effortlessly into the liquid black depths of sparkling space.

This strong response to a wall of textbook criticisms was one of many manifesting their presence across the world. Empirical researchers were already defying textbook restrictions on natural dynamics, producing phenomena which were anomalous by textbook declarations. Tom had a design. His goal was now to develop variations of the system. The goal was to produce the very greatest thrust in the smallest system volume.

First in his awareness was the need to develop electrical power. What compact system could provide the same currents as the welder's heavy arc transformers? No engine could have that much mass and fly aloft. Was there a way to store heavy currents and release them in bursts to provide an equivalent arc welding current? Yes. He would couple small high voltage induction coils to mica capacitors and let the arc explosively burn across a spark gap.

Gas would be injected into the arc space where it would literally explode into electrical plasma, being accelerated out of the reactive area. A more continuous acceleration along the arc channel would provide the most "complete" thrust. Small garage bench experiments with heavy battery discharges proved the spark's ability to "shoot" little pieces of tinfoil across the table. Here was a tiny demonstration of the effect which he sought.

The one problem which yet bothered his aesthetic sense, dealt with the actual power source for the engine. In chemical rockets, the fuels provide both the reactive explosion and the gaseous mass simultaneously. This was their simple beauty and essential advantage. An electrical rocket, a plasma rocket, relied on electrical sources which the gases did not produce. The ideal situation would require a gas, or gaseous mixture, which could produce electrical current for him. Was there such a mixture which he could find?

Here was no impasse. Here was an opportunity. The young dreamer was a fine student. He juggled the variables. Chemical rockets had most of their thrust coefficient on the mass side of the equation, producing heavy showers of molecules at high temperatures. Plasma rockets had most of their thrust coefficients on the velocity side of the equation. When considered from their volume and total unit mass, the preference seemed to fall toward the electrical rocket principle again.

As the increased need for thrust in chemical rockets increased, their total mass increased. With electrical rockets, the power generator reached a "fixed" unit mass. At a certain point in the size ratios, the electrical rocket would win out in terms of efficiency. The thrust equation swung back and forth toward each system in an intriguing way!

He now thought only of reaction velocities. How fast did electrically charged molecules actually travel? Much faster than the molecules impelled by chemical reactions. But, what would be the very highest achievable velocity when using the plasma rocket principle? It would be the velocity of light. No data table even given by J.J. Thomson gave values that high. Molecular velocities in chemical explosions ranged to about 3 kilometers per second, while those in explosive electrical discharges could range up to 3000 kilometers per second.

If high enough electrical velocities could be obtained, would mass cease being a necessary part of the thrust equation? Would a craft which powerfully discharges an electrical impulse just lift itself on its own field? Now this made him think beyond the ordinary concept of rocketry. And here is where our story really begins.

Tom began gathering information in order to embellish his thoughts in this avenue. The information came in the many "collections" left by Victorian electrical experimenters. Were there yet other, unthought "electrical" means by which to propel ships through the air and space? Why stop at using electrically fired chemicals or gases? Could electricity indeed propel a craft by some strange interaction between fields?

Tom had seen high voltage static machines in operation. Surmounted by bent metal pinwheels, these produced violet flames while propelling the wheels around at rapid rates. Here was a real "electrical thrust". Could pure electrical discharge be used to make a ship move without any other propulsive mass?

The real moment, where dream and reality were about to merge came in school. Tom witnessed a very high vacuum electrical discharge tube in operation during a physics class demonstration. Very high voltage impulses of direct current were applied to the X-Ray tube. The heavy wire lines connecting the tube to the induction coil were loose enough to move. Whenever the electricity was applied to the tube, Tom noticed that both wires "jumped" up. In addition, each time a spark suddenly discharged to the tube, the wires "jumped". What was this? When the wires moved did the tube also move? He asked the instructor to do it again.

Tom was not looking at the tube at all. He was focussing his attentions on the free mass, on the wires. The wires jumped each time the impulse was applied to the tube. The sudden jump ceased after he current continued. When the current was removed suddenly, he also noticed a slight wire-jump. This latter jump was not as strong as the initial one, but it was there nonetheless. For an instant, his mind soared out toward black radiant space. Was this exactly what he had been thinking about, right there in front of class?

The electrical discharge inside this very high vacuum tube was pure...a discharge of pure cathode rays. This was pure electricity without any gas molecules to contaminate its progress. The cathode rays were travelling at the very highest velocity to which any particle could be accelerated in such a short space. And here, outside the tube, was a propulsion effect. It was happening exactly where he thought it would be found, in direct line with the free vacuum discharge.

Tom thought deeply on this idea. If electrical discharges of this kind could be made to impulse into free space from a special "gun", then the entire projecting system would move "upon" the extending electrical field. The reaction would be a combined thrust produced, not by mass explosions, but by a charge explosion into free space. The reaction would be a propulsive reaction based on electrical field interactions. Cathode rays held the key...the key to space travel!

ELECTRO-REACTIONS

So...the jumping effect only worked when impulses were first applied. Was the motion simply magnetic in action? Texts said they were, but only an experiment would solve the noise in his head. He was desperate to try this out for himself, at home in his shop. He needed an induction coil of some strength. This was easy. Truck ignition coils were plentiful. Complete with vibrating interruptive switches, these units could produce quite a spark for any young experimenter. He also needed a high vacuum discharge tube, an X-Ray tube.

There might be many reasons for the "jumping wire" phenomenon which he observed. Applied wire current might be straightening out because of magnetic field lines. But, how could these be strong enough to move the heavy wires, being small. The high voltage wires of the induction coil were barely conducting current at all. In fact, within the X-Ray tube, there was practically no current!

Maybe this was some little-known electrostatic effect; something which happened when wires were charged against a dielectric. Vacuum was a dielectric. Maybe the effect happened with greater power in more perfect vacua. Maybe the effect was a simple electric rocket effect in which light, ultraviolet, X-Rays, or some unknown particles were flying in the opposite direction. Maybe the wires jumped because the tube was propelled for an instant by reactive particles flying from the wire to the air as invisible sparklets.

There were more thoughts on which his mind took defined paths. Like the invisible cathode rays shooting along invisible lines, young Tom's thoughts soared through space all around the experiment he had seen the day before. The glorious power in these new thoughts lifted him to such a height of inspiration that it was difficult for him to do anything else. Just imagine! A rocket effect in metal conductors! Would the whole unit, tube, spark coil, and wires, move through space? So many possibilities. So many thoughts. There was only one way to find out for sure.

Electrical discharge tubes could be obtained commercially in various forms, being the popular science artforms of the century's turn. Geissler Tubes were low pressure gas tubes of sparkling beauty. These could be ordered and obtained from hardware stores. These bulbous glassmaker's wonders were curiously bent and coiled, containing various phosphorescent chemicals. High voltage electric streams produced brilliant colorations in these; a true wonder to watch.

Another variety of the Geissler Tube was truly marvelous to see in the dark. These tubes were the miniature electrical "textbooks" of a more elegant age. These tubes were often used as curious centerpieces, the variety being truly Victorian and exquisite in design. Specially treated flowers were poised on a single wire electrode. Each flower was treated with special phosphorescent chemicals. When voltage was applied the flowers glowed in haunting reds, violets, blues, yellows, and oranges. The stems and leaves sparkled with the most wonderful green, extending their thorny sparks to the blue-grey glass tube walls.

While both of these varieties were beautiful phosphorescent display tubes, what Tom wanted was something which matched that which he witnessed in class: a "hard vacuum" tube. The higher the vacuum, the stronger the effect. He was now sure of this. The effect did not occur in Geissler tubes or neon lights. As dangerous as it sounds (and it is EXTREMELY dangerous!), young Brown obtained a small "Coolidge-type" X-Ray tube.

It must be understood that, during this time period, small X-Ray bulbs were not considered as dangerous as they are. One could order and obtain both Geissler tubes from hardware stores and small X-Ray tubes from local pharmacies. Tom was fortunate, using a small capacity induction coil to activate the tube. This was bad enough, but at higher voltages, the effects would have been deadly.

His first experiment would be to try the wire-jump effect for himself...at home. He duplicated the arrangement which he first saw in his school. Applying the high voltage suddenly, the wires jumped. He was amazed, but not satisfied. He had already thought through much of the possible reasons why this might not be what he wanted it to be. Now he wanted to see whether the whole tube moved, wires and all. If it did, just the slightest amount, a new world would be born. A new technology, a new science, and a new transportation potential for humanity.

He allowed the tube a certain degree of friction-free movement on the shop table. The impulse was applied. Wires jumped...and the tube jiggled. In an absolutely incredible manner, the phenomenon seemed to be gaining strength. He was in a sweat now. He knew the next step already. Preparing for this, his mind reeled. It was going to happen. He knew it! The next experiment would clarify both the phenomenon and his future. He expected to see propulsive motion.

The tube had to be suspended in a free swinging manner. Hung in a small pendulum arrangement, the electrically impulsed tube clearly moved in a single direction. He was fascinated, thrilled, awestruck. Unable to contain himself, he yelled, laughed aloud. Dancing around the garage bench, he was the lone dancer in a victory which signalled the birth of an age.

When he finally composed himself, he began thinking with much greater relief than he had during the last several weeks. For here, here was a real mystery indeed. The tube was demonstrating real translation through a fixed distance, without ANY visible reactants. He counted X-Rays out of the puzzle, they had no mass at all. There was no mistake, no "unknown combination of known forces" here. No. His intuitive skills were sharp. What he was seeing was some kind of new field reaction. It involved the longitudinal extension of electricity through a space. Somehow this was the key to releasing the propulsive effects of electricity. Here was propulsion with no mass at all!

The tube was now mounted on the end of a thin wooden rod, counterbalanced, and suspended from a strong ceiling position. Once again, as before, he electrically impulsed the tube. The wonderstruck teenager saw a dream before his very eyes, as the tube began to rotate the entire suspension rod! Each time he impulsed the tube with a sudden jolt, the tube gained speed. The power was cumulative. Each successive jolt drove the rig around with increasing speed.

The tube always moved in a specific direction, always with the electropositive side forward. The new phenomenon inherently contradicted all existing electrical theory in too many ways. He focussed his thoughts on the dielectric nature of the vacuum. After all, the vacuum was a special kind of dielectric. It provided an expansion space for the force lines along the charge flow...a longitudinal expansion. Perhaps the appearance of X-Rays was exactly what Nikola Tesla suggested so many years before. Perhaps X-Rays were the particulate release of pure electricity, even more fundamental than electrons.

The vacuum tube was acting as a "release valve" for some forgotten feature of the electric force, without which no propulsive effect would result. If the instantaneous charging of the plates inside the tube produced expansive force lines to the environment, the effect should have preferential directions with respect to geography. He examined this possibility by noting the strength of each impulse and its propulsive result with respect to compass directions. No difference in the motional effect could be seen despite direction. The whole tube always moved, electropositive plate forward, regardless of compass direction.

The phenomenon circumvented Newton's third law in some mysterious manner. Perhaps the release of high voltage electrical impulses in hard vacuum tubes broke the laws which bind objects together. Did electrical impulses somehow disrupt the natural order in some way? Could electrical impulses in vacuum tube "expanders" be modifying gravity itself? In the mind of Thomas Brown, an entirely different way to propel a spaceship through the deep was now being designed.

The phenomenon which he had just demonstrated to his own satisfaction had no conventional equal. It was far from the engine with which he had begun. Nevertheless, he was by no means disappointed. On the contrary, what he discovered greatly outstripped each of his initial proposals for a compact rocket engine. The phenomenon brought him into a truly alien realm of technology.

There were simply no previous examples or analogies on which to base his theoretics. The closest to these effects were those fleeting recollections made by an elusive Nikola Tesla. When discussing electrical impulse, Tesla spoke of "special reactive forces". At the time in which Tesla made these remarks he was not at liberty to discuss the phenomenon or the technologies which he had developed.

Certain facts now presented themselves to young Thomas Brown. First, his engine needed no reactant nozzles at all. There was no mass in this thrust. All he needed to supply was a steady barrage of high voltage direct current impulses. Current was not even required here. This made the requirements even more simple and elegant. A manifold of high vacuum tubes could be coupled together to form multiple reactions and far greater thrust. Perhaps he could redesign the tubes to better focus and release the longitudinal thrust. There were new thoughts and new technologies to develop — original technologies. The dream space rocket engine was his now.

The young and pensive high school physics student had pushed back an insurmountable wall of conventional objections and academic restrictions, imagination providing the final thrust. His thrilling observations became the heart of a revolution in electrical science; one which seized the world of physics in its day. We have not heard of the phenomena associated or the conditions of initial observation only because theoreticians now consider it "impossible".

His thoughts now turned toward the future, his future. While applying for colleges, he busied himself with meditations on gravitation. He wondered if he had not, in fact, discovered one great and mysteriously hidden gravity secret. Such a secret could be cultivated into a world revolution. Such a secret could reach toward the stars. Summoned by a pilot of the future, his engines could lift an entire crew into the deep reaches of other worlds. He was sure that his system would successfully lift a spaceship through the heavy mantle of gravity out to the sapphire edge of space.

CONTACT

Thomas Townsend Brown entered the California Institute of Technology in 1922. He was a brilliant seventeen year old. With deliberate intentions, he tried desperately to gain the attention of certain notable professors on staff. Robert Millikan was one for whom he held great admiration. He believed that sharing his experimental observations with Millikan would be fruitful. By this time he had developed a substantial base of observations through his home research to inspire others who were "far more capable than himself" of studying his new electric force effect.

Passionate dreamers can never contain their minds. Dreamers seem as impulsive as the energies with which they are involved. Tom simply wished for the development of that spacedrive engine so that humanity could explore the great, unfathomable depths of radiant black space. Had he known the ill manner in which Millikan behaved toward the great Nikola Tesla, he would ever have wasted his time or heart's hope.

Millikan scoffed at Tesla when the latter claimed to have discovered cosmic rays. When a few others corroborated some of these claims which Tesla made, Millikan would not yield. In fact, academe to the bitter end, he made several refutations of the notion altogether. When at last he could not maintain his stubborn ground, he himself yielded, claiming to have discovered "new cosmic rays" of his own.

Therefore, when Tom was rejected by Millikan and several other staff members, he was crushed. The biography of heralded academes usually indicates that, when their dreams were rejected by their superiors, they yielded under pressure. After such a period of ruthless "hazing", they refuted and recanted all their deepest hearts dreams. "Cleansed" of their obstructive romanticism, they find a sudden and peculiar affection being extended toward them once again. Occupying the remainder of their lives mocking other dreamers, they successfully perform highly profiled technical minutiae, and ending in highly honored vacuousness.

For Tom Brown such rejection was difficult to bear. Not one academe, including the illustrious Robert Millikan, accepted Tom's ideas or research work. The anomalous new electric force was, for them, nonsense. The phenomenon had no place in the accepted academic lexicon. He would not yield the romanticism. In this he saw the life of true and noble science. Without passion there is nothing. The small town boy dreamed the world of science. It was his whole life and esteem. The famed persons, the romantic ideals of science and its glory...all of these formed his heart. Things did not go well for him in California after these crushing rejections.

In this first collision with rigid academia, Tom received a precious guidance. What he had was real. He knew it. Why would they not know it? Were they afraid of truth, or just unworthy of its secrets? Exposed to the cold, he now knew where to find the warmth. He would never again give his honor toward those who would not take the time to listen to or learn of his secret. Observing and feeling that frigidity of response from academicians would trigger his ever cautious reflexes. He soon learned that the "rejection" pattern was strangely predominant in places where he would least expect it: in universities and research laboratories.

Tom Brown himself managed to remain insulated from the poison of this contagious academic illness long enough to make legendary advancements in research, taking gradual strength from the first brush against the glacier, he suddenly realized what had occurred. Perhaps Millikan was too busy having lunch those first few times. The others were probably steered by Millikan's opinion of young Brown, preferring the security of their poise to a casual conversation with young students of great promise.

Whatever the case, Tom knew beyond all doubt that his discovery would shake the foundations of science itself. With his simple little garage experiment, he introduced enough imbalance into the accepted lexicon to repel the greatest physicists. Tom witnessed this curiously annoying, but amusing response among those whose prejudices could not simply accept scientific truth.

Tom joyously transferred (1923) to Kenyon College in Gambier, Ohio. Closer to the warmth of home, he no longer felt that greatness was identified with great publicity. He continued his research at higher levels of perfection with funds of his own. Tom, no longer in need of any attention or honor, gradually "opened up" to his physics professor after the latter learned of his private research. Describing his home laboratory experiments and the phenomenon which he had proven beyond all doubt, Dr. Paul Biefield was greatly intrigued. And encouraging. Tom's quiet dignity and confidence said everything which Dr. Biefield needed to see. Serious, sober, sincere. The description of this strange effect greatly aroused curiosity. He wished to see the effect for himself. There was a reason.

Dr. Biefield was once a classmate of Albert Einstein in Switzerland, remaining a close colleague and friend throughout the years. It was obvious to Dr. Biefield that Tom's electromotional effect was something profound. The phenomenon could not find conventional explanation because no conventional expression had been provided for its manifestation. The effect was a gravitational one. Dr. Biefield and Tom both discussed how the momentary blurring of electric and gravitational forces might have provoked an electrogravitic effect. With this discussion came a long standing friendship which would forever change the life of Thomas Townsend Brown.

Dr. Biefield now contended on behalf of Tom's discovery, maintaining that he be encouraged to bring his work to the professional research facilities which the College could provide. Tom was deeply touched, the warmth cautiously returning toward the academic world once more.

Dr. Biefield believed that the "distinct gap" between electricity and gravitation, which academia had fixed for so long, was apparently violated in Brown's small tabletop experiment. Further research evidenced consistencies in similar effects which had been observed in isolated incidents since the turn of the century. The motional force effect was observed through the use of different electrical apparatus. These observations were chronicled throughout the years in several different journals.

Edward S. Farrow (November 1911) wrote an extensive report on his own findings concerning gravity reduction. He observed that ignition coils, attached to aerial wires and plates, lost weight. When placed on accurate, nonmetallic scales and "fired", the entire apparatus lost one-sixth of its weight. Mr. Farrow performed these demonstrations willingly and openly, allowing every portion of his experiment to be dismantled. Furthermore, he encouraged others to attempt reproducing the result. Although with weaker force, the effect was observed in these instances. The difference for the strength of his demonstration had to do with a specially developed a rapidly rotating "spark wheel" which he kept enclosed. Mr. Farrow believed that his apparatus was nullifying the local gravity field which he termed the "vertical component". He considered gravitation to be a special electrical effect which acted within neutral matter. At about this same time, a small antigravitational device was independently developed in Paris. In this, a highly charged mica disc spun at high rate and levitated when electrostatically charged (Ducretet).

Dr. Francis Nipher (March 1918) conducted extensive research on a modified Cavendish Experiment. In the classical reproduction of the experiment, he arranged for the gravitational attraction of free-swinging masses to a large fixed mass. Dr. Nipher's modification included the electrification of the fixed mass. When the fixed mass was highly charged by an electrostatic machine and shielded in a cage, the free-swinging masses yielded unexpected and inexplicable motional effects.

The free-swinging masses first showed reductions in their gravitational attraction when the fixed mass was slightly charged. At a certain charged stage, the free-swinging masses were not attracted at all. Beyond a critical charge limit, Dr. Nipher showed the complete reversal of gravitational attractions. Therefore, shielded electrostatic fields were demonstrably effecting gravitational modifications in controlled experiments, Dr. Nipher considering that electrostatic force and gravitation were absolutely linked. Dr. Nipher's reports were thorough and extensive, forming a real research base on which to perform further research.

George S. Piggot (July 1920) designed, built, and utilized a fantastically potent electrostatic machine with which he observed powerful electrogravitic effects. The device was heavily encased and "dried out" with high pressure carbon dioxide gas. With this dramatically dehumified static generator, Mr. Piggot observed a strange electro-gravitational effect. It was first seen, the result of accidental occurrences while performing unrelated electrical experiments.

Mr. Piggot was able to suspend heavy silver beads (1/2 inch in diameter) and other materials in the air space between a charged sphere and a concave ground plate when his generator was fully charged at 500,000 electrostatic volts. The levitational feat was only observed when the charged sphere was electropositive.

The Piggot effect was clearly not a purely electrical phenomenon. If it were, then the presence of the grounded plate would have destroyed the effect. The very instant in which a discharged passed to ground, every suspended object would have come crashing down. But, without the ground counterpoise, the levitational effect was not observed. Mr. Piggot believed that he was modifying the local gravitational field in some inexplicable manner, the effect being the result of interaction between the static field generator and some other agency the ground.

Piggot further stated that heated metal marbles fell further away from the field center than cold ones. These suspended marbles remained in the flotation space for at least 1.25 seconds even after the static generator ceased rotating. The marbles fell very slowly after the field was completely removed; a noticeable departure from normal gravitational behavior.

Mr. Piggot stated that suspended objects were surrounded by a radiant "black belt". The surrounding space was filled with the ephemeral electric blue lumination common with very powerful electrostatic machines. Many academicians explained such phenomena away. Employing electro-induction theories, it was stated that the effects were "simple outcomes of highly charged conditions in conductive media". The suspension of matter in Piggot's experiment was explained by academes to be the simple result of charge attraction and gravitational balance. Accordingly, charged metal balls would achieve their own balancing positions as long as the field was operating.

Piggot stated that tiny blue spots could be seen running all over the suspended metal marbles, evidence of electrical discharging into the air. This being the case, no net attractive charge could ever develop, simply leaking away with every second into the surrounding air. Considering that the intense field was "grounded" to a concave electrode plate, no consistent charge condition could develop in such a space. Obvious similarities are noted when considering all these cases, the electrogravitic action being stimulated by intense electrostatic fields. Effects developed by Piggot were entirely similar to those observed by Nikola Tesla, who employed high voltage electrostatic impulses.

The Piggot device certainly discharged its tremendous charge in a rapid staccato-like fashion to the ground plate. The rate of this disruptive unidirectional field would be determined by considering the parameters of the sphere and the concave ground plate. Judging from the actual capacities involved, and the sizable free air space, certainly it was a very rapid impulse rate.

Nikola Tesla observed and described the action of such staccato electrostatic impulses on matter in Colorado Springs; particularly on the levitation of dust particles. He later described a heavier than air ship which he said was entirely driven by electrical energies, lacking propellers or jets. There are those legendary reports by those who claimed to have seen this device in operation which have surfaced. An elderly gentleman, the son of a local rancher, described what his father claimed to have seen one night several miles from Tesla's experimental station in Colorado Springs.

Tesla was seen standing on a platform, surrounded by a purplish corona, some thirty feet above the ground. The contrivance had a small coil aft, and was entirely covered underneath with a smooth surface of sheet copper. The platform was perhaps two feet in total depth, being crammed with components. Tesla strode over to the platform, stood before a control panel, and whisked aloft in a crown of white sparks. The excessive sparks subsided with increased distance form the ground, often arcing to metal fencing. Tesla went out of his way to avoid the numerous metallic ranch fencing beneath his aerial course.

What originally attracted the rancher out into the night air was a stallion who had become spirited by the strange buzzing craft. It was said the Tesla often delighted in soaring through the night air for hours each night. He was dressed in his characteristic garb, sans top hat. Tesla became enthralled with the operation of his flying platform, travelling to great distances. Tapping energy directly from his Magnifying Transmitter, the device had an unlimited range. Others had witnessed these strange midnight journeys across the ranchlands.

Dr. Biefeld was familiar with all of the experimental papers, and many of the legends. The growing electrogravitic effects bibliography provided ample reference material by which it was possible to focus in upon the effect which Tom Brown discovered...the "Brown Effect".

Einstein utilized the concept of distorted space to explain the appearance of all forces. According to this theory, there should be various means by which to bridge each of the forces. Could this effect which Tom Brown had discovered not be the "bridge" between electrical force and gravitational force? Einstein said this bridge must exist. Now Tom had a good theoretical model with which to work. If gravitation was truly the result of a distorted space, then high voltage electric shock was somehow further modifying that distortion.

Through Dr. Biefeld's kind and generous support, Tom performed hundreds of experiments with various highly charged tubes and capacitors in stringent laboratory conditions. These experiments were intent on measuring the force exerted by high voltage charged tubes, capacitors, and solids in free space. Charged objects, rather than vacuum tubes, were suspended just as Brown had done in his smaller garage experiments. Professional laboratory instruments measuring every aspect of the motional effect and its epiphenomena, it was now possible to document and film the effect for all to see.

With the new instrumentation and enhanced laboratory access, several details in his strange electric force effect now became apparent. In 1924, he mounted two spheres of lead on a glass rod and suspended them by two strong insulating supports, forming a swing-like pendulum. When each sphere was oppositely and highly charged with sudden impulses of 120 Kilovolts the entire pendulum swung sideways to a maximum point...and very slowly came back to rest. The electropositive sphere led the motion once again.

What Tom now saw was truly astonishing. The pendulum literally remained suspended in the space for a long time. There were two clearly observable phases in the whole action. The "excitation phase" took less than five seconds. The "relaxation phase" required thirty to eighty seconds, coming back to rest in a series of "fixed steps". Astounding! Here was visible proof that a distorted space had power over matter, constraining its movements as if it were solid matter. In fact, Tom now recognized something of his original thoughts concerning the vacuum tube rotor.

As a student so many years before, he had accurately sensed that cathodic expulsions were producing a thrust without mass. Now he understood that the "missing mass" surrounded the device. It was space itself, distorted by electrostatic means. In this condition, space behaved better than any chemical fuel rocket. Space was everywhere available! All one needed to do would be to distort or "warp" that space. Thrust would simply be the result of creating the distorted condition.

GRAVITATORS

Completing university work in 1926, Tom Brown became a staff member of Swasey Observatory (Ohio). There he remained four years. During this time he was married. Maintaining his experimental passions, Tom continued his work on electrogravity while teaching and performing the sundry duties of instructor.

Research never far from his heart, he continued his work privately. His next amazing development came when he replaced his "classical" rod-connected double spheres with a more compact and commercial capacitor stack. This capacitor consisted of alternating layers of aluminum and paraffin-saturated cloth. Unlike conventional electrical capacitors, these aluminum foil layers did not interdigitate. Only the end metal plates were charged during the experiment. The entire stack was coated with asphalt and housed in bakelite, the end terminals protruding as large binding posts.

He made several of these for continuous laboratory testing. When these capacitors were electrically impulsed, the space distortion effect spread throughout their interior. This spreading effect produced greater pendulum swings than he had ever seen. These capacitors absorbed the applied impulse for a much longer time. They also required a much longer recovery time before reaching their "rest point". These devices stayed suspended longer during and after their electrifications. Tom Brown observed persistent space distortion with this apparatus. Five minutes were required before these capacitors reached complete relaxation.

The space distortion effect was now clarifying itself. It actually blended both the Coulomb electric force law and Newton's gravitational force law in a most curious manner. He measured the tantalizing variables which were required to produce maximum movements. He found that longer impulse durations required longer relaxation times. Greater dielectric mass in the capacitors amplified the thrusts. Increased voltages amplified the thrust. He also verified that electrical current had nothing to do with the distortion of space at all. Tom estimated the current in these gravitator cells at 3.7 microamps: virtually a "zero" value. It was the electrostatic impulse which effected the space "warp".

His analysis of the actual effect was quite simple. The high voltage impulse was directed through the body of the entire capacitor stack much like a shockfront, from the electropositive end to the negative end. The dielectric material in which the plates were cast was mildly electroconductive. The spacewarping effect spreading faster than electrical currents, riding the electrostatic shockfront through the capacitor. The result was a space which remained distorted for nearly one minute. The entire capacitor mass evidenced the subsequent motional effect in that region of distorted space.

Tom Brown saw the manner in which space distortion effects could actually move matter. The idea that a strong electrostatic shockwave could actually "warp" space was penetrating. Clearly, the space warp effect suffused the entire capacitor, continuing to collapse rigid space for several seconds. In some cases, with the right dielectric mixture, the collapse continued for several minutes.

During this time, any matter in or near the collapsing space was drawn into and through the collapsing warp. The warp had peculiar boundaries, extending around the capacitor to some extent. In this collapsing space, matter was moved. Here was the "missing momentum" which was evidenced as a thrust. These improvements were so totally different and new, that he renamed the components. He called them "cellular gravitators".

Gravitator effectiveness was related to voltage impulse. Like a saturation, rippling through the gravitator, the high voltage produced a continuously distorted space. It was a saturating shock wave effect, which continuously expanded throughout the mass of the gravitator for a good long while after the impulse had ceased. A prolonged space distortion dragged the gravitator. The space distorting wave front reached a maximum state until total saturation was achieved. Once the gravitator had absorbed the distortion, it stopped accelerating.

No amount of additionally applied voltage had any motional effect on the gravitator after this point. There was a defined reliance on dielectric substance. Space dynamically interacted in the dielectric with the electrostatic shock. This is exactly what he had understood when first examining his X-Ray tube apparatus. The effect was indeed entirely reliant on the nature of the dielectric. The dielectric substance provided the "release mechanism" for the space warping effect.

Eventually Tom Brown developed a simple mathematical relationship which accurately described the successive actions taking place in the gravitator. He then undertook the improvement of gravitators by first formulating a new dielectric "mash" of litharge (lead oxide) and bakelite. Multiple aluminum plates were set into molds, the dielectric matter being melted and poured in over them. The whole assembly was coated with asphalt and housed in bakelite. These units could be mass produced and operated with greater efficiency than his early models. Once cooled, the improved dielectric block was electrified with impulses of specific duration. These gravitators successfully produced a remarkably strong kinetic drive.

Hundreds of these gravitators were now produced for experiments and demonstrations. Gravitators developed terrific and continuous thrusts while the impulse was being absorbed. Placed on opposed ends of a balanced insulator, they formed a rotor. Equipped with slip-ring commutators, they spun with remarkable power when electrified. These results were photographed. The blurred image of the gravitators and their out-bowing support wires is an unprecedented wonder to see.

Tom utilized several gravitators "in tandem". Together, these maintained the drive effect. "Active" gravitators provided thrust while the "spent" gravitators relaxed. It was possible to configure a gravitator cadence in order to produce a constant thrust. Similar to a multiple piston engine, a continuous thrust of great force was demonstrated. The attainment of his original dream had been in part realized for earth-bound service. He now had truly unique electrogravitic engines. The large impulse gravitators succeeded in driving model vehicles around his laboratory. Tom built large model trains and cars, outfitted with multiple gravitators. These powerful models operated on wheels. When properly impulsed in sequence, the gravitator provided continuous thrust, each pulling quite a load across the room.

Dr. Brown designed, built, and tested special linear motors for maritime applications. His ship models worked well in water, mass not being an objectional feature in ships. Large model ship engines demonstrated a terrific drive effect through water. In addition, Tom discovered that distorted spaces also produced "ripple free" water movements, a fringe-benefit of using gravitator drive. He calculated that larger multi-ton gravitators could silently drive ships across the seas with minimal electrical power requirements.

In addition, such drive engines could more efficiently use fuel, since the space warp eliminated all seawater drag. The entire engine, along with specified parts of the hull, was enveloped in the spacewarp. Thrust was produced throughout the interior of the gravitator as designed. But at stronger intensities and with specific space-shaping designs, the space warp could be projected outside of the engines so that even the seawater moved with the ship. In this case, the ship simply slid through a frictionless water environment. The thrust effect flowed along with the ship's motion.

Whether in water or free space, the general thrust effect was curious when closely examined. It relied on the distortion of space, not the medium in which motion was produced. Everything in the distorted space moved along with that distortion. These devices worked with great effectiveness on land and sea. The thrust of these gravitators contended with other engine designs. He hoped to develop drives for airplanes and, eventually, his originally dream. Spaceships!

Next came a series of "rotary gravitators". Mounted on long axles, these massive cylinders were so configured to continually saturate, move, and relax with every applied electrostatic cycle. Dr. Brown described the kinetic efficiency of these space warping "self-excited" motors. The thrust provided by these rotary units vastly exceeded the electrostatic impulse power required to initiate and maintain their rotation. These motors developed one million times the kinetic force of the stimulating electrical input.

They were "self-exciting" because they relied on the collapse of space itself to move their mass forward.

Critics again, playing the old conventional electric game, insisted that his motional effects were the result of "electric wind". He submerged each of his designs in large tanks of oil, only to find that the gravitators worked with increased power! The oil, acting as dielectric, actually magnified the space warp effect beyond the gravitator volume.

When all such tests on his primary engines were completed, it was decided that he obtain an American Patent. Trundling off all of his meticulously prepared applications to the Registry, his hopes "ran high". Unfortunately, his claim was immediately rejected. The official cause of rejection had to do with "improper terminology".

The inability or purposeful unwillingness to recognize world shattering technology makes its strongest stand at the United States Patent Office. The screening process which discerns and separates "dangerous" technologies from "consumer" technologies is very much in place. Those familiar faces who are seen searching the daily flood of patent applications, report back to nameless superiors. The paper chase does so at the behest of the old established money. Inventors of the early Twentieth Century learned all too late that the world markets which exist beyond American shores are far more interested in revolutionary technology than those here having agendas of their own.

Those old elites who coerce national policies and declarations of war to protect their foreign investments do not waste time destroying new technologies which potentially usurp their pedestals of power. Nevertheless, inventors take note....both foreign investors and markets are far more ready and able to implement every new developments which emerge from the sea of dreams.

No doubt the experimental results of Dr. Brown were sufficient evidence that a revolution had indeed arrived; one which certain highly stabilized dynasties did not wish to see proliferated. Disappointed but undaunted, he applied in England for a patent. Was it not curious that there, he was immediately granted patent 300.311 on 15 November, 1928!

In this wonderful disclosure, he recounts his original Coolidge X-Ray tube experiment. Each of his experimental arrangements are preserved in developmental stages. This patent is a true "textbook" on electrogravity, Dr. Brown openly stating that electrified dielectrics moved as whole units through space without measurable reaction. He also indicates that, in this condition, Newton's Third Law of motion is apparently violated. He was granted an American patent thereafter for his rotary "electrostatic motor" in 1930 (1.974.483).

SPACE WARPS

While working with the gravitator, Dr. Brown discovered that its behavior as a pendulum varied literally "with the phases of the moon". In addition, there were startling effects which the sun evidently impressed on the gravitator during its charge-discharge cycles. Whether solar or lunar, it was clear that natural gravitational field conditions were observably affecting local space conditions right before his eyes. The peak maxima and minima of the gravitator varied so much during full moon phases, that he was able to chart the performance against the celestial activities with great precision. After acquiring so much data, he was able to predict what celestial conditions were occurring without visually sighting them. This is when the military became intrigued with his work. A new phase of Dr. Brown's career here began.

Remember that gravitators, as pendulums, do not behave as ordinary masses. The mass of a gravitator is modified by the electrostatic impulse, forcing a new interaction with gravitation. The gravitator rises during the electrostatic excitation pulse, doing so rather rapidly and discontinuously. When carefully observed, the "rise" phase consists of several "graded steps". Once through this "stepped rise", the gravitator manifests a "tension" while held in its maximum position. The gravitator appears to be in a fluidic channel while suspended at an angle. In this levitated position, the gravitator "bobs" several times. After the shockwave has saturated its dielectric thoroughly, the gravitator begins its lengthy "fall" back to the rest point. Here, more than during the rise phase, one most clearly observes the "rest steps" which last for several minutes.

The discovery identified the number and position of spatially disposed "rest steps" with the positions of sun and moon. In more refined optical examinations, one could even discern the effect of certain planetary configurations on the gravitator. These fixed space "slots" became the most intriguing discovery since his original observation of the electrogravitic interaction.

In order to test the electrogravitic hypothesis, he was forced to go into "extreme" locations. He discovered that the celestial space warping effect appeared without loss of strength even in very deep caverns. Here was an effect which was decidedly not reducible to electrostatic effects alone. No, this was very obviously an electrogravitic effect. The only measurable and penetrating "force" in the cavern was gravitation. Enclosed on all sides by solid walls of conductive rock, the cave effectively formed a Faraday Cage. No net electrostatic field could be determined that far down. The combination of solar and lunar warped spaces was evidently "bathing" the earth.

Terrestrial space, being thus bathed in transient space warps, was constantly changing its "symmetry". It was not homogeneous. Along with this change in space symmetries, there were constant variations in gravitational potential in every locale. The effects could not be detected by inertial instruments alone. It could only be detected in electrostatically activated instruments. There, an interaction was taking place; one in which gravity and electrostatic charge produced strongly observed motional effects. On retrospect, Dr. Brown discovered that several professional observers had measured these "anomalies" and discounted them. Using special string torsion equipment, highly charged, their instruments gave "perplexing" and inconsistent results.

In 1930 he was employed in the Naval Research Laboratory. Between the years 1931 and 1933 his research in dielectrics became classified information. New effects were now observed in electrical systems. Immobile capacitors evidenced fluctuations in field strength throughout the day. He first observed that all charged capacitors revealed fluctuations which varied with both solar and lunar cycles. Relative space gravitational effects registered with strength in both capacitor pendulums and immobile capacitors. Extremely basic instrumentation were thus capable of delivering celestial information to observers, the electrostatic fluctuations evidencing sudden "events" of unknown significance. Critics claimed that these events were simple "internal noise" effects. Brown relocated to a bunker-like concrete test facility and immersed his instruments in refrigerants. The instruments worked with greater strength and finer precision. The signals were actually amplified. Silence is golden.

These were the first recorded instances where gravitational "waves" had been scientifically determined with great accuracy. These experiments were first performed in very heavily shielded ground level buildings. The Naval Research Laboratory funded his research between 1937 and 1939, establishing underground gravity wave measurement stations in caverns and mineshafts. These experimental stations, in Ohio and Pennsylvania, remained highly classified for years.

This information on gravity waves was considered by Brown as proof of Einstein's predictions concerning gravitational waves. The work was presumably classified because of the possible implementation of the sensors in warfare. In his underground stations, Brown measured daily changes in sun and moon. Despite the extreme depth of his apparatus, the signals continued manifesting in their strength. The results of this work yet remains classified, though Brown published certain of these on his own after the World War.

Identifying his secret underground systems as "gravity wave detectors", Brown was forging new scientific and technological ground. Greatly appreciated by the military, he remained in their secure employ for several years.

During this period of time, Dr. Brown also discovered several remarkable characteristics of matter in which was indicated previously unrecognized gravitic interactions.

He found that gravitational distortions actually altered the electrical resistance of matter. He made his own large carbon resistors by coating long porcelain cylinders with lampblack. These were scored with a rotary cutter, the resultant resistor being a fine carbon "ribbon" measuring over 500 Megohms. With these it was possible to show remarkably strong electrogravitic effects. These signal devices outperformed the capacitor detectors in signal strength and overall response characteristics.

In this wonderful interaction, Dr. Brown perceived that gravitational fluctuations and carbon were intimately linked. He later dared to speculate that certain physiological states might be caused by space warp effects. These, acting on the carbon of the body, produced symptomatic effects of nausea and malaise. In other more neurologically sensitive individuals, such interactions might intensify into perceptual distortions and anxieties. It was the possible "organismic interaction" between physiological states, perceptual states, and gravitational warps which caused Naval researchers to study Dr. Brown's reports with now greater intent. The NRL had mysteriously pursued every aspect of perceptual science since this time period.

Experimenters developed highly complex equipment for measuring gravitational radiation during the 1960's. These large and highly funded academic installations, massive aluminum cylinders and ceramic strain gauges, never registered the definitive signals observed in Dr. Brown's elegantly simple apparatus. Since that time others have observed and confirmed Brown's findings (Hodowanec). The Brown gravity wave detectors represents a new astronomical tool which yet awaits academic implementation.

BLACKOUT

The increasing financial pressures of America's Great Depression forced Dr. Brown to leave the NRL, sign with the Naval Reserve, and join the Civilian Conservation Corps in Ohio. In 1939 Dr. Brown became a lieutenant in the Reserve and, after brief employment with the Glenn L.Martin Company, was directed toward the Bureau of Ships. There he worked on the magnetic and acoustic aspects of warships.

It was during this time that Dr. Brown was to embark on an adventure which would alter the path of his life forever. Many of the details and facts have been pieced together in a patchwork of intrigues. Gleaned from several reputable science sources, the incident reached public awareness as the "Philadelphia Experiment". What sequence of events triggered the Naval Research Laboratories to investigate the possibility of optically "cloaking" vessels of war?

It all began when several Naval researchers were asked to investigate a peculiar phenomenon which was plaguing a classified arc welding facility. This facility was classified because it protected a new Naval process for fabricating very durable armor plated hulls. The spot-welding process employed an incredibly intense, high amperage discharge. The process was similar to modern MIG welding, but was conducted on a titanic scale. Electrical power for this welding process was supplied by a massive capacitor bank charged to high voltage. Several steel plates could be thoroughly welded by this process, the metal seams absolutely interpenetrated at the weld points.

So intensely dangerous was this electrical discharge that personnel were restricted from the site when once the parts had been configured for the weld. Hazardous charge conditions being the least worrisome aspect of the process, X-Ray energies were released in the blinding blue-white arc. Applied by a heavily insulated mechanical arm, the arc was pressed to the plates by remote control as power was stored in the capacitor bank. The safety signal being given, a lightning-like discharge absolutely rocked the facility. Radiation counters measured the intense release of X-Rays. The process was a new advance in Naval technology.

Neither the extreme electrical or radiation hazards obstructed deployment of the system to other Naval facilities. Safety precautions were at maximum levels. Workmen faced no hazard outside the welding chamber. But another group of strange phenomena began plaguing the facility. Phenomena which had no reasonable explanation at all. Researchers examined the site, separately asked workmen to confirm the rumors they were hearing, and watched the process for themselves in the control booth.

What they saw was truly unprecedented. With the electrical blast came an equally intense "optical blackout". The sudden shock of the intense electric weld impulse was indeed producing a mysterious optical blackening of perceptual space, an effect which was thought to be ocular in nature. This peculiar "blackout" effect was believed to be a result of intense and complete retinal (rhodopsin) bleaching, a chemical response of the eye to intense "instantaneous" light impulse. This was the initial conventional answer. The more outrageous fact was that the effect permeated the control room, causing "retinal blackout" even when personnel were shielded by several protective walls.

Any effect which could permeate walls and render personnel incapable of sensation in this manner could be developed into a formidable weapon. The wall-permeating blackout was a neurological response which paralyzed the whole physiology, rendering it incapable of response to outside stimuli. So it was thought at this point. The research was earning and acquiring new levels of military classification by the day. Here was a possibly radiated phenomenon which temporarily neutralized neural sensation, transmission, and response.

The weapons experts knew that any electric radiance which could be substituted for a nerve gas would offer a new military advantage. An extraordinary means for deploying the effect, the horrid energy could be "beamed" to any site. If properly controlled, entire platoons could be rendered unconscious in a single "swiping flash". An unfortunate victim of such exposures was a certain William Shaver. Mr. Shaver worked as a Naval arc welder with much earlier and smaller hand-operated versions of this system. These systems employed intense impulses of low repetition frequency. After repeated exposures to this impulse energy, he began freely hallucinating. The result of neuronal damage, the centers of his will began shredding away.

This otherwise stable man ultimately lost his grips on reality, writing hundreds of pamphlets throughout his remaining years on the frightening topic of "beings from the underworld". It was subsequently discovered that exposure to sudden electrical impulses of intense potential and extremely low frequency produces a deadly nausea, in some cases even the neurological damage leading to eventual madness.

Careful examination of the effect before the NRL now proved perplexing. First, the "blackout effect" could be photographed as well as experienced. Therefore it was not a mere neurological response to some mysterious radiance. The blinding discharge was doing something to space itself. Researchers were now drawn into this project with a deep fascination. The "blackout" effect drew equally intense interest by Naval officials for obvious military reason. Careful study of research publications funded by NRL grants reveals an intense preoccupation with all such perception-related subjects.

But there were "other aspects" of the phenomenon which were chilling. Bizarre rumors were being shared by certain of the original weld-site workmen. Remember, these men were on the site throughout the period which proceeded the project's classification. They were privy to certain other phenomena which had no rational explanations.

Personnel hoisted hull materials and braced the pieces in composite arrangements for the discharge operation to commence. The warning alarm sounding, all workers and inspection teams promptly left the site, frequently dropping tools and other implements where they stood. Capacitor charging required several minutes. The switch thrown, a tremendous rocking explosion shook the site. The discharge produced the blackout effect, and when the room was declared officially "clear", workers returned to the chamber.

Workmen began noticing that tools and other weighty items, left on the floor or around the chamber, were somehow "misplaced" during the heavy arc discharge process. Imagining that these tools had been thrown into corners or possibly driven into walls by the room-rocking blasts, workers searched the entire welding facility. The tools and other materials could simply not be found (Puharich).

Now the mystery was intensifying to a degree which demanded a complete study of the phenomenon from its first observation. Workmen were called in to report what they had seen, felt, and experienced. Repeated stories matched to such a degree that the "rumors" were now taken as "personal testimony". The entire proceedings were so highly classified that military agents were not even aware of the study. What workmen told examiners was that their tools and other site materials were "disappearing", and disappearing "for good".

Foremen had scolded and ridiculed them repeatedly about this loss of materials and tools until experiencing it for themselves. One fact was clear, when the alarm blew and the discharge exploded, objects disappeared. Where they went, none could say. High speed films proved that the effect was real. Objects were placed on pedestals near the discharge arc. On discharge, the objects dematerialized. The films proved it. They were certainly not "thrust away" at high speeds, or even "impacted" into walls by the intense arc blast.

At first again, the conventional answers came forth. The blackout effect was seen as a mystifying radiant energy, possibly a specific variety of X-Rays. These rays had power to both neutralize human neurological response and disintegrate matter in its immediate vicinity. Here was the possible "death ray" for which the military had long been searching. The Second World War was raging, a possible second "theater" was developing in the Pacific, and this sort of fundamental discovery was enormous in military potential. To end the war was the aim. The only aim.

If this effect could be developed into a weapon, it would be deployed instantly thereafter. A weapons program of this kind would require the nation's most eminent scientists, and levels of secrecy which demanded the very highest stringency. Several Naval personnel were summoned for this study. Dr. Brown was requested to examine "the phenomenon". His knowledge of "dielectric stress" phenomena and the activities associated with arc discharges made him a perfect candidate. Keeping him "in the blind" concerning their ultimate hopes for this new discovery would not be easy. He "had a name" for being the dreamer.

When Dr. Brown reviewed the material, his conclusions were strikingly different from those which others gave. While academes adamantly insisted that the observed dematerializations were the result of "irradiation" and subsequent vaporization, no such evidence for the "vaporizations" could ever be found. Careful analysis of weld-chamber atmospheres proved negative in these regards. No gasified metals were detected in the room air throughout the discharge event. Truly mystifying. The NRL had to know more.

Dr. Brown was sure he knew what was happening here. Despite the fact that he had never observed these effects, his intuition taught him well. Though in his early experimentation, he never experienced any of these blackout effects, but Sir William Crookes had seen this very thing. Within the action space of his now famed high vacuum tube, Sir William beheld a curious sight. There, suspended over the cathode, was a black space which was actually radiant. The radiance extended beyond the tube walls in certain special instances. Sir William had no difficulty accepting the fact that this was a "space-permeating" blackness, a radiance having far greater significance than a mere physical phenomenon. Crookes believed this radiance was a spiritual gateway, a juncture between this world and another dimension.

In the blackout phenomenon Dr. Brown yet recognized the signs that space distortions were taking place. What was the upper limit in strength of these space distortions? What other bizarre anomalies would they manifest? His own small gravitators operated through high voltages now considered "small". When compared with those used in the new welding site, they were minuscule. Nevertheless, his experimentation proved the effects of small space warps. Material dragging was one such anomaly. In short, he believed that every anomalous inertial behavior could be traced to such space distortions.

In studying the entire effect, no single part was unimportant. Dr. Brown knew that the massive hulls played their part. They were somehow "spreading out" the electrical field and giving it a specific shape. The electric arc, focussed onto the hull by the mechanical applicator was the formidable field source. But something "extra" was occurring. Another realm of realities was entering the scene where arc discharges blasted through the welding chamber. He was the only one, perhaps with two others in the nation, who would propose that the phenomenon was the result of an interaction which was "electrogravitic" in nature. These were electrogravitic phenomena.

EVENTS

Associates reviled his thoughts and rejected his analysis of the problem. But the military needed answers. If Dr. Brown could bring them closer to their weapons goal, his explanations would take precedent. Acquiring the complete attention and respect of very highest military specialists, he was asked to formally address their small and elite corps.

Dr. Brown very casually described what he strongly believed was happening, citing his own work and familiarity with such phenomena. While his own experimental apparatus never produced spatial distortions of this extreme intensity and focus, he nevertheless observed similar effects which had power to move matter. Having no conventional electrical explanation, the only resolution was found in the Einsteinian proposals concerning electrical and gravitational force unity.

Einstein had already predicted that intense gravitational fields would produce optical blackouts. But his theory involved the huge masses of collapsing stars. Total blackout phenomena are theoretically achieved only when extremely dense matter is compressed or pinched in stars, producing black holes. Masses alone could not be whirled into producing the intense effect which intense electrical impulses had seemingly achieved. But was it a true blackhole effect, or something entirely distinct?

Einstein's work toward unifying electrical and gravitational fields through space geometries never found completion on paper. But the Naval experiment had proven the essential correctness of his thoughts on the topic. Perhaps tiny blackhole phenomena might have been achieved in miniature when stimulated by intense electrical fields. This mystifying effect was the most intense electrogravitic interaction purposely, though accidentally, ever produced in technical history. The military assembly was absolutely in awe.

Dr. Brown continued to describe what was occurring in and around the arc channel. The channel itself was producing its own "hard" vacuum in stages. Though occurring in atmospheric pressure, the explosive force of the plasma arc had thrust all atmospheric gases out of the arc in its first few microseconds of formation. The full force of the blast was now occurring across a vacuum dielectric. The vacuum actually hindered the complete discharge of the capacitor bank for a few more microseconds, allowing the potential to build beyond those effects observed in weak lightning channels.

It was in a sudden avalanche that the entire discharge occurred across this vacuum space, warping space through an electrogravitic interaction. The interaction was directly related to the voltages, the dielectric volume, and the brevity of the impulse. The normal density of inertial space was being instantaneously pierced, the arc literally "punching a hole" through the continuum.

The explosive vacuum arc set the stage for "uncommon" observations. Surrounding the intense electrical impulse, space itself was collapsing; space and everything within that space. The strange blackout effect would be expected if all available light was being bent into the arc channel. Incapable of escaping the distortion of space, the blackout effect spread outward. Provided the distortion was intense enough, a specific large volume of space would be "drawn" in toward the arc channel. The interaction took a few microseconds to effect. There was no escaping its presence.

More volumes of available light would vanish into that growing distortion until a maximum blackout volume manifested itself. Walls could not stop the effect because it was not a radiance, not a radiant electrical phenomenon. Space itself was literally being "warped" by the arc discharge. Dr. Brown cited instances when extremely intense lightning channels appeared "black" to witnesses. The phenomenon had also been photographed by professional researchers. Each had erroneously assumed the effect to be a bleaching action; one which was both ocular and photochemical. No one recognized the real significance of what had been recorded on these photographic plates.

Only one researcher successfully glimpsed and accurately identified the possible cause of these blackout effects.

George Piggot mentioned the mysterious "black band" which appeared around his highly charged suspended metal marbles. Light seemed to disappear into these zones. But it was Nikola Tesla, whose forgotten and ignored testimony on the perceptual effects of high voltage electrical systems took first place. Tesla produced such intense electrical arcs that the same strange blackout effects were repeatedly observed. In the case of Tesla's famed Colorado Springs Experiments, the blackout effect produced a lingering state which Tesla accurately described as a perceptual-spatial distortion.

Noted in his published diary, the results followed the intense activity of his Magnifying Transformer. Visual distortions, clarifications, black shadows, black streamers, black waves, lingered for hours all around his plateau laboratory, whereby he stated that:

"These phenomena are so striking that they cannot be satisfactorily explained by any plausible hypothesis, and I am led to believe that possibly the strong electrification of the air, which is often noted to an extraordinary degree, may be more or less responsible for their occurrence."

Dr. Brown suggested that space had been warped to a degree where all the available light had been completely refracted into the channel of distortion. No other force symmetry could have effected this manifestation. Furthermore, the blackout would produce various effects in "successive stages". At weak levels, one could maintain the blackout effect without noticing any effects on nearby matter. There would be an intensity at which significant "modifications" of matter would be noticed. These would include internal material strains and spontaneous electrical discharges. Provided the blackout effect was "slow enough", these material modifications could tear matter apart in an explosion of electrical brilliance.

With increasing arc intensities, a gradual series of space distorting effects would be expected. Dematerializations would follow light disappearances. Matter so significantly distorted, would follow the ever-extinguishing light into the arc. This would occur only if the discharge was "quick" enough to strain a given volume of mass. The tools which had disappeared were too short for such explosive displays to be observed.

Dr. Brown confirmed that the welding arc, used in the Naval welding facility, had reached the theoretical intensity where nearby matter was being drawn into the arc at a rapid enough speed to vanish. Mere vaporization was a possible "escape route" for the materials, for the steel tools. But the problem was compounded by the absence of metal trace gases after the effect. If metal tools were vaporized, one would expect some trace of this in the site atmosphere after the event. But there were no detectable vaporization products in the chamber following the blackout. This inferred that, were the metals vaporized, they had been rapidly drawn into the space warp. No by-products. Complete dematerialization!

The films confirmed the suggestion. Materials enveloped in the expanding blackness completely vanished. The mystifying disappearance of the materials was thus, a space distortion effect. Equations were summoned in proof of these speculations. The vanishing of matter proceeded, from surfaces nearest the arc directly through the masses to their outer peripheries. The blackout proceeded as an enveloping wave; expanding to a limit, and then retracting. After the blackout wave, materials were "erased" from our space.

Summarizing his keynotes, Dr. Brown emphasized that the effect in question was clearly an electrogravitic one; where sufficiently intense electrical fields had inadvertently been focussed to a degree which collapsed space into a thready distortion. Concluding his talk, he added that his descriptions of electrogravitic interactions were the only ones in existence, seeing that his precious research in these avenues had been conducted amid a great deal of academic doubt and derision.

He would not have been surprised if this small cluster of military superiors and engineers rejected his approach to the problem. The group was spellbound. Never had anyone presented such a complete analysis of the phenomenon. The otherwise bizarre event found a most satisfactory explanation. Imagine, they had the very fabric of space in their grasp! What things could now be done.

Dr. Brown also predicted that spontaneous levitations and other gravitational anomalies would follow the event. Gravitometers could never accurately register the brevity of these impulses, being interpreted as "anomalies" and instrumental "failures". Military use of the effect would not hide itself from his own electrogravitic detectors. These detection systems would necessarily be as highly classified as the effect itself.

Little more than his description existed as bibliography. Questions and conversation followed his formal presentation. Space warps having the focus and intensity of the new Naval welding facility had never been encountered by human agencies. Data of events taking place within the arc channel simply did not exist. Unknowns filled the discussion. Where was the dissolved matter going? In what form was it going? The questions were endless. The technological potentials, just as endless. One could not be sure exactly what was occurring within the arc channel until closer examinations were made possible. This was dangerous work. Once matter entered into the warp, all discussion became theoretical.

The speed at which the blackout swept space and matter also determined the completeness of its effect. The timing of its manifestation was critical.

The "event" spread through space, a shockwave of deadly effect. Once within the "event horizon" one would be removed from local space correspondence. Annihilated.

How materials were being drawn into the arc channel was simply not known. The arc seemed to be a thready channel, but whether or not matter was reduced to a thready vapor while entering that channel was also an unknown. What if matter was being wholly transferred into another space? What space was it then? These considerations brought science fiction dreams to mind. Could objects be teleported by this means? Were there other worlds into which voyagers could go? One thing was clear: dense matter from "local space" was being "translated" to another space. Where "that space" was remained unknowable.

There was danger in these experiments. Once initiated, the effect could "grow" against all control factors. One simply did not know what would happen when once the process was triggered. Suppose the very act of initiating the disturbance created space "drafts"? How long before they would cease? Would they cease at all? Would a space warp, once started, ever stop drawing light and matter into its ever growing event horizon? The frightening possibility that the blackout wave could become haphazard or uncontrolled was a possible threat. If the intensity of arc focus reached a critical unknown "universal constant", a complete local space collapse into another space could occur.

The space distortion "flashed" in its radiant and striated blackness, the rays reaching out in all directions. Those who had been in the control room were fortunate not to have been "drawn away" in the event horizon. This early discussion was rich in terms by which "black holes" are described. All agreed that careful examinations of the effect would only be possible with a modified welding system; one in which very precise adjustments of discharge power could be arranged. Only then would researchers be able to "grade" the effect, watching for the successive phases of phenomena manifesting near the explosive arc.

After he had concluded his presentation, the verbal display of congratulations was overwhelming. The shy and reserved man remained silently smiling amid the applause. But something troubled him about these proceedings. Even as he made his delivery, a strange heart-tug would not permit him to disclose the other and more deeply mysterious aspects of the spacewarp blackout phenomenon. He refused to discuss all the features of the spacewarp because of its possible use in weaponry.

It was obvious that "they" were not intrigued with the phenomenon from its more spiritual viewpoint. This phenomenon represented something of far greater importance than even he was able to perceive. Without such knowledge, without such wisdom, the technological application of this technology would become the terrible weaponry of future wars.

RECALL

True to form, military discussion continued into the night. The scientific applications of this effect would be endless. This new technology would improve their methods of warfare. Devastating. Such a device could eradicate every material in a specific circumference with no evidence that anything was once in existence! Better than an atomic bomb! They brimmed from ear to ear at the prospect. This accidental observation could be the "new power" which the world was awaiting.

Numerous phases of the effect could be used for various functions in warfare. Derivatives of the effect could be selected and deployed. By the time the night was over, whiskey and cigar smoke having thoroughly saturated the room, several schemes were proposed. First, it was determined that an experiment be conducted with the "blackout" effect. An attempt to "cloak" a selected region of ground space would be their first goal.

Once this experiment had proven its usefulness, then other selected portions of the space distorting effect would be used. More dramatic and complete destructions could therefore be deployed with each successive research success. Wonderful destructions. This first phase could not be designated by any explicit terminology. The "blackout effect" was itself a priority. The working title of this first project would necessarily be an elusive one. The removal of light being their goal, an oppositional term, one describing vivid color, would be used. The now infamous "Project Rainbow" was initiated.

The obvious preeminence of Dr. Brown's work on electrogravitic effects, together with his working knowledge of "space warps" made him the most reasonable advisor in the NRL's newly proposed "Project Rainbow". This period of Dr. Brown's biography remains shrouded in silence. By this time he had achieved the rank of Lieutenant Commander. His presence during the early phases of the now infamous experiment have been confirmed. Thereafter, he seemingly vanished from the team. Why this occurred had everything to do with the nature of the man himself.

Brown was willing to speed the war effort, saving lives and bringing a swift conclusion to all the horrible suffering. After carefully considering all the experimental evidence, it was agreed that a sustained "black out", capable of suffusing a region of ground, could be achieved. But others would have to design the main parts of the apparatus. His mind firmly against the deployment of spacewarp technology as weapons, Dr. Brown would only provide the initial components for a cloaking device. Already his heart and mind were striving with him to depart from the work. Something in the official treatment of his analysis troubled him deeply. What were they really after? A weapon? Or were they really only interested, as he was, in the wonder of an invisibility apparatus? A new facility was established toward these ends. The weldsite apparatus was modified for experimental research now. Heavy duty switches, arc-discharge chambers, and precise control components were developed.

By carefully monitoring the arc discharges, with precise attention on intensity and discharge "speed", the effect could theoretically be safely maintained. But capacitor fired arcs proved to be unstable and intermittent in operation. Therefore, judging from the thesis which Dr. Brown proposed, special generators would be needed to sustain the effects at highly controllable levels. The continuous generation of the warp was now sought.

A rotating high voltage machine could be mechanically governed, its momentum regulating the necessary high voltage for sustained periods of time. It was also recognized that, since a unidirectional electrical impulse produced the warp, then a reversed impulse could disrupt the warp. The suppression of any possible "runaway" reactions could be immediately switched and reversed on this principle. So, with these safety factors at their disposal, the research team strove into the unknown.

The pressure to achieve the goals of this project were unbearable. More important were the rapid calculations and visionary penetrations with which Dr. Brown viewed the true purpose of the Project. How far would the military go with this new power which they were about to unleash? On whom would these most dreadful forces be unleashed? Had he played his part as a puppet in a much larger theater where only the malevolent were entertained? Suddenly he knew what they were going to do with the technology, and where the final outcome would lead. This Project, this Rainbow quest was not going to end with mere "protective" technology. The very term now came through to his conscience like a mocking lie. He knew what the next phase of research would demand. He could not give it. He would not participate. It would have to stop here and now.

In his tortured thoughts he saw the empty cities, emptied in a black flash. He saw the twisted, distorted faces when torment was summoned; the black flash slowly pulling them apart. He heard the cries, the screams of innocents enveloped in the black and not appearing once the inky blackness passed. He could not easily withdraw from the experiments now, though moral obligation impelled that movement. Any such declarations would be declared acts of cowardice, or even of treason. He was an officer with officer's duties and oaths to uphold. The war was on, and he was helping his own nation develop a power more loathsome and morally abhorrent than the atomic bomb itself. He had to leave now or forever live with his conscience. What was he to do?

What we know of Dr. Brown's "official" disposition after this time period was that he was in a state of "complete nervous collapse". The extremity of his condition forced him out of the research project. His position in the Project now permanently "retired", his classification level "demoted", Dr. brown returned home to rest and wait out the time. There are those who accept the story of his "complete nervous collapse" without question. There are those who speculate on the true nature of his withdrawal. Had he sacrificed his own rank and prestige in order to block progress on a most horrid application of his technology? Was his "condition" the only logical recourse left to him in order to be prematurely "retired"? Had he obtained his purpose in this coverture?

RAINBOWS

His own integrity and moral fibre were intact, an admirable reward for so dubious a military venture. The decision made, his future would not be an easy one. It would be almost impossible to obtain work as a Naval consultant ever again. Credibility not harmed, he would have a difficult time with the outside engineering community if it was learned that he "failed under pressure". Furthermore, if the NRL wished to "pursue" him as they had done in the past with their other "defectors", he would stand no chance at all. Would they reach around him with a "ring of power", making him invisible?

Loved, respected, and highly appraised by all of his superiors and associates, the outcome was not nearly as bad as he feared it would be. He was treated with genuine kindness and respect. By now he was thinking about the NRL and what they were about to do. His departure did not stop the Project. They continued without him. New experts were brought in. Technicians and designers acquired the various Project tasks assigned them. Lacking the intuitive insight which comes with inspiration, their assessments did not match those of Dr. Brown.

In this technology, the smallest error of judgement in theoretics, design, or operation could become a dangerous and lethal situation. A newly designed dynamo provided the initiating energy. This rotating machine generated a very high voltage field. Remember, Dr. Brown had discovered that the spacewarp required only an intensely high voltage. Current was not the required feature in the process. Also, the employment of suitable high dielectric strength capacitors, better than vacuum, would allow the emergence of blackout effects at much lower voltages. In addition, it was imperative that the blackout effect, which was a dangerous spacewarp, be "shaped" to specified geometries. One did not wish to be enveloped in that blackness.

Some stated that the optical blackout was not a desirable state in which to cloak any tactically vital object. Any enemy could see a blackout". Enemies could simply fire shells into the heart of a blackout to reach their mark. What was desired was a more subtle application of the spacewarp effect; something far more refined and controlled. It was then hypothetically proposed that complete blackness might not be the first optical distortion effect in the "sequence of effects" which Dr. Brown had described in his report.

Perhaps it might be possible to so "tune" the effect that, prior to "black optical states", a condition of transparency or invisibility might actually be achieved. The idea was to attempt bringing the space warp to an intensity in which any material objects surrounded by "the effect" would vanish. Vanish "from sight", but not from local space. Some superiors stated that they would be happy even if the spacewarp achieved "radar invisibility" for sustained intervals. Optical invisibility would then be an "extra treat". With this proposal, Project Rainbow found its true inspiration.

Years earlier in Hungary (1936), Stephen Pribil demonstrated an "invisibility system". His system utilized special heterodyned light beams to render objects transparent and even invisible. Under the beams produced by his special lamps, opaque objects gradually became transparent. The effect could be controlled, optical transparencies sustained at specific "intensities". Radio cabinets, exposed to his lamps, faded from view. Astounded witnesses saw through the cabinets, while the interiors stood out with amazing clarity. Metal parts, tubes, and chassis could be seen darkly shining through the wooden enclosures.

Allowed to thoroughly handle each part of his apparatus as well as the objects made transparent, it was also obvious that the system actually worked. He rendered transparent any object they wished by his extraordinary method. What became of Mr. Pribil, his device, and his theories remains a mystery. In what manner spaces are distorted by very ordinary energies remains part of the scientific record. Numerous instances in which these strange phenomena make themselves apparent have been noted. The sharp black edges which surround grounded iron materials and evergreen trees seem to be related to the more extreme electrogravitic phenomena in outward appearance at the very least.

The first series of Project Rainbow experiments was first performed with the aim of "cloaking" tanks and heavy gun installations, that phase of the experiment which principally had stimulated the participation of Dr. Brown. The high voltage impulses would necessarily be applied to a ring of capacitors which surrounded the object. It was critical that these capacitors be oriented in such a manner that the spacewarp would not engulf anyone within its perimeter.

In the first accidental observations, the arc discharge engulfed all surrounding space. With its singular axis and vertical disposition, the warp radiated from arc to periphery. The designers needed to find a way for the radiant warp to move in a confined zone. Any such system in which personnel were involved would necessarily contain a "safety zone". In this safety zone, one could observe and yet remain unaffected by the warp. Could an array of extremely high voltage capacitance actually bend space around a fixed perimeter? Safety from the warp was the item of concern.

Was there no shield from this potentially deadly effect? Were there no materials or field configurations which could shape and "guide" the spacewarp away from human participants and, eventually, crew members? Could the effect be beamed to a site? How far could such a "beam" be projected? Already the deliberated weapons research phase was meshing with the Project. Technicians attempted redesigning the symmetry in which the warp, the "black flash", would expand into its surrounding space. The capacitor axes acted as the radiant points for the expanding warp.

The system was remarkably and ruggedly simple. It is unknown what impulse durations and repetition rates were actually employed when applying the high voltage to the capacitor ring, but their repetitive frequency most certainly were very low. As Dr. Brown had empirically discovered with his gravitators years before, large capacitor values required longer warp "saturation" times.

The capacitor "axes" were radially disposed at a great distance from a central point. From above the installation, the power cables formed radial lines out to the periphery of a circle. Each radial cable was connected to a terminal of a capacitor. There were many of these. The capacitor axes pointed "in line" with the radial cables. Their precise alignment was absolutely critical, since any deviation from this radial symmetry would bring disaster. The outer terminals of all the capacitors were joined by a single circular loop. Voltage was thus applied from center to periphery. These capacitors, electrified by the high voltage dynamo and its ancillary impulsing components, were collectively a very large capacitance.

With the capacitors themselves at that circumference, the spacewarp would expand as a sheath of specific "thickness". The thickness depended entirely on the capacitor thickness. Their dielectric material was a special composite, probably of barium titanate powders. This dielectric permitted the concentration of very high voltage electrical fields without leakage and subsequent field loss. This material also permitted the use of significantly diminished voltages and impulse rates. Each capacitor projected a warp. Warps meshed between capacitors. The cloaking effect would be induced as a ringlike wall surrounding an interior space, a safety zone. Or so they hoped.

In this manner, any object placed in the center of the "cloaking ring" would be surrounded by an optically indistinct zone. According to the analysis which technicians now provided, no blacking effect would be experienced within the ring, the warp being confined to a specified outer periphery. The system would provide a convenient wall, shrouding one's presence from any enemy's gaze. Questions now were directed toward the perceptual distortions which the crew might experience. How one perceived the world "beyond the ring" would require another series of experiments.

The voltage output of newly designed dynamos was extreme. In fact, they operated more like electrostatic generators than current generators. Ratings of several million volts would not be a conservative estimate. The initial experiments apparently succeeded, the Project moving into its progressive developmental stages. Problems of communication through the warp cloak was going to be a problem. How would the crew remain in touch with superiors? In wartime, communications would be the necessary and critical item of concern.

Could one use ordinary radiowaves across the threshold of a spacewarp? Optical blackout would block all communications on both sides of the cloak wall. But transparency might allow specific radio frequencies through the wall. At sufficiently high radio frequencies and power levels, there would be no problems in this application of the warp. Special UHF antennas appeared throughout these experiments, reported as "Christmas tree-like" structures.

The Naval Research Laboratories now undertook the first deployment of the system for use on a much grander scale: to "cloak" and render a ship invisible. The chosen ship, USS Eldridge (DE 173), was now outfitted with the experimental system in two sections. The first was loaded on board. This included the dynamos and a modified capacitor ring assembly. The second section, presumably controlling devices and data readout systems, was located off shore. The moored ship was connected from ship to shore by a great system of very thick black cables.

The capacitor ring surrounded the ship's hull in the same symmetry as was used to surround grounded objects. Several pairs of wire loops were placed completely around the ship's hull. Between these loops, a great number of the specialized capacitors were placed in radial orientations. Since the capacitor "ring" assembly saturated the warp for such a long time, it would be possible to again lower the impulse rate. Continuous transparency could be arranged by employing a cadenced series of such assemblies.

It was vital that the electrical axes be absolutely radial with respect to the ship's deck, and that no unexpected diffraction effects be encountered by the expanding warp. Such diffractions would destroy the warp profile, bringing its deadly effects into the safety zone. Crew members would have to stay deep in the safety zone interior or risk being destroyed by the expanding warp.

The assembly produced an intense space distortion which surrounded the ship. In unmanned tests, the effects appeared benign, invisibility actually having been achieved. It has been rumored that the aged Tesla was called from his secure Manhattan retreat to act as consultant to this project. The father of electrical impulse science and inventor of electrostatic impulse generators and transformers, his expertise added certain finishing touches to the massive equipment. In addition, his own experience with the blackout phenomenon added special credence to the project. No doubt, he himself was enthralled at the discovery and its amazing properties.

The development and proposal to deploy the system by the military did not rest well with Tesla however, his sharpened intuition recognizing their ultimate goals. It had been reported that Tesla left the Project abruptly after recognizing the potentially harmful effects on crew members and the natural environment. The safety of crew members or other military personnel who would be exposed to such an energetic effect would indeed be the critical factor.

Could human beings operate near such a powerful warp and remain completely unaffected? The early exposures of control room personnel produced mild and temporary cessation of conscious process. Physiologically, no reports of after-effects were ever mentioned in connection with the blackout. It was critical that the effect be closely monitored and isolated. When deployed on a battleship or aircraft carrier, the capacitor rings would produce an enormous warp volume. Required to bend space around the ship, the system would be taxed to its design limits. Until standardized, the system itself would continue to remain "in the design stage".

The "black out" state would not be engaged, it previously being determined to achieve a mild state of transparency. This cloaking system would provide a camouflage resembling a "blurry sea horizon". After tests with laboratory animals proved safe, a crew was selected. Once aboard the ship, crew members were asked to simply "go about their business as usual". Warned to stay well within the interior of the safety zone, it was stressed that they treat the hull as they would a deadly electrified fence. The crew was told that there might be possible "strange optical effects" when looking outside, but that this was part of the experiment. If they were successful, they would be "war heroes". The Navy would have obtained a new weapon for the war and they would be first to try it.

The systems were initiated. Power was applied. The ship acquired a "swimmy" appearance against the blue sky. Increased power brought the effect toward the theoretical mark. The ship becoming more blurry. Power to the system was raised a few more degrees, and to the amazement of all who watched, the blur became a foggy transparency. The ship could not distinctly be seen at all! The effect was sustained for a few minutes, and then removed gradually. All seemed well.

What the NRL did not know was what had happened in those few minutes. It was impossible for them to receive any distress calls, the entire crew having been incapacitated. There must have been a leakage effect somewhere in the hull, the "ill-formed" spacewarp moving into the safety zone. It has been thought that the excessive steel of the ship itself acted in some strange uncalculated manner to draw the warp effect into the ship. Furthermore, in the few seconds after the power was applied, many of the crew members attempted diving off the ship. They never made it to the warp edge, being saturated in the radiant warp field.

The experiment had horrible side-effects. Each exposed crew member was neurologically damaged beyond all hope of healing. Having become known as the "Philadelphia" experiment, it remains but one highly classified chapter in military science. What we know of this event has come through significant and credible sources. The disastrous results of this experiment were not easily forgotten. It has been "presumed" that the NRL relinquished these developments. Others offer more striking evidence that none of these systems have escaped continual examination.

SPACEWARD

Having subsequently lost his security clearance, and being retired from active service on a small pension had its advantages. By 1944, those who sought him found him working as a consultant on radar systems for Lockheed. Radar, the leading conventional postwar science responsible for saving lives and virtually winning the war, was a safe and moral place to land. Playing it "conventional" after the war meant earning a steady income. Nevertheless, but his dream of spaceflight continued weaving its lovely and intricate patterns through his mind. Dr. Brown was well aware of the horrid outcomes obtained in Project Rainbow. Though his moral obligations were fulfilled in these regards, he could not recall the Project without the deepest sadness. The designers were simply not equipped to properly engage the power and fabric of space in such a cavalier manner. His examination of their proposals revealed the numerous flaws in their configuration. It was amazing to him that no one had even considered the diffractive effect of the hulls when raising the warp. Yet, had he been actively involved in the Project, his superiors would have demanded complete cooperation in deploying the weapons aspect of the warp phenomenon.

There had been ramifications of the NRL experimental nightmare which disturbed his own plans for the future. For a time he very well imagined that spacewarp drive was deadly. If used in the intensities demanded by his own calculations, the unshielded effects would do to a flight crew exactly what it did to the men of the USS Eldridge. For a time it seemed that spacewarp would kill voyagers if employed in aviation or space travel.

Patient study and intuitive vision finally brought forth a solution to the deadly problem. The answer was contained in considering the nature of dielectrics. Dielectrics of proper composition could guide the mobile spacewarp in a smooth and confined flow. Unlike the NRL, tests on these notions would require sufficient investigations prior to deployment. His crushing suspended thoughts on the wonderful vision of his childhood finally returned with renewed strength. Never would he reveal it to those in the NRL. He knew how to shape and isolate the warp with absolute safety.

The War was over. And his own war was over. Once again, the dream sea began surging in human minds. The skies deepened and people began seeing things, new things, in the deep blue outer reaches. A sudden wave of reports flooded the media with observations of "flying discs". Saucers. The vague and elusive forms travelled at speeds which rivalled every known military aircraft. As with the mystery aeroships and ghost-rockets, these aerial sightings remained true to the archetypal pattern of manifestations. Their first appearances were vague and dreamy, cloudy and indistinct. With time, their forms seemed to solidify and acquire sharply detailed surface features.

Each dream manifestation in the historical series brought totally new behaviors to human attention. Saucers were no different. By their very shape and activity they comprised an utterly exotic and unexpected event. Lights, search beacons, colors, flashings, color changes, aerial caprices, right-angle turns, sudden disappearances, simultaneous radar visibility and visual invisibility...the flying saucers exhibited their new behaviors as each previous aerial dream display had done.

Their appearance came on the tail of the war in Europe, pilots reporting strange "foo fighters" and mysterious "fireballs". While some investigators cited the development of special electrically activated robot aircraft in Nazi Germany, most could not explain the incredible variety of flying saucer sightings in every other part of the world.

Having ability to out-maneuver the best Allied bombers, these strangely animated objects were more than a public curiosity by this point in time. World national security being the principle theme of every government, flying saucers were military "objectives". Every aspect of these flying craft was carefully scrutinized. Who were they? Were they a private group of highly advanced engineers? A maverick or rogue community of technologists at war with the world of nations? Were they aliens? If aliens, then which planet? What were their motives? Why were they playing with military facilities in such a nonchalant and arrogant manner?

Memos, reports, flaps, and aerial skirmishes. News releases, press releases, cover-stories, cover-ups. Diversions, debunkers. Files exist, files do not exist. Crafts retrieved, crafts shot down, crafts do not exist, bodies retrieved. Hangars, weather balloons, radar balloons...the amusing charades yet continue. While flying saucers behaved in comic fashion, the Air Force was burning the sky down. Helpless. Powerless. While attempting to unravel the puzzle and appear in total control, the simultaneous juggling act of desperate data acquisition, cover-ups, and denials proved to be a far more entertaining theater than the saucers themselves.

The ridiculous spectacle of earthly power and dominion in the act of square-dancing at high speed, was an rich amusement of derision. The intended effect. The dramatic and focussed concern of military agencies for these flying saucers betrayed a singular impotence. A demand for control. Power, the deception of control and dominations, was being mocked by a playful energy which analysts yet do not recognize.

Nevertheless, the obvious superiority of flying saucer crafts, if they were indeed crafts, left military superiors at a total loss. How did they fly? How were they able to execute inertia-defying maneuvers without shearing to pieces? How were they able to accelerate complete out of visual contact in a millisecond without incinerating? What was their mode of propulsion? What was their metallurgy?

Every vaguely related terrestrial saucer technology was probed. Piecing through the Patent Registry, the military people were directed by experts to examine those propulsion devices which, though deigned "impossible" by major aircraft corporations, nevertheless actually worked.

Neither the Air Force, however, nor the NRL communicated in a mutually cooperative manner. The "tradition" among military groups forbad this. Now, especially after the war, these "traditions" were strengthened beyond measure; seeing that each group was in possession of new and potentially domineering technologies gleaned from the war years.

Suddenly, the NRL was "interested" in Dr. Brown again! In 1949, against all principles of normal military security, Dr. Brown was brought into the NRL again. Living in Hawaii, his relaxed manner was noted; a defined and dramatic change from the days during the now-infamous Rainbow Project. Highly cloistered military engaged the dialogue. "Brown" was the closest they would ever get to solving a particular "problem" which they had.

Dr. Brown decided that a formal proposal be made for research funding in 1952. After several months of research, a small NRL group was convened. Dr. Brown spoke once more to NRL examiners. He imagined at first that their "problem" had something to do with the recent sightings of UFO phenomena. He mentioned that he could surprise them in these regards if they were patient enough.

He began his talk by attempting to prove that aerial flight with gravitator engines was the new high frontier. He named his experimental foray "Project Winterhaven". Hoping to lead the military gaze toward true space travel itself, he made his appeal to the NRL. Now done with the last war, he encouraged them to seek out some new technological venture. They smiled and listened on. The time was indeed ripe. The race for new spaceward investigations had captured every mind. The dream sea surging. Spaceflight was the obvious new challenge. Dr. Brown reminded and warned his colleagues of their former "negligence" with pre-war technology.

The Air Force was painfully aware of their ignorance concerning rocketry, until both V-1 and V-2 packages began raining death and ruin upon England. Here again was their time, ripe for new technology, ripe for the experimental implementation of his astounding discoveries. Most of the NRL examiners agreed, almost laughing among themselves at an obvious "inside joke".

Now was the time. Now. Everything was moving spaceward now. Now was the time when space travel could really come to pass. He then mentioned his careful study of early European Rocket Club periodicals which revealed forgotten fancies in both space travel and space engineering. Some designs and hypothetical plans were bizarre fantasies in mathematical form, but they had practical merit.

Taking their lead from Robert Goddard, the Europeans developed elaborate and elegant solutions to the real problems of rocket-driven spaceflight. As with so many wonderful scientific developments, the mathematical work on various aspects of spaceflight had all been published and forgotten during the War. He elucidated on these, highlighting the fact that while some were fanciful, most were inspirational, visionary, and instructive. Russian engineers designed a ground-boarded space cable car system. Some imagined "space carrousels" which could extend capture cables earthward, towing groundlings skyward. It was the mathematical precision which most thoroughly engaged professional readers and students of these journals.

Certain of the more serious schemes, proposed independently by both Willy Ley and Arthur C. Clarke, considered the stepwise approach of moon landings and moon colonization. These topics provoked great interest and the writing of numerous notations. The methodic deployment of "clock orbit" stations would certainly establish the practical means for landing on the moon. Space islands, permitting short trips with long rest intervals, would be positioned at critical "L points". In such points the stations would be space stable. Fuel, rest, food, supplies. All of these requirements could be shuttled between successive L stations before embarking toward the lunar surface.

Such a system of "space stepping stones" would not necessitate the titanic and impractical rocket needed for a straight earth-to-moon voyage. They shifted as he went on, seemingly uninterested. When once the L stations were in their sequenced positions, only small escape velocity shuttles would be required for the lunar journey. When he mentioned his plan for a "sortie vehicle", one which could launch a spacecraft from ground to L station number one, he suddenly acquired their interest. He described his new engines. They listened intently to everything he said now.

The electro-impulsed gravity engines would operate in space at peak efficiency. Utilizing electrical power alone, the spacecraft designs required no liquid fuel. The gravitators needed an initial lift before achieving true free flight. Dr. Brown reported his development and successful testing of small ion thrust jets, the electric arc rockets which he designed in his youth. These would provide the necessary takeoff momentum. Coupled together, the system would gracefully propel a disc-shaped vehicle out to space.

Implementation of his ion thrusters alone would revolutionize the art of flying, ultimately permitting rapid deployment of small interplanetary crafts. He patented several of these ionic thrust engines (3.018.394, 3.022.430, 3.187.206) throughout the 1950's and early 1960's. These devices produced both thrust and electrical voltages of unheard strengths. They were early examples of MHD generators. Magneto-hydrodynamic generators produce remarkable volumes of electrical power directly from flame.

Photographs which he presented to the group looked less like his early "Flash Gordon" rocket forms, and more like the flying saucers which had been recently reported on the worldwide scene. NRL examiners remained completely silent, while writing what they saw. There were those who came to inspect his experimental facility. His designs had taken on a distinctly otherworldly appearance; being large area, beautifully smooth surfaced "discoids". Convex metallic forms. Flying saucers which actually flew. Their obvious link to the UFO rage of the early 1950's was a compelling aspect which attracted many curious academes and Air Force military advisors later.

He had designed an overall shape which combined all the features required for a "flying wing". The forward craft section was charged electropositive, tension being established with section aft, charged electronegative. The Brown warp thrust effect would move the entire craft forward when sufficient electrical impulses were delivered to the aerofoil. These saucer shaped designs were tethered to a central support rod, having perpendicular extensions of ten feet. On these were fixed two "discoid aerofoils", each some two feet in diameter.

Supplied with long electrical impulses of 50 Kilovolts electrostatic potential, the discs raced around the sixty foot circumference, achieving an angular velocity of nearly twenty feet per second! These discs operated on pure spacewarp power. They were dragged along with the distorted space which they electrically projected.

In quick steps, he had increased both the size and speed of these tethered discoid capacitors. The discs were now three feet in diameter, running around a fifty foot course at higher speeds. Their movement was swift and silent, a military dream. These experiments had been classified by NRL examiners who had previously come to observe his progress. Classified, but not accepted...a strange contradiction!

The NRL was interested, but something in their manner alerted Dr. Brown to some deeper motive. What really was it that brought them here? And why here and now? Could it be that their interest was in pursuing some other feature of his work? He nevertheless continued the dialogue. The discoidal aerofoil was to be equipped with several propulsion systems. Each would be engaged with each stage of the aerial ascent. Dr. Brown's ion rockets were designed to provide initial lift for his crafts, while generating enough electrical field strength to power the gravitator cluster.

At orbital heights, the entire craft would achieve stable gravitator function, ion engines being disengaged. The ion rocket could be used for controlled re-entry after a spaceflight reached home. This plan not only sounded professionally impressive, its demonstration was impressive. Dr. Brown had succeeded in achieving the self-sustained flight of his aerofoil models without much NRL assistance. His frugal use of their recent funding was not wasted. They asked how he proposed to shield the crew from the possibly deadly aspects of the warp drive. And this is when he stopped sharing. For now he recognized their true motive for both reinstating his security clearance, funding his research, and for now being in his facility.

In his private researches, Dr. Brown discovered that dielectric material could be molded to any specification. Any geometry could be accommodated. In these secrets, he had encrypted his elegant solution to the problem of isolating space warps from crew members. The NRL experts desired this information. This is why they were there.

He attempted to engage and elude them now, to obtain his own information. The behavioral patterns and actions of one's opponents betray their intentions, acquisitions, and fund of knowledge. Dr. Brown watched them closer now. He demonstrated his devices repeatedly for them. Now, the NRL was drawing their examination to a close. Suddenly, after all they had heard and seen, none agreed that the effect was "truly electrogravitic" in nature. This was designed to "throw him off" the trail which he was pursuing.

Typically, NRL examiners accused Dr. Brown of "poor science", stating that his effect was produced by "nothing more" than a high voltage ionic wind emitted from the gravitator poles. The gravitators would therefore be useless in space environments. This decoy, this derision, was the commonly employed tactic of those who wish to conceal. In this case, the NRL wished to conceal the fact that they had already perfected the warp drive engines. Dr. Brown knew exactly what they were doing in his laboratory now. They were desperate. They needed a piece of information which he alone had. What was it though?

The frail hearted would have been dejected, their morale destroyed. Believing the "rumor" of their own failure, in the mouth of military "experts", such individuals had been known to drop their work completely. The mind game. Nations turn in upon the wellsprings of their own creativity, secluding discovery to a cloistered elite. The power game. On behalf of whom? The old money. Who are they? The old families remain anonymous. The nouveau riche are pushed to the foreground.

The actions and behavior of the NRL told him everything he needed to know. One or two more surprises would give him a piece of information which he now desired. He provoked them with an impromptu demonstration. It was a wonderful, but old experiment. He had performed this test the very day after he discovered the thrusting effect in his Zanesville garage as a child. In true scientific bravado, Dr. Brown completely submerged his gravitators in oil. The oil-submerged devices again outperformed their own atmospheric tests! He was enthralled at this last face to face "victory". This meant that, with increasing altitude, the gravitators would take an increasing role in overall thrust! He thanked them for their "original" objections.

For the meanwhile, his critics were silenced...but not thrilled at the findings. This response was shockingly conspicuous. They had already seen such demonstrations elsewhere, had they? But, by whom? And when? It was clear that they had already reached this technological plateau. Quite obviously, these examiners were not "now" interested in his system, because they were never really interested in his system from the start. They had already taken the original project beyond the bounds which he was displaying for them, having developed superior thrusters of their own.

The examiners were there specifically to discover whether or not his system incorporated a "warp shield". It was simple. They knew that his knowledge of the Rainbow experiment was the leading theme. They knew his awareness of their failure. He knew the dangers of the warp. He knew their failure to shield from its deadly effect. Yet, here he was developing a spacewarp drive. No one with his expertise would undertake such an experiment. There would be no future in it at a certain point in the developmental stage. Hence, himself quite aware of the potential dangers of deploying a large scale warp drive, he had to have solved the "shielding problem".

This was the very problem which their "experts" had originally, and disastrously failed to produce in Philadelphia. Therefore, they had approached and "caressed" him for favors. Technological answers to a new project, one in which he would never play a part. It was also obvious that the internal functions of the NRL had become the property of post-war trusts. The chronological facts and behavior of NRL examiners yet remain part of the historical record.

FREE FLIGHT

The "successful" results of his NRL encounter further encouraged him to reach his originally envisioned goal of spaceflight. But he would do so on a non-military level. Numerous military personnel were now commandeering aircraft industries. Such individuals would be useless to him if co-opted by establishment concerns. In that case, their every endeavor would be marred and marked by a simple concern to resist new technology for the protection of investments.

He would take his system to privateers. Commercialization of space crafts would revolutionize the very manner in which citizens envisioned themselves. The gravitators would indeed work in space with far greater efficiency than they did in his earthbound lab experiments. The engines did not require the military excess common with ordinary rocketry.

It was so strikingly clear that Dr. Brown was not concerned at all with the rejections of NRL or military developers. Their own stealth had betrayed them. He knew there were crafts already being deployed in test facilities. He knew that the future of his technology had a future. The manner in which the NRL had first approached, inquired, acquired, and then rejected his work told all.

The commercialization of spacewarp systems was predicated on the practical simplicity and relative inexpense of their development and industrial deployment. Dr. Brown was not going to be stopped this time in achieving his spaceward dream. It was not money he needed. His was the mind on which they had built. He had the dream and the knowledge. No one funded his work henceforth. The NRL maintained their cautious and cloistered distance. It was his own fortune which was consumed in perfecting these experiments and making scientific history. But costly equipment and gigantic facilities were not the necessary resource. The dream was the only necessary resource.

When American investors failed to join his project, he answered queries made by foreign companies. Travelling abroad to Europe in 1955, where he believed more willing hearts and competitive minds were listening to new ideas, he began his presentations. He first demonstrated his system in England, but received no commitment of funding. Investors there were thoroughly embroiled in duties of the empire, unable to disentangle themselves from their first loyalties.

But, the National Aeronautic Society in France greatly desired to examine these designs for testing and development. His demonstrations captured them completely. Enthralled with the potential access to spaceflight, their enthusiastic plans for his technology took off at once. Calculated proposals indicated that larger disc aerofoils would become increasingly more efficient than the models. Upscaling the voltage supply required a new type of engine.

While French engineers were temporarily obstructed by this impasse, he reintroduced his ideas about ion rockets. He developed his most famous and powerful MHD jet engine in France. The lightweight MHD powerpack could be supported by the craft itself, removing the noisome connection to ground. In addition to his oil tank demonstrations, he successfully performed the gravitator experiments in Paris at the highest obtainable vacua. These tests, made in 1955, proved conclusively that "ionic wind" activities were not the cause of thrust. In fact, not only was it impossible to detect any sort of ionic winds, but it was found that the gravitator became ever more powerful with increased vacuum.

The only liquid fuel would be the ionic "seed" materials with which to provide charge. Inert liquid gases, seeded with salts would provide the powerful fuel for his electric arc jet. In addition, permanent magnets would split the ionic jet into electropositive and electronegative charges of stupendous potential. This lightweight system would produce over three million volts, simultaneously supplying both initial thrust and gravitator action potential. Now the systems carried their own power supplies and could support sizable payloads.

A quick merger between rival factions of the National Aeronautic Society brought Brown's dreams down. Thoroughly disappointed and out of funds, Brown returned to the States in 1956. Nevertheless, he had reached toward his dreams with success. He had the system and the patents to prove it. In 1957, he was invited to work on antigravity designs with Agnew Bahnson, a wealthy businessman. Brown travelled to North Carolina to establish a privately funded antigravity research laboratory.

While Bahnson and King were intent on developing and patenting designs of their own, Brown was simply becoming a consultant. This disappointing poise caused him to refrain from sharing much of his technology with them. Nevertheless, he did attempt to urge them into considering his discoveries, and the superiority of his patents. There are films of some of the experiments in which an older Dr. Brown may be seen standing alongside Mr. Bahnson and Frank King. Smiling, warm, and friendly, he is seen demonstrating the Brown Effect on a small (but weighty) metal disc. A momentary impulse appears, and the disc noticeably rocks to and fro. It does not seem that they were at all intrigued with this prospect.

In a freak accident of the strangest kind, Bahnson was killed. His private plane ran directly into a high tension line. The research lab continued for a time with Bahnson's associates, who patented several inefficient electric wind-lifted devices. During this latter period, Dr. Brown spoke with Frank King of certain theories which he had apparently kept to himself for years. His experimental research had convinced him that inertial mass differed from gravitational mass. He taught that these differences could be elicited only when mass was electrified.

Positive charging at high potentials would increase inertial mass while decreasing gravitational mass. Negative charging at high potential would increase gravitational mass while decreasing inertial mass. Dr. Brown had proven this in a great number of strictly controlled demonstrations.

The development of rocketplanes emerged from the commercial drawing boards as the "best means" for reaching space. The corporate venture, co-opted by both established money and military entanglements, was destined for disappointment. In the typical "nationalization" policy, and on the very eve of its success, the X-15 venture was twisted out of designers' hands and placed entirely in military control. The obvious reasons for a military "first" assault on space need no further discussions. No commercial space ventures would be "permitted".

The remarkably effective superguns of Dr. Gerald Bull were already proving that an inexpensive means for launching unmanned commercial payloads, satellites and scientific instruments, was available and ready to meet customer demands. How they pursued Dr. Bull, provoking him to seek investors elsewhere, is another matter of the historical record. Each of these dreams were methodically eradicated, not because they were "ineffectual", but because other powers and dominions demanded total control of space. It is amusing, and pathetic to recognize how ignorance best blinds those who promote it. The industrial-military complex now considers themselves in possession of the "only means for achieving space". Remember, however, Dr. Brown did not design an expensive and inaccessible system. Since that time, others received viable patents for devices which demonstrate modifications of gravitation (Wallace).

PETROS

Between the years 1958 and 1962, Brown acted as consultant in a number of corporate ventures. Years after being declined by the NRL, Tom Brown was yet considered "too knowledgeable to be left alone". Government investigators continually infiltrated his demonstrations and meetings with deliberate intent. The infiltrators each observed his experimental work, accurately reporting what they saw to their superiors.

He continually claimed that his electrogravitic drive system would power ships to the stars, representing his system to potential buyers and developers. It seemed that all had fallen victim to some mass lethargy with the exception of himself and a few other inventors. Investors, on the other hand, were simply incapable of entering sound business ventures any longer. They had not the zeal and adventure of their forefathers who made their financial principle on new technology. Those originators of dynasties placed money on the technology by which America's steely face was forged. These younger inheritors seemed intent only on amassing and securing their capital.

Brown retired, but privately continued his research for the remainder of his life. It seemed that his dream of flying off into space was going to fall to others. He therefore decided to serve the future best, by supplying an adequate fund of knowledge for future developers and dreamers. His later experimental work intended to prove the Unified Field Theory by showing that connections exist between electrical and gravitational forces.

Years before, he had discovered that electrified carbon resistances and dielectric masses emitted spurious noise signals. These noise signals were correlated with cosmic changes. Dr. Brown discovered that specific rocks, notably granites and basalts, manifested strangely "spontaneous voltages". Dr. Brown claimed that sudden gravitational fluctuations were being transformed into electrostatic charges when encountering special materials. They were named "petrovoltaic" charges, the phenomenon referred to as "petroelectricity".

These petrovoltaic charges could be measured, their outputs having sizable strengths. In very specific silicates, these outputs were significant. Here was evidence of a reaction process in which space transient warps were stimulating electrical generation! A large enough arrangement of the proper rock could be an eternal generator of gravito-electricity.

Dr. Brown recalled the work of Dr. Charles Brush, the American physical chemist who investigated gravitational phenomena during the Victorian Era. Dr. Brush claimed that certain rocks actually fell "slower" than other materials by a slight, but significant degree. On further examination, Dr. Brush found that these rocks were possessed of a peculiar "excess heat".

Dr. Brown obtained samples of these rocks, Lintz Basalts, finding that they indeed gave very high spontaneous voltages when instrumented. Wire contacts touched the materials on their surface, yielding several millivolts. The rocks could be sliced, and placed in multiple contact to produce a net voltage which was sizable. Rock batteries! Dr. Brown's petrovoltaic effect produced nearly one volt with larger specimens.

Rock voltages are strangely current free, increasing their output at 6 PM each evening and decreasing at 7 AM each day. The effective output increased with increasing elevation, but lost the smooth tracings which strip-recorders revealed for rock specimens measured on the ground. Others duplicated and corroborated these findings with surprising results (Hodowanec). When certain researchers travelled to the Andes to test Brown's claims, they received up to 1.8 volt surges from a single rock specimen! Brown discovered that increased graphite content in the rocks also increased the output voltage: a reminder of his findings with carbon resistors.

Voltages derived from rock specimens contained two electrical components. The first, a steady DC bias, contained a second superimposed fluctuating signal. This latter signal varied with solar-lunar conditions, positions, and configurations. Minute electrical bursts were correlated with distant gravitic pulse sources in space. He now discovered that other silica-rich rocks spontaneously produced these electrical charges.

This interaction of gravitic fluctuations with crystal lattices had other applications. With this phenomenon as his astrophysical tool, he was able to chart stellar effects which normally required huge radio telescopes. Dr. Brown sighted pulsar action and supernova events long before they were formally announced by radio astronomers. Dr. Brown measured solar flares with these instruments, all of which were contained in deep vaults once again. The instruments were shielded from radioactive, thermal, and optical energies. The sensors continually registered these same signals, though maintained in the isolated environment.

SIGNALS

The alliance of military and business communities have invested their monies and efforts on ordinary rocketry. The supposedly dependable military rocket systems were deployed with furious intent. Huge, dangerous, unwieldy, and flawed, they caused the military great concern in 1956. Sputnik rattled the smug military stronghold with deliberation. When called upon to launch the small grapefruit sized Vanguard satellite into orbit, each industrially mass-produced rocket failed.

Later in the decade, Dr. Gerald Bull would lose his original direction and become a formidable arms dealer. Had Dr. Bull patiently born the time out, he might have made a fortune in the business of privatized satellite launches. The military chose its single investment in rocketry, there planting their fortress of commitment. They did so on behalf of others, not on behalf of discovery. Space was to be a frontier only for governments and military units.

The outlandish deferment of the space travel dream has turned our nation inward. Inward and downward. National confusion and chaos is defined by its true origin and source. Social confusion and social derangements derive form the engineered deferment of dreams and expectations. National campaigns propagandize one issue, and then produce another. Often, these reversals are calculated against the younger generation, since the maintenance of dynastic fortunes requires the exertion of deliberate social controls with each new generation.

Nevertheless, discovery and technology are never at rest. They are now alive in little basement laboratories and garage research facilities. An armada of indefinite proportions, from which will emerge the surgings of the dream sea. New discoveries will come and make their forays into the world. Despite the established systemology which methodically harvests and burns these discoveries, they will yet come. No one suppresses discovery. Discovery, with the apparitional manifestation which appears in certain night skies, is THE relentless force. It destroys those who seek its harm.

Others will bear the technological crown away when nations seek and hunt to destroy providentially supplied knowledge. The nation which resurrects Brown's research will rule the air and space. Dreams are never destroyed. They permeate and haunt societies whose leaders have rejected their materialization with a special vengeance all their own. Providence is kind to the dreamer. Visions come to the devoted heart.

Dr. Thomas Townsend Brown loved warm and sunny climates, spending the remainder of his life in Umatillo, Florida. His scientific achievements remain legendary among those who honor and preserve his memory. Imaginative vision itself is the true spaceflight. I believe he recognized in his own dreams the true power, acceptance, and glory which he rightly deserved in life. His wonderful childhood dream of the stars remains.

CHAPTER 8

DEADLY SOUNDS: DR. VLADIMIR GAVREAU

SIGNALS

He listened and closed his eyes as the rolling waves of sound poured over and through his being. Thrilling, intoxicating, the hysteria of heaven, the enthralled and frightening flight of angels. Electrifying. Messaien's organ music signalled messages of meaning, titanic foghorns ululating among dimly perceived near-worlds. Olivier Messaien, master composer of musical expressionism, used the ground thrumming tones of great Parisian cathedral organs to evoke sensations which may only be called otherworldly. Masterfully macabre. Black foundations, blue pillars, and rainbow ceilings.

Sound, rhythm, and space. Ultra-chromatic chord frames, rising like rock walls from the black depths. And immense stellar crystallizations, radiating tonal perfumes through deep and black radiant space. Lovely and lyrically swooping melodies, the flight of birds through delicate limbs. And melodic lines, reaching up toward unknown depths of space, each had their foundation in ultrabass tones of rooted depth. The basso profunda of Messaien are the critical foundations, the strong vertical pillars of an immense architecture which extends beyond performance walls. He scoured the deep and unreachable roots of worlds to hold his musical cathedrals together. Such majesty and grandeur of sound! Rich in the intelligence which flooded and made the world, the musical currents and the atmosphere of tones. Fluidic music and meaning.

The most fundamental signals which permeate this world are inaudible. They not only surpass our hearing, but they undergird our being. Natural infrasounds rumble through experience daily. There manifestations are fortunately infrequent and incoherent. Infrasound is inaudible to human hearing, being of pitch below 15 cycles per second. The bottom human limit. The plynth. The foundation. Infrasound is not heard, it is felt. Infrasound holds a terrible secret in its silent roar.

Infrasound produces varied physiological sensations which begin as vague "irritations". At certain pitch, infrasound produces physical pressure. At specific low intensity, fear and disorientation. Nazi propaganda engineers methodically used infrasound to stir up the hostilities of crowds who were gathered to hear their madman. The results are historical nightmares.

At a very specific pitch, infrasound explodes matter. At others, infrasound incapacitates and kills. Organisms rupture in its blast. Sea creatures use this power to stun and kill prey.

The swelling bass tones of the cathedral seem as though they can burst the very pillars which uphold the ancient vaults. Stained glass windows have been known to erupt in a shower of colored fragments from the organ's basso profunda. Impulsed ultrabass tones...thunder. Somewhere in the almost inaudible roll of these basement sounds there was a devastating and fearful power.

THUNDER

As thunderous tones deepen, their power seemingly intensifies over frail barriers such as glass windows. Certain abrupt thunder peals often shatter windows into tiny fragments. In the apparent absence of thunderous tones we may observe the strong and continuous vibration of glass window panes during storms. A sudden eerie silence, and the window is shattered before our eyes.

Natural phenomena are prodigious generators of infrasound. The potent distal effects produced when natural explosions occur produce legendary effects. When Krakatoa exploded, windows were shattered hundreds of miles away by the infrasonic wave. Wind was not the causative agent of these occurrences, as no wind was felt or detected. Seismographic stations registered the blast, and barometers measured the shockwaves. The "ringing" of both earth and atmosphere continued for hours. It is believed that infrasound actually formed the upper pitch of this natural volcanic explosion, tones unmeasurably deep forming the actual "central harmonic" of the event. The island of Krakatoa was literally lifted into orbit in the fatal blast. Brilliant sunsets followed for many years thereafter, the sad memorial of all the souls who perished.

The power of explosives, in shattering and devastating property, lies in two zones. The first zone is that with which we are principally familiar; the actual blast site, where chemically released gases and metal fragments push back everything in their perimeter. The second less familiar zone extends very much further from the blast site than can be imagined. It is in the powerful sonic wave which expands outward that an equally destructive danger lies. Thick pressure walls of incredible momentum, interspaced with equally thick walls of reduced air pressure, travel far away from the blast site. The blast site is the small destructive zone by comparison. Few objects can survive this destructive tide.

Analysts contend that infrasound is composed of a very broad band of pitches. These tones of immense pressure and duration "accommodate" themselves when encountering resonant cavities. All such resonant cavities are "found and destroyed" when the proper pressure waves flow into their resonances. Rooms, halls, alleys, spaces among buildings, courtyard areas, cellars, subways, sewer chambers; all these burst open into flying fragments when infrasonic waves flood them. Infrasound is the cruel tonal giant, tearing open whatever it finds in its path.

Study reveals that the sudden shock wave of an explosion propels a complex infrasonic signal far beyond the shattered perimeter. Incoherent though such shockwaves may be, their destructive influence dissolves distant walls and windows seconds after the shrapnel has done its deadly work. Objects of all shapes, sizes, and compositions explode when the infrasonic impulse passes through their space. No shield can block infrasound. Physicists have studied the refuse which remains after an explosive charge has been detonated. Few materials can maintain their integrity. Those objects which manage to survive explosions are noteworthy as infrasonic "resistors". Screen reinforced concrete does not easily succumb to the infrasonic blasts of explosive charges.

EARTHQUAKE

The sound of Krakatoa exploding up into space, a vertical excess of one hundred miles, succeeded in blasting out windows at a thousand mile radius from the epicenter. Certain earthquake activities produce large and virtually insensate vertical displacements of the ground surface, in extreme instances amounting to a few feet per pulse. In this case, the ground becomes the surface of a drum, ringing out its deadly cadence at infrasonic pitch hours before the event. The ground undulates with infrasonic tones, an elasticity that eventually cracks under the heaving stress.

Ultralow pitch earthquake sounds are keenly felt by animals and sensitive humans. Quakes occur in distinct stages. Long before the final breaking release of built up earth tensions, there are numerous and succinct precursory shocks. Deep shocks produce strong infrasonic impulses up to the surface, the result of massive heaving ground strata. Certain animals (fish) actually can hear infrasonic precursors. Precursory shocks are silent, being inaudible in humans. Animals, however, react strongly to the sudden surface assault of infrasonic shocks by attempting escape from the area. Animals cannot locate the source and center of these infrasonic impulses, behaving in a pitiful display of circular frenzies. The careening motion of wild horses and other domestic animals indicates their fear and anxiety. Poor creatures, neither they nor we can escape the infrasonic source. Encounters with natural infrasound reveal their vast extent, covering hundreds of square miles of surface area.

Certain animals employ infrasound as weaponry. It has been known that certain whales are able to stun their prey with powerful blasts of inaudible sounds. Called "gunshots", whales focus these powerful blasts at large squid and other fish to paralyze and catch them. In some instances, they have been known to burst their prey apart by tonal projection alone. Human experience with these inaudible blasts have been reported. The distress calls emitted by little beached whales was sufficient to push a veterinarian back several feet in the water. Others have experienced these pressure waves, reporting that their hands could not be brought close to the sinal area of small whales because of their inaudible acoustic projections.

Infrasonic shocks produce characteristic pressure effects on structures and organisms alike. The sensation flattens the body. It is as if one were struck with a solid invisible wall from which there is no escape. There are physiological effects as well. Anxiety, fear, extreme emotional distress, and mental incapacitation are all part of the unpleasant phenomenon. Notable among human exposures to quake-correlated infrasound is the precursory nausea which many report. This strong sensation leaves its more sensitive victims helpless. Feeling the momentary deep motion of the ground strata beneath them, numerous individuals have been used to report these sensations in a bizarre earthquake "alarm system". Unfortunately, physiological reaction to infrasound remains continuous, long after their irritating presence has ceased. The harmfully stimulating influence of infrasound renders physiology permeable and ultrasensitive to every available environmental sensation. The extreme irritability of infrasound victims has been noted.

Earthquake infrasound manifests only at intermittent intervals, producing drastic and sustained negative modifications of consciousness. The human organism continues to reel under intermittent infrasonic assault for numerous reasons. After less than a five minute exposure to low intensity infrasound of 10 cycles per second, dizziness will last for hours. Infrasound of 12 cycles per second produces severe and long lasting nausea after a brief low intensity exposure.

FLOOD

Surf pounds the shore, producing shocks of 16 cycles per second, just short of the true infrasound range. Ocean waves which pound the atmosphere across huge ocean areas produce an acoustic energy with a mean pitch of 16 cycles per second. The phenomenon of the "barisal guns, fog guns, lake guns" form a well documented bibliography of anomalous acoustic phenomena. These "booming" anomalous tonal phenomena are not isolated to one area or nation. Every nation has accounts of these sea-related mystery sounds. Some regions call them "bay detonations", since they come as abruptly explosive intonations from certain bay areas.

Some of these tones manifest their shocking tones at haphazard intervals. But there are those water-related booms which are periodic, residents near these sites being accustomed to their mysterious occurrence. The strange "explosive" sounds come at certain times of the day, at certain times of the month, and at certain times of the year. There are certain other related anomalous natural tones which ring, hoot, and buzz. Some have been likened to organ tones, tuba blasts, and the deep intonings of very large bells. Bay sizes, wave sizes, and geologic compositions of bays and shores have been woven into complex mechanistic attempts at explaining how these mystery sounds are being generated in certain environments.

The detonation may be caused by a sudden "slapping" of bay water by a singular wave having the "right" breadth and momentum, matching the natural resonant pitch of a bay. The underlying bay rock matrix may resound in the manner of a bell, gong, or cymbal. The geological composition of the bay plays the greater part of the effect, sudden winds or water surges knocking the natural "sounding board". These natural bay tones have great infrasonic content.

The infrasonic outputs of the mystery tones are significant. Each of these phenomena produce a range of very low pitch tones. These booming sounds have rattled windows and rocked some small towns. Animals are startled by their inaudible precursors, and humans are often dizzied after their manifestation for hours. In several areas, people are hospitalized by the "boom" related illnesses.

Waterfalls are notorious generators of infrasound. Numerous susceptible visitors at Niagara experience a peculiar nausea which is not associated with the normal fear of heights. Thundering cataracts produce strong infrasonic shocks to which mile exposure stimulates the common malady. Lake ice and glacial ice produce deep booming sounds which ring for hours, behaving as large tympanic surfaces. The thunderous sounds associated with these occurrences produces infrasound of pitch related to ice surface mass, breadth, and length alone. Antarctic research experienced nausea in relation to ice related sounds.

Tidal waves and other sudden variations of water surfaces produce large magnitude "seiche" waves. These have been sighted by ocean going ships, where oceanic surfaces have drastically changed elevation in an incredibly short time. Ships "drop" into such huge ocean troughs and rise again after the wave passes. In dropping, some have crashed to the very rock bottom of their bays, only to be lifted in pieces when the wave resurged. Film footage of the great and horrid Alaskan Earthquake (1964) reveals this devastating sea "drop-out".

Upon such lethal seiche tides, even in the fortunate absence of earthquakes, comes nausea and other coastal related illnesses. Large intensity infrasonic sea shocks have their powerful effect on the overlying atmosphere of their regions. These infrasonic shockwaves travel for long distances. Certain bays are known for the high incidence of such illnesses, the result of resonant baywater "heavings" which occur daily. Their sickening effects are seemingly "stored up" in physiology, lasting for hours. While these phenomena proceed from deep in the heart of earth, and on its surface, there are phenomena which generate infrasonic sources...from space.

Aerial earthquake sounds have been reported by observers. Such rushing, thrumming sounds seem to come from "everywhere" above the affected locale. Typical of infrasound, the sources cannot be accurately located. When Krakatoa exploded, barometers fluctuated rapidly in short time intervals. It was recognized that a new means for detecting earthquakes and other earth movements had been found. The horrifying destruction of Krakatoa prompted the emergence of a new science. The rapid development of sensitive barometric instruments provoked the discovery of a whole new "infrasonic world".

The opposing nations of the Cold War years used barometers and seismographs in determining the relative explosive yield of periodic underground atomic blasts. Sensitive barometric detection gradually began searching the atmosphere and earth for infrasonic "events". Mysterious and sudden barometric variations indicates that natural infrasonic generation has a much wider source than the subterranean earth. Infrasounds associated with the Aurora Borealis are too numerous to mention, a well chronicled occurrence.

The aurora borealis is heard to "swish, crackle, sizzle, and...thunder". Quantitative analysts "cannot understand" how these sounds can be "heard" but not recorded. It is obvious that certain auroral sounds stimulate physiological responses which will never register in biologically unmodified electronic systems (Lawrence). Though debates continue when referring to higher auroral tones, the "thunder" of the aurora stimulates aerial infrasounds which can be measured.

Low level auroras have been actually seen and felt. The sounds and odors associated with this rare phenomenon are unmistakable. In one instance a chemist was fortunate enough to have lived, after witnessing the effect of auroral grounding throughout his laboratory. The incredible luminescence produced in several platinocyanides, electro-phosphorescent chemicals, were duly noted and reported. Another such incident involved the grounding of the aurora into an elevated radio tower. The radio engineer suddenly heard a crackling sound "from everywhere", was unable to transmit any signal power, felt completely electrified, smelled ozone everywhere, and heard the "crackling" sound. Numerous witnesses who saw the event, describing the colored column of light which suffused the tower, feared he might have been killed by its power.

The auroral high pitched sizzling sounds are augmented by deep and ominous thrumming. These deep tones sweep through the bodies of listeners who are fortunate enough to survive the dangerous encounter. These permeations produce an irritability and a dizzying nausea. These sounds were always equated with evil by the Eskimo. Their legends of the aurora are always fear-filled. The trademark of anxiety and dread highly characteristic of infrasonic influence, their tales also recount the "taking" of souls by the "ground walking" aurora. English observers reported that the aurora actually "swept along the ground" like a column descending from the sky. It took a multicolored appearance all along its meandering path.

Blasting through interplanetary space, solar flares assault the earth with a barrage of stupendous proportion. Their disturbing effect on electrical systems is historically noted. The appearance of electrical power surges during solar flare events has amounted to many hundreds or even thousands of amperes line-induced current. Northern lands design their power systems to accommodate these periodic manifestations of great power. Oil lines in northern lands must be carefully grounded and insulated to prevent the continuous induction of such harmful electrical surges. Standing arcs of brilliant blue current had been observed upon the surface of loose pipe joints during solar flares and strong auroral episodes.

Few writers have discussed the intermittent effects of solar flares on atmospheric pressure. The sudden changes noted in air pressure, which cover many thousands of square miles, are obviously sourced in the solar wind. The effect of this natural atmospheric assault has defined and disturbing influence on both the weather and human behavioral patterns. Some 139 solar flares were recorded between 1980 and 1983. There is a statistical 155 day periodicity in solar flares, a rhythm often violated by several interstitial flares. Principally used for predicting their expectable effect on radio communications, specific military observers monitor solar flares with continued concern. Correlations of flares with jet stream behavior is strong.

Jet stream behavior, in its meanderings and undulations across vast geographic regions, is not mysterious when considering the intermittent effect of solar flares and the normal "background" bursting of the solar wind. Travelling at thousands of miles per hour, flare pressures aperiodically barrage the neutral atmosphere. The explosive influence of vast power shocks the entire weather system, electrically active flare disturbances violently disrupting all atmospheric processes. The very obvious outlines of flare contacts with the atmosphere can be traced as major pressure changes on weather maps. Solar flare impacts strike the earth like a bell. Auroras result, and have been correlated with thunderstorm activities.

But continual minor background disruptions also exist, propelled by the sun. In its normal process, solar expulsions do not arrive at the atmospheric boundaries as a homogeneous pressure wave. The arrival of solar products comes as a pressure wave of inconsistent density. This intermittent barrage induces harmonic atmospheric disturbances which continually modify and chaoticize emerging weather patterns. The effect is exactly like "thrumming" an evenly sanded drumhead with innumerable impacts. This imprint of "background" pressure waves, rattling daily upon the atmosphere from solar winds, can be seen as "Chladni" patterns on weather maps.

Both solar flares and the normal thrummings of the solar wind generate infrasonic pulses throughout the atmosphere. The infrasonic shockwaves of the aurora are normally not heard, but definitely sensed. Measurements have registered a continual infrasonic background noise level. This pressure energy emanates "from above" atmospheric strata, radiating downward in large patterns. Atmospheric infrasound is most strongly measured during daytime hours, a clear indication of their source in the intermittent expulsions of solar wind. Atmospheric infrasounds arrive at measuring stations with pitch between .67 and 1.5 cycles per second. Their pitch continuously oscillates between .67 and .83 cycles per second. These solar sourced infrasonic impacts very definitely correlate with sudden swings in human behavior, having very obvious sociological implications. The energetic content of atmospheric infrasound represents a vast and untapped potential.

WIND

Infrasound moves, unaffected, through and across both winds and storms. But wind and storm can generate infrasound. The powerful harmonic rotations of storms shears the atmosphere, radiating a cyclonic series of expanding infrasounds. The sense of impending fear which proceeds hurricanes is due to infrasonic emissions. The infrasound of seasonal winds and weather patterns produces illness in certain persons. Some individuals can hear the jetstream and its thunderous pitch, peaking between 30 and 40 cycles per second. More and more populations are reporting the persistence of ultralow pitch sounds which render them weak and fatigued. Having often unexplained sources, we find the bibliography flooded with cases of persistent "atmospheric...and underground sounds".

Victims of such infrasonic assaults report severe stomach upsets associated with such infrasounds. Persons who report these persistent "underground" rumbles often live in a very localized region. These loci have been as small as six miles in diameter. Wind shear action between the rapid jet stream (.75 miles per second) and more quiescent lower air strata might continuously generate this persistent infrasound. Natural infrasonic generation is difficult to determine in all cases where it has been detected. Clashing winds may produce such sustained low pitches by misunderstood "shearing" actions, similar to Von Karrman vortices.

Wind shearing may be modified by local topology. Mountain ranges of specific geometry may offer the most plausible explanation for infrasound in certain areas. Their obstructive presence among regionally prevailing winds may produced sustained aerial vortices from which infrasounds continuously radiate.

Why however does the infrasound focus on certain ground points? Some theorists claim that wind enters caverns, producing an immense artificial whistle of infrasonic resonant pitch. Careful examinations of these caverns reveals infrasonic pitch of 20 to 30 cycles per second which does not "register" on tape recordings. Some have suggested that these infrasounds are only sensed in physiology, being "electrostatic" in nature. They also claim that the incidence of ground-focussing infrasound is an electrical manifestation, the result of emerging terrestrial charges in highly localized regions.

How does sustained infrasound affect manmade structures? Gusting wind has often applied such instantaneous pressure to manmade artifice that strong rock walls fall flat in tiny pieces. These sudden events often occur when winds seemingly ceased for an instance. During that brief interval, windows are often blown through, and walls are toppled by infrasonic impulse.

What is the sustained influence of infrasound on humans and human behavior? Mysterious desert humming sounds fill the night of nomads with superstitious dread. Deep, buzzing, and threatening, these continuous humming tones have produced anxiety and fear among bedouins for centuries. The "ghost wails" appear in the mythology and folktales of the desert people.

But these deep and virtually inaudible humming tones are not confined to the desert plains, where they thrash among themselves across sandy dunes. The Mistral, the northward winds of the African continent, sweeps over the southern Mediterranean coastlands during late fall. These familiar hot winds emerge from their desert journeys with a strange power, lasting throughout the winter. These winds leave an indelible trace among exposed communities, a phenomenon which has been misunderstood for centuries.

The Mistral, weak in infrasonic intensity, does not wreak havoc with material structures. But the Mistral works its permeating harm nonetheless. For the inhabitants of certain coastal areas, the low intensity infrasound of the Mistral brings with it a peculiar seasonal anxiety and depression. In certain locations across the Mediterranean coastland there are individuals who suffer from "seasonal nervous exhaustion" and other "neurophysical maladies". It is known that whenever the Mistral blows, there will be increased emotional tension, depression, and irritability. The Mistral, in numerous cases, has produced fatalities.

Infrasound travels long distances, often exceeding one thousand miles, with virtually no attenuation. Its pressures thus arrive at great distances with the same force and intensity as when generated. A deadly pressure. The atmosphere sustains prolonged and powerful infrasonic vibrations. How natural conditions can systematically modify human behavior for protracted seasonal periods is frightening. How natural conditions can systematically modify large-scale social behavior for protracted seasonal periods is equally frightening. Not much acoustic power is required for infrasound to produce such extreme and sustained physiological symptoms.

Fohn winds are dry and warm southerly winds which traverse the Alpine regions of Europe. Fohn weather is characterized by clear skies, high visibility, and dry atmosphere. Studies of "Fohn weather" and the Mistral alike have revealed some intriguing and frightening statistical correlations. The biological effects of both Mistral and Fohn weather have been well documented. These include extreme irritability, accident-prone loss of objective judgement, slight disorientation, mild nausea, and diarrhea.

It is an established fact that sustained low intensity infrasound alters human behavior and health. Higher accident rates are correlated with pre-Fohn weather onset. This high accident rate rises until the establishment of Fohn weather, having been attributed to the infrasonic content of the winds.

ENGINES

Vibrating manmade structures stimulate the artificial generation of dangerous infrasound. When turns are made at 60 miles per hour, car chassis vibrations produce a peak infrasonic emission. Travel sickness can be associated with prolonged infrasonic exposure to any vibrating chassis. Cars, buses, trains, motorcycles, and jets alike each register hazardous intensities of infrasound. Each transportation mode has its characteristic infrasonic pitch, the necessary outcome of mechanical frictions and inertial resistances.

There is difficulty in recording and reproducing ultradeep tones for study and analysis. They have to be generated on site for experimental purposes. Theater-sized sound systems can never completely transmit all of the sensations associated with naturally occurring infrasound. But there have been instances where audiences have become frighteningly ill because of the accidental generation of infrasound in a theater space.

Of critical importance is the comprehension of human tolerances to infrasound. Military medical teams have long studied the effect of machine vibration on human judgement and behavior out of necessity. If jet pilots and rocket pilots alike evidence even minor errors in judgement through their exposure to infrasound, disaster can result. Certain critical errors in judgement and accuracy have in fact been noted during short flight times.

The powerful infrasonic vibrations of jet chassis absolutely saturate the bodies of pilots. Continually saturated with these infrasonic energies throughout their flight time, pilot reflexes are severely diminished. Military procedure recognizes this factor, and routinely limits flight time. It is known that excess infrasonic exposures endanger pilots and their flight missions. Pilot damaging effects include decrements in vision, speech, intelligence, orientation, equilibrium, ability to accurately discern situations, and make reasonable decisions.

THE ABYSS

The Cold War was on. The United States alone held the dread secret. The most terrible weapon yet developed was the private property of one government. The mere existence of the atomic bomb was threat to nations whose motives were not entirely altruistic. Motivated, aggressive, and imperialistic, obtaining atomic bomb data was a priority for several nations. The only manner in which some nations obtained the secret was by stealing it. When Stalin's science officers finally developed an atomic duplicate of the American bomb, pressure suddenly was placed upon every other European nation to achieve an equivalent or better device.

When one seeks to defend one's borders, the consequences of releasing weapons of devastation to the world do not seem important. Weaponry is death-oriented by nature. But there are moral differences between weapons of defense and weapons of offense. Previous to this atomic proliferation, competing nations concentrated their weapons research on truly bizarre and equally deadly means for defending their national boundaries. A great variety of such deadly weapons were perfected in rapid succession. This included deadly variations and combinations of gas weaponry, pathogenic agents, and radiant weaponry. Stalin's research teams investigated psychic powers as a possible means for destroying an enemy. Psychotronic warfare was developed among numerous groups, both private and national, with measurable success. Information on some simpler psychotronic weapons have recently been obtained through an increasing process of Soviet disclosure.

In truth, the larger the weaponry the less safe the national boundaries truly were. While the superpowers concentrated their weapons development programs on mass-destructive nuclear weaponry, others focussed on more practical conventions. The limited tactical warfare of small battlefields seemed a more immediate need. While developing their own atomic device, France sought defensive tactical weaponry on every possible technological front. Short range weapons would best defend against a conventional national assault. But other systems were also sought; systems which, though non-nuclear, were equally invincible. As the great Frankish Knight, Charles "the Hammer" Martel repelled ruthless invaders from the medieval east, so a new hammer would be sought to defend France against possible new enemies from the east. Even as Charles Martel arose from obscurity, so this strange new "hammer" would arise in equal obscurity.

GAVREAU

The central research theme of Dr. Vladimir Gavreau was the development of remote controlled automatons and robotic devices. To this end he assembled a group of scientists in 1957. The group, including Marcel Miane, Henri Saul, and Raymond Comdat, successfully developed a great variety of robotic devices for industrial and military purposes. In the course of developing mobile robots for use in battlefields and industrial fields, Dr. Gavreau and his staff made a strange and astounding observation which, not only interrupted their work, but became their major research theme.

Housed in a large concrete building, the entire group periodically experienced a disconcerting nausea which flooded the research facility. Day after day, for weeks at a time, the symptoms plagued the researchers. Called to inspect the situation, industrial examiners also fell victim to the malady. It was thought that the condition was caused by pathogens, a "building sickness". No such agencies were ever biologically detected. Yet the condition prevailed. Research schedules now seriously interrupted, a complete examination of the building was called.

The researchers noticed that the mysterious nauseations ceased when certain laboratory windows were blocked. It was then assumed that "chemical gas emissions" of some kind were responsible for the malady, and so a thorough search of the building was undertaken. While no noxious fumes could be detected by any technical means, the source was finally traced by building engineers to an improperly installed motor-driven ventilator. The engineers at first thought that this motor might be emitting noxious fumes, possibly evaporated oils and lubricants. But no evaporated products were ever detected. It was found that the loosely poised low speed motor, poised in its cavernous duct of several stories, was developing "nauseating vibrations".

The mystery magnified for Dr. Gavreau and his team, when they tried to measure the sound intensity and pitch. Failing to register any acoustic readings at all, the team doubted the assessment of the building engineers. Nevertheless, closing the windows blocked the sense of nausea. In a step of brilliant scientific reasoning, Gavreau and his colleagues realized that the sound with which they were dealing was so low in pitch that it could not register on any available microphonic detector. The data was costly to the crew.

They could not pursue the "search" for long time periods. During the very course of tracking the sound down, an accidental direct exposure rendered them all extremely ill for hours. When finally measured, it was found that a low intensity pitch of a fundamental 7 cycles per second was being produced. Furthermore, this infrasonic pitch was not one of great intensity either. It became obvious that the slow vibrating motor was activating an infrasonic resonant mode in the large concrete duct. Operating as the vibrating "tongue" of an immense "organ pipe", the rattling motor produced nauseating infrasound. Coupled with the rest of the concrete building, a cavernous industrial enclosure, the vibrating air column formed a bizarre infrasonic "amplifier".

Knowledge of this infrasonic configuration also explained why shutting the windows was mildly effective in "blocking the malady". The windows altered the total resonant profile of the building, shifting the infrasonic pitch and intensity. Since this time, others have noted the personally damaging effects of such infrasonic generation in office buildings and industrial facilities. The nauseating effects of exposure to a low intensity natural or manmade infrasonic source is now well appreciated.

It has become a routine architectural procedure to seek out and alter any possible such resonant cavities. The sources often appear in older buildings, the result of construction rendered faulty by previous lack of this knowledge. All such "improper" architectural formats are modified by the additions of sound-blocking materials.

WHISTLES

Dr. Gavreau and his research team now carefully investigated the effects of their "infrasonic organ" at various intensity levels and pitch. Changing the spring tension on shock mounts which held the fan motor, it was possible to change the pitch. Various infrasonic resonances were established throughout the large research building. Shutting the windows blocked most of the symptoms. When the window was again opened, however weak as the source was made, the team felt the nauseating effects once again. In the business of military research, Dr. Gavreau believed he had discovered a new and previously "unknown weapon" in these infrasounds. Aware of the natural explosives by which infrasonics are generated, Dr. Gavreau began to speculate on the application of infrasonics as a defense initiative. The haphazard explosive effects of natural infrasound in thunderclaps were quite effective in demonstrating what an artificial "thunder-maker" could do. But, how could a thunderclap be artificially generated in a compact system? These thoughts stimulated theoretical discussions on the possibility of producing coherent infrasound: an infrasonic "laser".

The first devices Dr. Gavreau implemented were designed to imitate the "accident" which first made his research group aware of infrasonics. They designed real organ pipes of exceedingly great width and length. The first of these was six feet in diameter and seventy five feet long. These designs were tested outdoors, securely propped against protective sound-absorbent walls. The investigators stood at a great distance. Two forms of these infrasonic organ pipes were built. The first utilized a drive piston which pulsed the pipe output. The second utilized compressed air in a more conventional manner.

The main resonant frequency of these pipes occurred in the "range of death", found to lie between three and seven cycles per second. These sounds could not be humanly heard, a distinct advantage for a defense system. The effects were felt however. The symptoms come on rapidly and unexpectedly, though the pipes were operating for a few seconds. Their pressure waves impacted against the entire body in a terrible and inescapable grip. The grip was a pressure which came in on one from all sides simultaneously, an envelope of death.

Next came the pain, dull infrasonic pressure against the eyes and ears. Then came a frightening manifestation on the material supports of the device itself. With sustained operation of the pipe, a sudden rumble rocked the area, nearly destroying the test building. Every pillar and joint of the massive structure bolted and moved. One of the technicians managed to ignore the pain enough to shut down the power supply.

These experiments with infrasonics were as dangerous as those early investigations of nuclear energy. Dr. Gavreau and his associates were dangerously ill for nearly a day after these preliminary tests. These maladies were sustained for hours after the device was turned off. Infrasonic assaults on the body are the more lethal because they come with dreadful silence. The eyesight of Dr. Gavreau and his fellow workers were affected for days. More dangerously were their internal organs affected: the heart, lungs, stomach, intestinal cavity were filled with continual painful spasms for an equal time period.

Musculature convulses, torques, and tears were the symptoms of infrasonic exposure. All the resonant body cavities absorbed the self-destructive acoustic energy, and would have been torn apart had the power not been extinguished at that precise moment. The effectiveness of infrasound as a defense weapon of frightening power having been demonstrated "to satisfaction", more questions were asked. After this dreadful accident, approaching the equipment once again was almost a fearful exercise. How powerful could the output of an infrasonic device be raised before even the operating engineers were affected?

With greatest caution and respect for the power with which they worked, Dr. Gavreau began recalculating all of his design parameters. He had grossly misjudged the power released by the pipes. He had, in fact, greatly lowered those calculated outputs for diagnostic purposes. Never had he imagined that these figures were actually far too great in the world of infrasound!

Empirical data being the only way to determine how infrasonic energy correlated with both biological and material effect, the tests were again attempted with a miniature power supply. First, the dimensions of these devices had to be greatly reduced. Their extreme length was objectionable. In order to provide absolutely safe control of the deadly blasts, several emergency cutoff switches were provided. These responded to the radiated infrasonic pressure wave. The intensity could be absolutely limited by use of automated barometric switches.

In an attempt to achieve more compact and controllable infrasound generators, Dr. Gavreau designed and tested special horns and "whistles" of various volumes. These were each remarkably simple flat circular resonant cavities, having a side output duct. They were simply the large analogues of foghorns and police whistles. These flat forms were volumetrically reduced in successive design stages because it was found that their output was far too great. The infrasonic foghorns could produce a frightening two kilowatts of infrasonic energy, at a pitch of one hundred fifty cycles per second.

The flat "police whistles" were more easily designed to required specifications. Their overall characteristics were quite simple to determine, a mathematical formula being devised for the purpose. The whistle's resonant pitch was found by dividing its diameter into a numerical constant of 51. Increasing the depth of the whistle effectively increased its amplitude. A whistle 1.3 meters in diameter produced an infrasonic pitch of 37 cycles per second. This form violently shook the walls of the entire laboratory complex, though its intensity was less than 2 watts infrasonic power.

DANGER

Not much amplitude is required for infrasound to produce physiological malady. Several researchers accidentally did themselves great harm when, by deliberate intent or accident, they succeeded in generating infrasonic vibrations. Tesla used vibrating platforms as an aid to vitality. He delighted in "toning the body" with vibrational platforms of his own design. Mounted on heavy rubber pads, these platforms were vibrated by simple motorized "eccentric" wheels.

Their mild use, for a minute, could be pleasantly stimulating. The effects invigorating the whole body for hours thereafter. Excessive use would produce grave illness however, excessive aggravations of the heart being the most dangerous aspect of the stimulation. The entire body "rang" for hours with an elevated heart rate and greatly stimulated blood pressure. The effects could be deadly.

In one historic instance, Samuel Clemens, Tesla's close friend, refused to descend from the vibrating platform. Tesla was sorry he had allowed him to mount it. After repeated warnings, Tesla's concern was drowned out by both the vibrating machine and Clemens' jubilant exaltations and praises. Several more seconds and Clemens nearly soiled his white suit, the effects of infrasound being "duly recorded".

Tesla often went to great lengths in describing the effects of infrasounds to newspaper reporters who, behind his back, scoffed at the notion that a "little sound" could effect such devastations. Yet, it was precisely with such a "little sound" that Tesla nearly brought down his laboratory on Houston Street. His compact infrasonic impulsers were terribly efficient. Tesla later designed and tested infrasonic impulse weapons capable of wrecking buildings and whole cities on command.

Walt Disney and his artists were once made seriously ill when a sound effect, intended for a short cartoon scene, was slowed down several times on a tape machine and amplified through a theater sound system. The original sound source was a soldering iron, whose buzzing 60 cycle tone was lowered five times to 12 cycles. This tone produced a lingering nausea in the crew which lasted for days.

Physiology seems to remain paralyzed by infrasound. Infrasound stimulates middle ear disruptions, ruining organismic equilibrium. The effect is like severe and prolonged seasickness. Infrasound immobilizes its victims. Restoration to normal vitality requires several hours, or even days. Exposure to mild infrasound intensities produces illness, but increased intensities result in death. Alarming responses to infrasound have been accurately recorded by military medical experts. Tolerances from 40 to 100 cycles per second have been recorded by military examiners. The results are sobering ones. As infrasonic pitches decrease, the deadly symptoms increase. Altered cardiac rhythms, with pulse rates rising to 40 percent of their rest values, are the precursors to other pre-lethal states. Mild nausea, giddiness, skin flushing, and body tingling occur at 100 cycles per second. Vertigo, anxiety, extreme fatigue, throat pressure, and respiratory dysfunction follow. Coughing, severe sternal pressure, choking, excessive salivation, extreme swallowing pains, inability to breathe, headache, and abdominal pain occur between 60 and 73 cycles per second. Post exposure fatigue is marked. Certain subjects continued to cough for half an hour, while many continued the skin-flush manifestation for up to four hours.

Significant visual acuity decrements are noted when humans are exposed to infrasounds between 43 and 73 cycles per second. Intelligibility scores for persons exposed, fall to a low of 77 percent their normal scores. Spatial orientation becomes completely distorted. Muscular coordination and equilibrium falter considerably. Depressed manual dexterity and slurred speech have been noted before individuals blackout. Just before this point, a significant loss in intelligibility is noted.

The findings of Dr. Gavreau in the infrasonic range between 1 and 10 cycles per second are truly shocking. Lethal infrasonic pitch lies in the 7 cycle range. Small amplitude increases affect human behavior in this pitch range. Intellectual activity is first inhibited, blocked, and then destroyed. As the amplitude is increased, several disconcerting responses had been noted. These responses begin as complete neurological interference. The action of the medulla is physiologically blocked, its autonomic functions cease.

WATCHMEN

Infrasound clings to the ground, a phenomenon well known in the animal world. Female vocalizations and those of their young, take their traceable routes through the air. High pitched sounds are aerial in nature. This makes females and young natural targets for predators. Low pitched tones cling to the ground, being "guided" along the soil layers. Male vocalizations cannot be localized by predators. Male sounds "hug the ground", diffusing out from their source. Some males rumble the ground with voice and hooves. These are communications signals which they alone comprehend.

The fact that the ground draws and guides low frequency tones is a remarkable gift to the animal kingdom, enhancing the survival of male leaders. When herds are attacked by predators, the males can continue to give guidance to their companions, while remaining completely "invisible" and elusive. Predators cannot locate the voices and rumblings of male leaders because their low pitched signals are impossible to pinpoint. They are therefore also impossible to attack. Predators are often overtaken by the males who maintain their diffusive communications across and through the ground.

The same analogies would apply to an infrasonic defense system. First, infrasound does not lose its intensity when travelling very long distances across the ground. They remain at the same intensity as when released from their deadly sources. Also, because of the ground clinging effect, infrasonic sources cannot be located without special appliances. This would work well for those who used the weaponry of infrasound. But suppose some hostile force were themselves using infrasonics? Infrasonics are inaudible. The battle would be over before anyone knew it had begun. How would one know of an infrasonic attack? The first line of defense would therefore be the detection of the "unperceived enemy". The development of an adequate infrasonic weapons systems would first require an infrasound detector.

Dr. Gavreau first concentrated on developing infallible infrasonic detectors for the personal safety of his operators as well as for eventual tactical deployment. He experimented with several designs which followed the arcane analogues of old wireless detectors. One such design used enclosed flames to detect infrasonic pitches. They were reminiscent of those flame detectors developed by Lee De Forest just before his invention of the triode. The flame detectors of Gavreau employed variable resonant cavities. Flame amplitudes shifted with specific infrasonic pitches. He could calibrate the infrasonic intensity as well as the pitch with these detectors. But, flames are dangerous and fickle, not being very reliable in battle.

Dr. Gavreau next experimented with enhanced mechanical barometers. These coupled large resonant cavities with very fine barometer tubes. They displayed great sensitivity. Steady increases in barometric pressure were registered when large cavity bellows were compressed by infrasounds. The sensitivity of these barometers increased as the bellows capacity was increased. They were adequate, but frail.

Another embodiment resembled the early mechanical television designs of John Logie Baird. It utilized large tympani skins, mirrors, lights, and photocells. A mirror was fastened to the tympanum. A light beam flickered when infrasound struck the mirror. The photocell recorded these flickers as an electrical signal. This detector system was very reliable.

By far, the most advanced detectors which Gavreau designed and tested utilized an electrolytic process. In this analogue of systems developed by Fessenden to measure faint wireless signals, chemical solutions and fine wirepoint electrical contacts were used. Chemical solutions, separated by an osmotic barrier, were forced to migrate through the barriers whenever infrasound traversed the system. This chemical mixture was then measured as an increased electrical conductivity in a sensitive galvanometer. This system was reliable and accurate. All of these systems suffered from one possibility. The offensive use of an incredible infrasonic amplitude would burst them into vapor.

ARMOR

Claims were issued by french authorities, stating that Dr. Gavreau was not developing weapons at all. Several patents, however, betray this conspicuous smoke-screen. While it is impossible to retrieve the actual patents for the infrasonic generators, Dr. Gavreau is credited with extensive development of "infrasonic armor". Why would he "waste" such time and expense if not for an anti-weapons program?

Thus use of infrasonic weaponry necessitates the development and implementation of infrasonic shields. Dr. Gavreau spent more time developing infrasonic shields than on developing efficient infrasonic horns. Infrasound could not adequately be blocked, as Dr. Gavreau discovered early in his research. Infrasonic devices require extremely large baffles.

Furthermore, no one would dare initiate an infrasonic barrage on any invasive force without adequate protection. Infrasonic horns can project their sounds in a given direction, but natural environments "leak" portions of the sound in all directions. Infrasounds saturate their generators, flooding and permeating their sources in a few seconds. They "work their way back" toward those who dispatch their deadly signals. Infrasounds "hug the ground" and spread around their sources. Unfortunately, those who would release infrasonic energy would themselves be slaughtered in the very act.

The first method of Gavreau involves the conversion of infrasound into successively higher pitches, until the infrasonic pitch is "lost". This was achieved in his passive "structural" method, an enormous layered series of baffles and resonant cavities. This form is "passive" since it merely stands and waits for infrasonic barrages, absorbing and converting them into harmless audible tones.

The second method of Gavreau is more active and "aggressive". It actively engages and nullifies any offensive infrasonic power. The nullifier uses a well known physical principle for its operation. As an "active" shield, it transmits tones whose opposing wavefronts destructively interfere with incoming infrasound. Infrasonic attacks are nullified, or at least brought to much weaker levels.

This method requires high speed detection and response systems. The process involves determination of an attack pitch, generation of the same, and projection of the pitch "out of phase". The active nullifier method is not completely accurate or protective by any means. A highly modulated, mobile infrasonic source would be nearly impossible to successfully neutralize without extremely sophisticated electronics.

But an elegantly simple approach was imagined, one which would not require the defender to be exposed to his own infrasonic projections. While fixated on the old notion of gun installations and stations, Gavreau and the team had momentarily forgotten their first research endeavor. Robotics!

THE HAMMER

Let us recall that Dr. Gavreau and his team of pioneers were in the business of robotics. They developed industrial and military automaton systems. How difficult would it have been to couple his newfound weaponry with robotic applications? Dr. Gavreau combined the organ pipe and whistle format. The device was housed in a block of concrete. It was less than a cubic meter in volume. The primary whistle was poised within its interior. At its flared opening were placed several resonant pipes. The device was operated by highly compressed air. Its output was frightful. It was capable, in a conventional engagement, of utterly destroying an aggressor.

This infrasound whistle design was once sealed in an 880 pound concrete pier for tests, a concrete baffle placed over its projective end. Even with these precautions, the device succeeded in absolutely shaking a fan-shaped portion of Marseille. It broke through its supportive concrete pier and destroyed the baffle covering in an instant. Macabre. No sound was ever heard.

This design demonstrated great pitch selectivity, power, and directivity. In this last feature, Gavreau and his team achieved a safety factor of greatest value. Infrasonic defensive armaments could now be safely directed away from the operators against any foe. This weapon was a remarkably compact and efficient device. Its efficiency was gauged by the destructive output and the weapon volume.

A later embodiment of this terror disclosed another compact cube. The infrasonic whistle was presumably housed therein. Proceeding from the front plate were some sixty pipes, flared horns aimed in deadly forward array. It was said that this device alone, remotely guided into an arranged artificial battlefield, burst heavy battlements and tank interiors open with a hideous effortlessness. In addition, several other more frightening and unmentionable disruptions were observed with equal effectiveness. In each, not a sound was ever heard.

The device was mounted and mobilized. A robotic vehicle. Powered by diesel engines or compressed gas, the almost insignificant unit would be a bizarre foe for an army to engage. Preliminary experiments had proven the extreme danger of loosing infrasonic power among Gavreau and his workers. Without automatic remote control mechanisms each technician would succumb to the deadly sound and die, while the machine kept broadcasting its deadly sound. As defensive weaponry, such a device would be terrible and effective. The system would be a true deterrent for those who would be foolish enough to attempt ground assault on any nation so armed. Armies would fall flat. Once the infrasonic horns were unleashed against the foe, the battle would not even begin.

Such a war engine would be impossible to locate. None who saw its size would believe it to contain such a lethal power. Most would overlook the device completely. A flood of such devices, each emanating a peculiar highly modulated blend of infrasound, would be an unstoppable wall. Robotic tanks equipped with infrasonic generators could sweep an area with deadly infrasound, destroying all opponents to within a five mile radius. These terrible infrasonic weapons could easily be secured in drone jets, where aerial assaults could quickly and methodically waste any offensive approaching army.

Deterring would-be aerial attackers could be equally devastating for the offenders. Infrasonic beacons could sweep and scan the skies with a deadly accuracy. Infrasound passes through all matter with equal effectiveness, seeking out offenders with deadly consequence. The intensities which the Gavreau devices effectively broadcast into the environment are frightening. In these devices we see the perfection of phenomena which never naturally occur in such dangerous intensities. This is why these weapons must be deployed by remote control, operating as automatons at great distances from their operators.

Weapons are made to defend, not to offend. In Gavreau's own words: "There does not exist complete protection against infrasound. It is not absorbed by ordinary matter, walls and chambers do not suffice to arrest it". And so, once again, we stand at the cross-roads. We are called, summoned to appear before two pathways. On the one, we hear Messaien and the musical messages of peace. On the other, Gavreau and the musical messages of war. And again we choose. And again we must choose. Whose music will it be?

CHAPTER 9

THE FUSOR REACTOR: DR. PHILO FARNSWORTH

RAY OF DISCOVERY

There was once a dream of endless energy, of radiance without end. As dreams do, it found its material expressions in the discoveries of several truly gifted researchers. The dream symbols seek those who seek them. The discoveries were all the results of accidents. Fortunate accidents. Nevertheless, the world did see manifestations of dream-like energies whose sources were potent, mysterious, and eternal. It seemed then that these wonderful inventions, these strange power generators would fulfill so much of the dream and hope, the passion and desire of those who wait in silence.

These mysterious devices were received, gracious gifts from the providential source. Among those in whose hands the devices found their material expression, there was no question concerning their source. Here were anomalies of Nature. No one could have developed these things from existing lines of scientific inquiry. To do that would have required, in hindsight, another few centuries. No, the radiant energy receivers of Moray, Perrigo, Tomadelli, Hendershot, Hubbard, Coler, and so many others, were not the result of mere scientific persistence.

But, why were they given? Perhaps it was both because of an impending need and a humanly intended objective. Great revelations always precede potential world crisis. Great revelations precede human need. Perhaps they came to fulfill the world's need for energy. Clean, safe energy. They all appeared at around the same time period. Both sides of the Century's turn was witness to these wondrous discoveries. Perhaps also they came because the wrong use of certain natural energies would be proliferated on earth, endangering all life. There are those who will not argue with this understanding.

The discoveries arrive like a piercing ray. Where the discoveries appear makes no difference. Their benefits are intended for the whole race. They come like a ray in the night, striking into hearts who seek. The process is an active manifestation of external consciousness. Those who receive this blessed Ray make the discoveries which are sent to save the entire race from impending perils. They have each seen and spoken of that Ray of Discovery. What humanity chooses to do with such messages will determine the future of nations.

TECHNOPOLITICS

A recent announcement from plasma physicists promises new and thrilling advancements in the generation of electrical energy for society. The prize of this research is cheap, clean, limitless electrical power in vast, limitless quantities. The energy source of which they speak could theoretically last for millennia. For this they require new billions for continued research. And the billions are obtained.

The hot fusion advocates, mostly grant-greedy academes and their research teams, paint a lovely future before eager Senate hearing committees. The very fact that Senate committees entertain such proposals at all makes conspicuous several startling facts. It becomes at once apparent that large and radical policy shifts in technological funding have been called into action. Once upon a time, proposals for hot fusion and space program initiatives were squashed flat by committee members. Having their own ties with fossil fuel cartels, none were willing to risk the emergence of new and futuristic technologies at the risk of their own financial security.

But now we see a defined policy shift propelling toward the new technology, as if some new regulatory strategy were being deployed. It is difficult to assess these behaviors and fiscal movements. Traditionally, American Money is made by resisting discoveries and technologies, not by promoting them. But there may be new regulatory variables and fears which we must continue to assess.

Academes profess that the free social expression of hot fusion technology would trigger startling social revolutions. Like master behavioral modifiers, their words play the heartstrings well. The new improvement, they say, would mean the obsolescence of fossil fuels. Independence from every foreign oil source would be part of the "advantage". Along with this obsolescence would come the disappearance of every fossil fuel attendant environmental plague.

Next, they project the deployment of hot fusion power generating stations. Small industrial scale hot fusion reactors could supply all nationally needed electrical power, supplying it forever. Commercial hot fusion reactors will arise, they swear, if the government-backed research is licensed out to private concerns. According to the idealistic economists which they employ, the long-range effects of this hot fusion revolution will not destroy the fossil fuel interests. Their economic advisers suggest that fossil fuel industries gradually begin shifting their markets from fuel to petrochemical sales. While such altruism has not been the historical response of the coal and oil companies, the committee members strain to hear and absorb each word of promise. Some are still careful, since Standard Oil is an American based trust. Showing a vote for new technology could be dangerous.

As the utopian promises of clean, cheap energy are waved before the committees in yearly bids for extensive contracts, the facts are not so flowery. In fact, the facts are not promising at all. It is surprising that academes who pursue hot fusion research are never as caustic when reviewing their own data results as they are with the research proposals of others.

The contradictory note in this scientific fund-raising symphony is very dissonant to those familiar with Hot Fusion's past history. Yet, the bid-making proposals go on, the yearly show which brings them salaries. The hot fusion "prize", sought through modern methods is, at best, elusive. Analysts who have followed the shaky course of current hot fusion research know the disappointments, failures, and practical limits of the entire hot fusion program. And there are limits to which real success in the hot fusion endeavor is also limited. But these are all very costly disappointments. Physicists who periodically announce minor "success" in the hot fusion race usually employ these announcements, obscure technical diversions, in their continuing effort to obtain continued support. But the dreams they wave are figments, the patchwork fragments of a former dream which was fulfilled in numerous energy discoveries. Long before "atomic" or "nuclear" were termed, there were real dream fulfillments, whose material vessels matched their promise. And they were inexpensive to reproduce.

But perhaps we may cite a general ignorance of the scientific community to their own treasures once again. Those fusion researchers who made the most recent demands for federal funding, dared to project their promise of practical success out toward the year 2050! Never before has such an outrageous demand for monies been poised on such a precarious pinpoint. This bombastic announcement sounded more like a technological bribe, from individuals who wished to secure salaries and eventual pensions from government sources! Most skeptical individuals, themselves academes, would guess the obvious. While drawing heavy salaries, hot fusion researchers have nothing real to give. The modern charlatans at court. But, an historical shredding open of the technical evidence concerning hot fusion projects will prove most illuminating; the revelation being that controlled hot fusion was actually achieved...thirty years ago.

FUSION

Fusion Energy. What is "Fusion" energy? The atomic bomb operates when uranium atoms are split to release their binding energy. The controlled splitting of atoms in fission reactors produces heat and waste products. Fission reactors need highly toxic uranium or plutonium and pose environmentally problematic waste disposal.

In his theoretical approach to the "problem" of solar energy, Hans Bethe proposed that the sun uses another kind of atomic process. Dr. Bethe claimed that the fission process, but the fusion of nuclei results in the huge release of energy in the sun. According to Bethe, hydrogen is the fuel for the solar fusion process; a gas which almost entirely occupies the solar atmosphere. In the solar fusion reaction, hydrogen nuclei are drawn together in ever tightening collisions. The gravitational force of the solar mass providing the "compression" power.

Increased collision among hydrogen atoms first produces heating, ionization, and finally nuclear fusion. This process requires titanic extremes of compression which are unknown on earth. Nuclei, at a given radius, repel one another through electrostatic force. Within a specific radius however, the nuclear forces become prominent. With greater numbers of collisions, in ever dense collision radii, free nuclei begin to "merge" on collision. Energy is released when the nuclei weld. Many readers often question how this is possible. While it is easier to imagine the release of excess energy in fission reactions, many have difficulty understanding how an "excess" energy can be released through a "merger". In the early days of atomic theory, nuclei were not viewed as static structures of protons and neutrons. They were not seen as "locked" crystals without their own internal dynamics. In fact, nuclei were once viewed as very dynamic systems, having internal gaseous freedoms and particulate "currents" of extraordinary violence (Thomson, Lenard).

Nuclei did not have to move in order to acquire this energy. They were possessed of the violent internal motion as part of their own nature. As such, different nuclei interacted as whole systems, each having their own internal kinetic energies. When they approached each other, they did so with great individual powers. Pressures capable of squeezing them together were thus able to cause a blending of the stupendous nuclear dynamics. But, entering thus into nuclear combinations, the internal energies entered into an arena having structural "rules".

Despite the excessive independent dynamics of each nucleus, the rules of nuclear stability energetically constrain nuclei which merge together. It is here that we see the great world intelligence at work, evidencing naturally preferred states of structural stability at a nuclear level. The states exist throughout space, manifesting during nuclear reactions. The strong nuclear attraction which binds nuclei on collision cannot become a stable form without some changes. Energetic changes. Each nucleus in the meld contains its own internal vibrations. Its own nuclear vibrations and currents. Each enters the fusion state with its own vibrational energy, trying to form a new and more stable nuclear "structure". But there is a problem which occurs when the hydrogen nuclei fuse. There is a sudden excess nuclear vibrational energy which each independently brought. If the fusing nuclei do not eject their own vibrational energies, the new structure they have formed will shear apart. What forces them to do this?

Energy states seek spatial geometries which are "stable" and "secure". The stability of that new nuclear structure demands that a certain energetic excess be rejected. The vibrational energy which each nucleus once internally and independently possessed is thrown outside the new structure, and fusion has fulfilled its operation. Each "new structure" is an helium nucleus. When hydrogen nuclei "fuse" into helium nuclei, they necessarily release neutrons, heat, and light. The process continues until "environmental" changes occur. Changes in the amount of hydrogen nuclei, their collision speeds, the density in which their collisions occur, and the amount of products being ejected will each modify the continual fusion process occurring in the sun.

In nature, the process is controlled by solar mass, heat output, particle output, and subsequent solar volume. The sun acts as an immense "balloon" of gas, expanding when hot and contracting when cold. Too much heat, and it expands. This slows the fusion reaction to a moderate constancy. Too cold, and the mass is pulled back by gravitational forces until the heating and fusing process resumes. In short, the gravitational force is responsible for arranging, maintaining, and governing the solar fusion reaction.

Studying these reactions and the mode of their manifestations, physicists began wondering whether it might be possible to artificially generate "hot" fusion reactions on earth. The idea would be to compress hydrogen artificially, welding nuclei together in a tank. The controlled fusion of hydrogen atoms produces extreme heat, but no radioactive waste products. This latter fact is the most attractive feature of a potential "hot" fusion reactor.

Furthermore, the fuel for a hot fusion reactor would be two heavy hydrogen isotopes. Heavy hydrogens, deuterium and tritium, are both found in seawater. Seawater! There is a world of seawater from which to draw fusion fuel. Uranium industries would collapse overnight. A mixture of deuterium and tritium gases would be the fuel for a hot fusion reactor. A hot fusion reactor would have to heat this gas mixture and contain its colliding nuclei long enough for nuclear hot fusion to occur. And here is precisely where our story begins. The engineering problems accompanying this supposed "simple" reactor proved to be insurmountable for most researchers.

HOT FUSION

The practical research began with the notion that electrical power could replace the huge gravitational forces which the sun seems to exert on its own "body of hydrogen". This proved, for early developers of "hot fusion reactors", to be a thrilling mark toward which to reach. The essential differences between "neutral" gravitational forces and "polarized" electrical forces only became apparent with later experiments.

Electrically driven hot fusion reactor temperatures must be held at over a million degrees before the hot fusion reaction takes place. The first researchers made use of an effect which was discovered in Victorian electrical laboratories. Glowing electrical discharges which, otherwise filled their containers, could be made to "pull away" from their container walls completely by increasing the amount of applied electrical current. Further increased applications of current caused the glowing discharge to "pinch". Thus withdrawn to the container axis, the discharges intensified their brilliance beyond the ability of investigators to continue gazing on their light.

The "discharge pinch", a tightly constricted wriggling thread, held a strange energy secret. Early hot fusion researchers retraced the phenomenon, designing their own different pinch discharge systems. In some of these experiments, electrical power was applied in order to ionize and "pinch" deuterium gas at high pressure. Fritz Paneth conducted these investigations in the 1930's, discovering an incredible and anomalous release of excess heat when tungsten and bismuth electrodes were immersed in a high pressure deuterium atmosphere. The heat made his arc electrodes glow with a dull red heat. This red heat was sustained for a long time after the initiating current was withdrawn. An early fusion reactor.

In similar early experiments with deuterium arcs and palladium electrodes, Dr. Paneth demonstrated the practical and controlled release of fusion reaction heat. Although he was not aware of the true source of this heat, these reactions evidenced radioactive emissions equal to those of radium! Few researchers forgot his work until the Second World War. In typical manner, the scientific community forgets what treasures it possesses. After the War, new fusion experiments were engaged, but did not recall the successful high mark to which Dr. Paneth raised the art.

The first post-War designs were simple cylindrical discharge tubes having opposed electrodes. The more amperage applied, the more pinch was obtained. The more pinch, the closer the theoretical approach to hot fusion. Several early researchers measured neutron emissions, a sure sign that hot fusions were taking place. The problem was, of course, that a reactor would necessarily have to produce its own sustaining hot fusion reaction. This required a tremendous application of electrical power. Furthermore, the arc systems failed because electrode metals would melt into the arc, contaminating the reactions before hot fusion could occur. Contaminants blocked the reaction.

Others tried using a phenomenon discovered by Thomson. Developed by both Tesla and Lenard, the "electrodeless discharge" occurs in sealed glass bulbs when held near oscillating electrical or magnetic fields. In a powerfully oscillating magnetic field, gaseous discharges can "pinch" without metal electrodes. Replacing a transformer coil with a deuterium filled tube effectively approaches the design for a "practical" hot fusion reaction chamber. Magnetic energy induces electrical currents of high amperage in the tube-contained gas mixture. The magnetic pressure increases the applied power to a tremendous crescendo and the deuterium gas nuclei begin fusing together.

The goal of systems such as these was to reach "self-ignition" temperatures. Self ignition is the temperature at which it is possible to withdraw all of the externally applied electrical power. Once the initial ignition is achieved, one could simply supply the plasma with fuel. Obtaining electrical power from the reactor is then possible. But, there are significant problems in this prospect.

At the point of self-ignition, the deuterium-tritium mixture is hot. Super-hot. This is why the technology is referred to as "Hot Fusion". These super-hot ionized gases are dangerous. Containment of the heated hydrogen is the main problem with achieving hot fusion on earth. The ionized gas has to be kept from ever touching the container walls. If they touch, the results are devastating. A hot fusion reactor is far less than a controlled hydrogen bomb, but dangerous nonetheless. Photographs of the numerous failed projects reveals far more than ruptured containers and systems. They are miniature blast sites.

According to the academicians, the key to achieving controlled hot fusion reactions is still found in magnetism. A powerful magnetic field can both ionize and contain the gas away from the container walls. Magnetic containment systems were called "magnetic bottles". Ionized deuterium nuclei fused only as they absorbed energy from applied magnetic fields. The input magnetism acted simply as a "spark" to possibly ignite the nuclear reaction.

If all this energy could be sustained for a critical few seconds, hot fusion would begin. Once hot fusion has begun, the extra energy would appear as an electrical "blast" against the applied magnetic field energy. The problem of retrieving any of the released energy would be simply solved through the transformer principle. Hot pulsating fusion cores would induce pulsating electrical currents in field coils, externalizing the extra energy. The theoretical resulting output would dwarf input power by several electrical gigawatts! Power forever. Or so they say.

COLD WAR DREAMS

Nuclear hot fusion formed THE socio-scientific dream. Hot fusion was THE cold war quest. Chevrolets, rock and roll, space travel, and FUSION! Exciting fads, they were the external focal points of human social consciousness. The quest for "reaching fusion" began several years before the end of the Second World War. It occupied the minds of theoreticians, researchers, and scientific developers in every nation. Though forgetful of previous achievements in energy science, the promise of fusion research was limitless energy. Fusion, the single focus of social discussions on future energy, was interrupted one night by a solitary invader from space.

The Space Race abruptly appeared when the Soviet Sputnik was launched. All of the applicable American industrial resources were powerfully moved as by a strong wind when Sputnik appeared in the sky. Suddenly space, the politically necessary new venue, became a technological theme of major importance. The new and unexpected runner in the arena where science met politics, space industries were confined exclusively to the military funding process. The military owned the space race, and still does. This momentary shift in interest, from Fusion to Space, was not to last for very long. While the Space Race had its overt political overtones, the race for Hot Fusion had lasting and more permeating implications. Long after the satellites became capsules, and the capsules became lunar landers, fusion remained the single central technological quest.

Both were extensions of a wonderful dream which had emerged as social expressions from that mysterious source of all dreams. While every youngster had visionary hopes of becoming a space cadet, it was patently obvious that the "process of selection" was a military one. One had to be a military man to become a spaceman. Space belonged to the military, to men, and not to American Society at large. As the American Space program made its progressive forays against Soviet space technology, most Americans continued their lives with other futuristic visions in mind. The progressive movement of space endeavors, from social expectations to military achievements, was accompanied by a conjugate movement of social attentions from space to earth once again. It was a disappointment which has had no representatives, a subliminal hurt which altered the dream life of society forever.

This loss of social focus on space travel came back to haunt the space industries. It prompted the new and more socially accessible "space shuttle" project, which seemed to offer "all common citizens" a chance, however slim, of becoming "space cadets". But the race for Hot Fusion gripped the public with a more certain hope, an earth-fixed hope which was completely accessible and commercially valuable. Common citizens could relate more to a project which promised new and clean energy sources, and all the ancillary promises which accompany any such real technological revolution.

People knew that Fusion Energy would completely revolutionize our world. It would have in fact. They waited for its appearance, a lovely dawn on a clean and perfect horizon. Fusion was the Future. The word from researchers, and the promise which they held out to society, was that nothing could stop the forthcoming development of a fusion reactor system. Hot Fusion research had a counterpoise upon which it readily found funding and public support during this time period.

Fusion was the nuclear alternative to uranium technology. Atomic uranium. It was a disappointing phrase replete with dreaded associations. Disappointing because its very mention once filled the mind with imaginary force. The dream of atomic energy was once wonderful. Visionary. Uranium, an earth metal. A power which could take us to the stars.

The quest for atomic energy began long before the War, gaining dream strength in science gazettes and in a hundred small laboratories the world over. There were even private researchers who experimented with uranium ore, managing to obtain strange energy releases from the minerals. Some took uranium ores, right from the ground, placed them in low pressure gas chambers, and applied high voltage electrical currents to them. This released tremendous amounts of energy which greatly exceeded the applied currents.

In those days, the sense of the phrase "Atomic" was thoroughly different. And dream suffusions provide the atmospheres which flood phrases. Atomic energy had a "golden" radiant sense to its mention. Atomic Energy was a phrase first termed by Dr. Gustav Le Bon, the Belgium physicist who preceded all the familiar names having to do with conventional atomic physics. Dr. Le Bon designed small atomic reactors which engaged photonuclear reactions.

As Dr. Le Bon envisioned atomic energy, it was a force which could be used as desired with no deadly after effects. He had released tremendous amounts of photonuclear products in his simple reactors with no harmful after effects, and no waste products. He did not employ uranium or any of the heavy metals to release his energies. He did not employ fission reactions. The folklore surrounding atomic energy began with his writings. How the ideas were twisted into considerations of the heavy metals and of fission reactions was the result of work done by the Curies thereafter.

The words "Uranium" and "Radium" were interchangeably used in every science fiction serial of the 1930's. Popular fiction, operating entirely in the dream current of archetypes, had so glamorized atomic energy that none believed an atomic weapon possible. Heroic and fictional science figures utilized these elements in wondrous ways for the benefit of humanity. How they ever could have been used in a bomb was yet inconceivable by the vast majority of readers. Atomic energy had a golden aura, an enveloping halo of wonder. "Atomic" was the future hope. It was the means by which unlimited exploits could be secured by human effort. A new Golden Age would begin, one in which energy would play the major role.

Atomic Power would help humanity to tame the entire planet. Endless supplies of food and water could be obtained by the limitless energies derived from radium or uranium. Travel to any point in the world would never be problematic. Power to delve into the sea or ground to any depth would be secured. New alloys, new medicines, new energy applications would convert the world into the future. A golden future. After these wonders had been achieved, humanity would move out into the sea of space, a new Age of Discovery having begun.

Foreseen and told by H.G. Wells in his short masterpiece "Shapes of Things To Come", the reality was indeed manifesting among certain inventors the world over. The names Paneth, Hubbard, Winkelman, McElrath, and Burke were among those who actually obtained significant amounts of electrical energy from small amounts of radium and natural ores of uranium. These systems were safe. They utilized the natural radioactive decay process of their source matter, converting the energetic outflow of particles and fields into electricity. In numerous cases, the electricity obtained exhibited strange and uncommon characteristics. It was imagined by some that new charged particles might be responsible for the brilliant cold light obtained from these devices.

Progress in these primary atomic "generators" evidenced the dream materialization process. The new world was coming into being. But, simultaneous progress in another vein of study brought about instabilities in the technological expression. While attempting to accelerate radioactive decay, physicists discovered fission reactions. Hailed by academician as the "real atomic power", most recognized a fundamental danger in this research. Natural radioactive decay propelled enough energy, said private researchers, to supply the world for millennia. Atomic Generators simply required a simple means for extracting the available power. No other technique would be necessary.

Several writers, seeing ahead to the time where industrial scale fission systems might be employed everywhere, pointed out the dangers of forcing Nature to yield these energies. Some went so far as to predict a crisis of world proportions, should such energies accidentally be released beyond control. New words began emerging in the literature. CHAIN REACTION. WORLD DEVASTATION. These began permeating the human psyche with terrible consequence. Surface storms on the otherwise calm dream sea.

A few engineers calculated the problem of storing spent atomic fuel, stating that such fuel would be radioactive in lethal proportions. The storage of spent fuel would be the major problem of future generations. The lethal spent fuel cartridges would be strong enough to be used in their primary atomic generators anyway. Seeing this would be the result of fission generating systems, most researchers went back to their own experiments.

Deriving usable power from natural radioactivity would be the true means for utilizing this power. No purification process was ever necessary in these systems, as can be seen by the earliest atomic battery patents. In large enough devices, the output power could become enormous and constant, lasting for centuries. Certain of the atomic batteries were vacuum tubes. These systems employed radio frequencies to electronically "pump" the natural radioactive fuels, producing far more electrical output than has ever been seen today in atomic batteries. These systems remain, curiously and conspicuously, forgotten.

While these inventions were being patented and demonstrated in public arenas, fission was gaining new predominance in more military venues. The BOMB was a new "promise", by which present wars would be won, and future wars prevented. But there was a social expression which caught academicians quite unawares. It was a defined resistance to the development of destructive atom technologies. No one believed that atomic energy could become a weapon first, and then a social power. It was always assumed that atomic energy would be safe and benign. In the minds of informed society, "true atomic power" would have no repercussions. True atomic power was a gift, a benefit, a promise and potential of future safety. Viewed in the light of those forgotten atomic battery patents, one can understand the hope. The documents yet remain, proof that the dream sea expresses truth when willing hearts search and find.

Yet, the fission camp was far more powerful than the innumerable private atomic generator researchers. Intent on proving that Nature could be forced to yield her atomic energies, physicists began performing their deadly experiments under the cloak of military secrecy. Fueled by incredibly wealthy investors, who saw the future atomic industry in the hands of far more "academically accredited" personnel, fission research mounted.

The very term "atomic" had acquired a dirtied name. The sense in the word "Atomic" being reversed from one of golden hope to one of leaden death. Used in war, it was soiled. Guilt and shock. HIROSHIMA. NAGASAKI. Overkill. The thoughts radiated out like deadly rays. Guilt and shock. Even then, wherever and whenever the bombs were tested, fear and doom followed. MUSHROOM CLOUDS. The sense was cold dread. The fear real. The mere thought of an atomic accident made people go into a cold sweat. Cold War sweat and paranoia. Curious, how the fear of atomic fire brought cold sweat. Regulatory attempts to redirect consciousness, renaming it "NUCLEAR" energy, did not much matter.

The word ATOMIC was soiled. DIRTY bombs. Radioactive. FALLOUT. Atomic, Nuclear, terms did not matter. MEGATON. Nightmares in children betrayed the world-permeating sense. STRONTIUM-90. Duck and cover! Prayers directed to protect against WORLD WAR III. The world, ending in atomic fire. ARMAGEDDON. Children thought deeply. If high technology produced atomic energy at its apex, then the world was doomed. The social dream expression had been deferred, suppressed, and eradicated once again by the powerful social predators. The dreams of a few ruled the dream of the many. More hope deferred.

The cutting buzz words of the Cold Atomic War became a helplessness which eventually produced a strange social movement away from the curse of high-technology. It was as if the bombs had already done their deadly work in the heart, fallout covering the bodies of growing children and producing mutants. Running away. Running dreams. Running from the atomic light, from the Moloch which threatened to eat children, and parents, and worlds. Running within, rejecting every external thing. Rejecting and reforming. The inner movement among youth did not stop with low-tech. It ran down to no-tech. In a strange psychic plunge, American youth dove into the green psychic earth, re-emerging as Amerindians. PEYOTE. BEADS. LONG HAIR. GRAFFITI. EARTH DAY. New buzz words.

After the War, Uranium was a clean name only in the minds of those who sought it in deserts, with Geiger counters. Uranium was now a money making claim, not a science or dream questing technology. But Fusion was new. Fusion would be the perfection of the nuclear science which began with the Curies. Fusion would be clean. The process, though nuclear, was CLEAN. It was The FUTURE, a sleek and shining surface of blue metal. Streamlined, with glass spheres, cylinders, and dials. If fission split the world and dream and heart apart. Fusion would heal it back together again.

And so the dream wove itself anew. Smiling white-coated operators, walking through black marble halls and wavery glass windows. Pillars and gold filigree. Beautiful halls and Fusion Reactor sites set in the evergreens. Breathing clean air once again, the perfect vision again. Hydrogen gas, medically hissing from large blue-steel cylinders into the large central glass dome. The Fusion Reactor, a science fiction dream. It was a vision to drink in.

The power of a future civilization. It was where we could become our own future. We could be the advanced aliens from another world, the popular theme of science fiction folklore. We could become the placid Utopians once again. Internalized deep in the heart, where dreams are cultivated, where sensations formed the heart of the quest, we hoped for Fusion, and the promises which seemed never to stop.

Fusion technology would revolutionize every aspect of our technology. New metals. Bizarre alloys, stronger than steel, lighter than magnesium, and harder than diamond. New transportation. The power for reaching space, small and compact. One could ascend from any earth station and proceed directly into orbit. New medicine. Operations with light, anesthetic rays, healing rays. New communications. New architecture. New municipal works. Electrical cars buzzing along sleek unweathered multi-lane highways, whizzing past waterways carved into the Midwest by fusion powered earth movers.

New industries. All new and "fused". Like glass, fused, smoothened, lovely, and whole again. Nuclear Art. Fusion, the new and strange fire, would change our world. Le Bon, Moray, and the others were not even recalled now. The first dream forgotten, their dream images snapped apart like a cut necklace. Glass beads dripping off their string into a cup of cheap wine. The dream fragments plopped unwillingly into the new vessels.

How the material forms would actually fulfill themselves in real technological expressions did not much matter. That the quest would be fulfilled was enough for most. Healing the wounds and outrage of the former World War, most people simply wished for a peaceful, happy life in the here and now. All of them, working class people. These are the ones who are most touched by the regulators and those who design the working class perimeters. Aware of governmental machinations, now tired of having sacrificed their sons and futures in wars designed to protect foreign investments, most second generation working class people had reached their tolerant patriotic limit.

Fusion would become the accessible, expectable lifestyle of the American Nation. It was the promise which drove our consent. A dream for citizens, it prompted many to seek professions in related fields of endeavor. Hot fusion reactor designs are not new. The methods toward achieving the "hot fusion dream" have varied in form and principle over the last forty five years. It was a fractured dream however. It forgot the first discoveries which made it a redundant exercise.

Each of the Cold War hot fusion projects had wonderful sounding names. They were each worthy of the fractured dream which empowered them. Scylla, Project Zeta, The Astron Project, The Stellarator, the DCX, ALICE. The names retained their magick until a few more years. Projects appeared in every major industrial corporation. Each project site was enormous and covered up in secrecy. These were usually federally funded, and closely monitored by government agencies, especially whenever success seemed imminent — regulators had found something new to regulate. The race was on. Those who were first to reach the goal would bear away the technical prize, the admiration of generations, and the wonder of an age.

The public needed to educated concerning the forthcoming energy. General Electric even demonstrated a simple hot fusion reactor in their 1964 New York World's Fair pavilion. A public exhibition of Nuclear Fusion! Visitors waiting on lines outside the exhibition, awed by the explosive thunder and light display seen through the doors, waited in hope of glimpsing the future. Powered by a truck sized capacitor bank, the immense current discharged in a thunderous explosion through an enclosed tungsten spark gap.

Shocked witnesses were completely taken aback. The deuterium gas through which this lightning bolt surged was held under a futuristic looking dome of thick plexiglass. The sudden and blinding red-white discharge produced internal neutron counts which made neon number displays rise into the thousands. Each spectator left the windswept pavilion doors, looking into the lovely setting sun and dreaming of the sweeter day.

HOT REACTIONS

Several developmental stages appeared throughout the technological history of hot fusion devices. Arc discharges were first used because the power delivery to arcs was far greater and more direct than the systems which employed "electrodeless discharge". Electrodeless discharges produced weak arcs by electrostatic or magnetic induction of plasma states through glass walls.

Arc discharges were replaced by "magnetic mirror systems". Magnetic mirrors were a hybrid system which relied on arc discharges in deuterium gas. They employed large external magnetic wrappings in which arc terminals were enclosed. It was found that higher temperatures could effectively be attained with this system, the magnetic "mirrors" blocking the excessive loss of particles and heat from the plasma ends. Nevertheless, these were arc terminal systems. Contaminants from the electrodes poisoned and limited the upper thresholds of heat stored in the plasmas thus produced. Mirror systems failed.

Projects changed from arc research to electrodeless discharge research. Mirror systems were abandoned, replaced by transformer-like magnetic containment systems. Magnetic containment went through several developmental improvements. The most notable early transformer-like plasma system was the famous Project Zeta, a British endeavor. The toroidal glass vessel of Project Zeta was poised in a large transformer core. Pumped by huge electrical oscillations, the torus sprang to burning life. There seemed great promise in this new avenue.

With the deployment of information on magnetic containment systems came a new zest for the research. "Ioff" bars were developed by a Soviet plasma researcher of the same name. His thrilling successes with a magnetic system of his own design caught interest throughout the hot fusion world. The idea was the wrap the torus in various pitched coilings, as well as surround the torus with parallel bus bars.

The combined field symmetries stabilized and centralized the internally pinched plasma, reducing its own pulsations. But heat loss was still the main concern. No amount of applied energy could get the plasma up to ignition temperatures. It was not that size was the critical issue either. Most of these facilities were huge to begin with. The main engineering problem was not size. The main engineering problem in each of these hot fusion systems was the excessive heat loss. Inability to approach ignition thresholds kept the prize at a tantalizing distance.

Long before ignition could ever be attained. Hot fusion temperatures are enumerated in millions of degrees. No material on earth can withstand such a blast. There were materials problems as well as severe problems with containment. The ionized gas channels of these temperatures vaporized their containers, and research sites as well. In truth, none of the early systems could ever deliver the sufficient power needed to overcome the ion leakages. Moving through their containment systems like bees through nets, ions seeped fuel and heat away as fast as was supplied. As a result, all the potential ignition power simply seeped off in the heating process. This dangerous state often destroyed both the devices and the dreams of those who designed them.

After assessing the work of Dr. Loff, Western theoreticians proposed that a complexly wound "magnetic bottle" might hold the superhot plasma long enough for nuclear ignition to occur. Everyone who was not yet disappointed to tears jumped into the "magnetic containment game". Wrappings, symmetries, parallel Ioff bars. Bars wrapped at angles, bars wrapped with coils, coils wrapped with opposing coils...the magnetic combinations were endless. After all of these magnetic foibles, most of the American and British magnetic containment systems failed to achieve their theoretical results during their very first few trials.

New resistant plasma phenomena blocked all hopeful progress in this venture. Plasma instabilities. Thoroughly disgruntled magnetic containment researchers found that as power was applied to the gases, instabilities, ripples, and pulsating oscillations suddenly appeared throughout the plasma channel. It seemed as if nature simply did not want hot electrical fusion to happen! Plasma channels wriggled like unwilling snakes against the magnetic bottle, shaking off all of their energy in a single wriggle. In some cases, the superhot plasma columns burst through their metal chamber walls. These circumstances were extremely dangerous. Quasi-nuclear explosions did occur in test sites, radioactivity spilling out into facilities.

The score being a miserable record of defeats, none of the most serious and highly funded hot fusion ventures, unfortunately, succeeded. Each project failed to deliver the promise of sustained nuclear hot fusion power. Because of the numerous failed magnetic containment projects, most physicists simply left the controlled hot fusion race altogether. In methodic succession, and after several billion dollars were spent, it became apparent that magnetic containment systems would not succeed at all. There were those who preferred to collect grants, assuring themselves of bureaucratically "safe" positions. Rather than risk their yearly salaries on radically new scientific ventures, they chose alternate related routes of employment. Survivalism. There were those who used the fusion projects in "throw away" fashion just to survive.

There are those whose jaded personalities have taken a survivalistic stance, already accepting that hot fusion is a dead-end. A dead-end, but a steady income. Many researchers stepped down from the "performance risky" pedestal of success oriented hot fusion projects, assuming lower profile positions as hot fusion theoreticians. Books sell better and longer. The royalties continue after each unsuccessful project. It was easy to write papers and analytical discussions on the numerous failed projects, simply because there were so many failures from which to draw "copy". Survival. Library shelves became flooded with these fusion-related papers and texts. But still no fusion reactor.

DREAMS DEFERRED

The loss of the Hot fusion Dream became another lost social quest, another socially prolific disappointment. It seemed that, after World War II, every potential new technology of promise was methodically assessed by regulators as "impractical and impossible". Curiously and conspicuously, these assessments all coincided with the early war effort in Viet Nam. Obviously the re-alignment of national interests followed the re-alignment of old money in Indochina. And financial re-alignments have no problem with the loss of any socially vital developments. Whether or not a providential discovery becomes socially proliferated is no concern of theirs.

But the consequences did not become desperate for regulators. A disappointed society can be controlled. Each lost social dream becomes a new social malaise. National demoralizations manifest when social myths are deferred. Therefore, the synthetic manufacture of new and continual myths, of new "dreams", is actively sought. Government propaganda mills replace each deferred hope. When once having dissuaded the natural flow of discoveries for the sake of old money interests, government regulators sponsor synthetic dreams to replace what they remove. But even the smallest child knows when a dream, a real dream, has been taken away.

Unfortunately and tragically, it is the adults who discover that the deepest dreams have not only been deferred, but made impossible to achieve. The production of alternative social goals and other myths makes its continual appeal for our attentions. Such false dreams are drawn from a rich surplus of dreamers and schemers found throughout the surplus American intelligentsia. Choosing one synthetic dream out of many has become regulatory policy. Bread and circuses.

The dream visions and goals which a government hoists before the eyes of its citizens in state of the union messages promotes a social poise which can seem thrilling. Thrilling, for the spoken moment. But, if a dream vision has been truly obtained through the providential discovery source, then it alone succeeds. In the heart of every fantasy embellished dream, points a diamond truth. While the false prophets and their synthetic dream visions fell to the ground, like pretty paper kites in so many colored pieces, there was a truth to fusion. One man found it. The gross overfunding of new hot fusion projects represents a means for maintaining public morale at a very subliminal level by means of a synthetic dream. It, however, is a funding campaign based in abject ignorance. Fusion was achieved. A successful fusion reactor was designed and tested. Tested, and forgotten.

The stubborn development of magnetic containers yet continues. Even the Russian Tokamak Reactor, however gigantic, cannot achieve the short-range controlled hot fusion objective. This is why their Western protagonists demand federal funding for another fifty years! During the 1970's there were notable and alternative experimental systems which employed focussed laser light to trigger hot fusion in hydrogen gas. But these huge systems, however alternative in approach, proved grossly ineffectual. It was later discovered that Sandia Labs conducted these experimental baubles to cover a military project in which the hundred-yard laser played an essential role. Dreams deferred. Hopes disappointed.

If not for a single significant lost chapter in the history of hot fusion research, we would have grounds to accept the chatter of those who supply their own needs by publicizing the already fulfilled quest. All of the academic banter and poised public relations campaigns stand in conspicuous contradiction to an event which occurred in 1965, when a working controlled hot fusion reactor was both successfully demonstrated and discarded before 1966 under suspicious circumstances. Even as these newest magnetic containment projects are being designed and attempted, a controlled hot fusion reactor had already been routinely operated.

This piece of information twists the nature of the Hot Fusion research game in a new way. For if a working Hot Fusion Reactor already existed, then why were stupendous floods of money continually being supplied to newer and more complex hot fusion projects? Was this contradiction a "funding project", a means for deferring the instantaneous deployment of the successful reactor on behalf of threatened petroleum cartels?

The reality of anyone controlling hot fusion reactions as early as 1965 (reaching self-sustaining reaction) sounds truly bizarre to anyone familiar with the historical publications. But this single technological instance is by no means the only time this century that such a contradiction was nationally framed. The forgotten science which lies dormant in unstudied Patents and Victorian texts will revolutionize the mind of anyone who dares break out of the conventional thought frame.

Fleeing the implications of an ever regulated mind state, those who study in the sciences are often passionately forced into the Victorian literature where they rediscover true science. This is not a pleasant luxury for some. It is a necessity which determines their very lives.

But dreams are fulfilled by those who seek them out. Those who are sought and touched by the Ray of Discovery know its power and desire for humanity. Despite the forgetfulness of scientific researchers, there are those whose hearts are pure enough to receive new dreams. New visions. Answers to the unknowns which causes some to seek truth, as bearing solitary candles, in a windy night. Of all the venture projects which chased after the elusive hot fusion Grail, one researcher succeeded. How fortunate we are to have found this document. For it will become plainly evident that, if it were possible to have eradicated this document, those who now own it would have done so. But since the device stands as part of the public record, we can rest assured of its status. Here is a tale, a biography, best recounted with solid evidence from the Patent Registry itself. The device was patented!

DR. PHILO T. FARNSWORTH

The true father of electronic television, Dr. Philo T. Farnsworth, found the practical key toward achieving hot fusion. He demonstrated his hot fusion reactor before several highly qualified groups of analysts. They saw the system in operation and yet testify that it really worked.

In a brief biographic sketch, it must be recalled that Dr. Farnsworth is the true father of electronic television. He is one of the most conspicuously disregarded inventors of the Twentieth Century. In 1927, the young high school student received the entire working design for the electronic television system in a single insightful ray. Scrawled out on a scrap of notepaper and saved by his high school physics teacher, the very drawing became the document which sealed his name in later court proceedings.

His patents are remarkably advanced for their time. His ability to design electron tubes of extraordinary form allowed him to create incredibly new electronic components yet used by the military. No precedent had been set for electronic television. The system parts had to be invented. Proper design of the electronic television system required that every piece, every tube, every component be researched, tested, and implemented. In a rapid deployment of new designs, Dr. Farnsworth and his dedicated group of researchers designed, built, and implemented each part. During this arduous process, they learned how to manage the very production of their own parts, the television industry having developed among their members. This later enabled a small factory to be established for the manufacture of the various Farnsworth television systems.

At the time, no corporate enterprise was able to summon the genius in producing an adequate television system. Even R.C.A. relied on the old mechanical television systems as a primary base for developing a new system. These mechanical sets were the sparkling, whirling, multi-mirrored twirling wonders of an earlier Victorian time. Baird, Rosing, Jenkins, and other names come to mind when recalling those quaint and inspiring working designs. Others had brought mechanical television to its point of perfection. But the mechanical televisions, while being the perfection and wonder of their day, were little more than "flicker windows", producing vague and blurry shadows. Something utterly new was needed, some fresh departure into a new age of real television. That new departure was already patented and operational in Farnsworth's laboratories.

After several years and fortunes in failed attempts, RCA was forced to duplicate the results which Farnsworth had developed. Ultimately they had to use the Farnsworth System as their chief model from which to..."glean" their own components. They used Dr. Vladimir Zworykin to achieve this theft. Zworykin's legendary "photographic mind" was employed, through his numerous "visits" to the Farnsworth Research Laboratories, to permit the complete re-design of every Farnsworth component under the R.C.A. crest. David Sarnoff was thereafter able to issue Zworykin's patents for a television system without paying Dr. Farnsworth a single penny or a word of gratitude.

Nevertheless, the dream belonged to Dr. Farnsworth. Image dissectors, pulse transmitters, synchronizing oscillators, synchronous scanning, image analyzers, receivers, and special cathode ray tubes: Farnsworth conceived, designed, and hand-built each of them with his research team in 1926. Examination of the Farnsworth patents reveals nothing but novel tube designs which remain without contemporary equal. Dr. Farnsworth developed numerous unusual tubes to make his television oscillators, receivers, and transmitters more efficient. No existing technology could match the performance characteristics of his UHF oscillators, electron multipliers, and cold cathode signal amplifier tubes.

Notable among these designs were cold cathode vacuum tubes, some of which employed soft radioactive materials to achieve unheard electronic performances. He developed photomultipliers, multipactors, Infrared imaging tubes, image storage tubes, and image amplifiers. Military night-vision is a Farnsworth invention. ITT makes billions of dollars from this single Farnsworth patent.

LITTLE STARS

Throughout World War II, Dr. Farnsworth continued to explore new electronic alternatives, designing radically new species of electron tubes which became as famous as his earlier development of electronic television. The development of his "multipactor tube" was one such departure from convention. In this strange "cold" tube species, a photoelectric multiplying process saturates the vacuum with electrons. The simple application of a small direct current results in such an efficient avalanche of electronic charge that the tube bordered on "complete" efficiency. This meant that the input energy completely equalled the output energy, a condition not known in vacuum tube technology.

The multipactor tubes use opposed concave electrodes. In effect, they are concave electrostatic mirrors. These mirrors focus the ionized gases into tight little points, just as mirrors concentrate light. The concave mirrors permitted the re-discovery of electron optics; a phenomenon originally witnessed by Sir William Crookes in 1890 and "overlooked".

Dr. Farnsworth designed a great number of different multipactor tubes. His patent collection is enormous. Dr. Farnsworth noted very anomalous phenomena in several of his multipactor tubes. These included sharp energy surges on the output stages which seemingly appeared "from nowhere". The possibility is strong that he discovered an entirely new kind of energy source, having nothing to do with hot fusion. It has been suggested that these surges were of cosmic origin.

While testing his high power UHF multipactor tubes in 1935, Farnsworth discovered a strange phenomenon which caught his curiosity. Suspended in the tube center, he sighted a tiny brilliant blue starlike point. The little starlike point of light became more brilliant with increasing application of voltage. The little starlike points never touched the walls of the container, remaining fixed in the space where first sighted. Farnsworth recognized this feature as a control characteristic which might somehow be employed in the future.

Farnsworth multipactor tubes can be small enough to be hand-held. The larger models are the size and volume of a thermos bottle. Used as UHF oscillators, they produce enormous outputs of power. The optically focussed little stars are instantly formed within the multipactor tube, exhibiting all the control-response characteristics later sought desperately by hot fusion reactor designers.

Farnsworth realized that hot ionized gases could be bound into these small starlike points, their rare stability managing any applied power load. The little stars could absorb and hold tremendous amounts of applied energy, an aspect which deeply impressed Dr. Farnsworth. His original notion was to utilize the principle in high power UHF transmitter tubes. For metallurgical purposes Farnsworth thought the process would have industrial applications. The star points could be directed into any material surface. Melting tiny holes in metals would be no problem for the intense freely floating little ionic star. Soon his mind turned toward nuclear energy. The starlike "plasmoids" could be loaded with any amount of electrical power and be maintained away from the container walls. They were stable, could absorb fresh gas and electrical power with theoretically no limit to the attainable temperatures. The notion of using the principle to construct a nuclear furnace deeply intrigued him.

THE "FUSOR"

Thermonuclear energy was used in the hydrogen bomb technology. The scientific community was astir with talk of hydrogen energy. Farnsworth also studied the problem of controlled thermonuclear energy. The gaseous temperatures had to be immensely high, and safely contained. By as early as 1953 he had conceived of a means for using the star like phenomenon to produce controlled nuclear hot fusion reactions. He published his theoretical research on usable hot fusion energy.

In 1959 H.S. Geneen (Raytheon) invited Dr. Farnsworth to address the ITT board of directors on controlled nuclear hot fusion. It was against the verbalized misgivings of the AEC that this lecture was presented. Farnsworth was then formally approached by ITT after announcing his plans to investigate hot fusion reactions. Farnsworth designed a new and dramatically original tube which he named "The Fusor". In this new tube, the starlike plasmoids of deuterium were isolated, shaped, confined, treated, balanced, and moved absolutely without the need for magnetic confinement. He conducted the first tests in his own home laboratory space, the deuterium tanks and electrical cables running throughout the living room to the cellar. Shortly thereafter, preliminary tests on the first Farnsworth "Fusor" was performed in a small ITT basement laboratory. His first design for a hot fusion reactor system was realized late in 1958. ITT monitored all the research and brought its own supervisors into Farnsworth's team.

The "Fusor" is a device which produces controllable hot fusion reactions and does not utilize magnetic confinement. The design is a radical departure from all the designs of its time, a simple optical electronic system. The Fusor is no larger than a softball. In its center is the electron-radiating cathode. This cathode is surrounded by a spherical anode. A group of deuterium guns are symmetrically mounted about the anode ball. Their beam axes face each other and intersect at the tube center, firing ionized fuel into the reactive focus. It is perhaps the most advanced electron power tube ever designed.

Deuterium gas particles are propelled and focussed into the center of the tube, establishing the star-like plasma at the focus. Magnets are never needed to contain the gas. Nuclei which are trapped in the starpoint can never escape the focus. They are maintained in place by their own inertia and the incoming barrage. Deuterium nuclei are literally hammered into the required density in the central region by the process of "inertial containment", a term which Farnsworth first coined. Potentially escaping nuclei are stopped by layers of surrounding charge until they are forced back into their center point.

Ionic shells are held in the vise-like grip of applied power. Confinement power can be poured into this center almost indefinitely because the trapped nuclei cannot escape the field energy. Nuclei which "fall" into the centermost virtual electrode have fusion energies, and are contained at a density sufficient to produce controlled fusion reactions. With sufficiently high power applications, the hot fusion reaction can be sustained and controlled at will in the Fusor. Dr. Farnsworth worked out an elegant means for extracting the energies of fusion, energies which remain electronic in their nature. Developed fusion energy produces an electronic pressure blast against the applied energy field. This experimentally appeared as a dramatic back-surge in power. This electrical blast may be directly harnessed and used in external loads.

Even as stars govern their own output by expanding and reducing plasma density, the little stars were found to be remarkably resilient and resistive to instabilities. In fact, the only instabilities seen in the Fusor were those which came from the outside. Tube external power instabilities required new safety systems to be developed. Maintaining the constancy of application required "pure" electrical inputs.

Magnetic containment never reached this degree of success. Farnsworth's system was compact, simple, elegant, and inexpensive. He solved the particle confinement and energy conversion problems in one simple design. On October 8, 1960, the Mark I Fusor produced a steady-state neutron count when deuterium was admitted into the device with very low power application. This meant that fusion was happening. What Farnsworth sought in these first tests lay in the control of fusion reactions under increasing power application. The self-sustaining reaction would be gradually approached in steps.

Farnsworth established and charted increasing neutron counts with increasing application of electrostatic power. His methodic experimental method was necessary in the uncharted fusion territory. He repeatedly tackled the possibility of a "runaway" reaction, designing newer electron restraining guns to prevent this horror. It is suggested that the reader obtain and study copies of the Fusor patent for further understanding of this design aspect.

SELF-SUSTAINED FUSION

Farnsworth had to learn the operating parameters of a practical fusion reactor. Being a brilliant mathematician, his theoretical work was published along with the design patents. He established several criteria for testing the reality of achieving nuclear fusion in his system. The entire assembly was submerged in oil and was confined behind thick lead-concrete walls. The experiment took on a decidedly ominous tone after this procedure reconfigured the system. The entire Fusor reactor occupied the volume of a very small lecture hall. This volume included the power sources, tanks, shields, and monitoring devices. It was a rare miniature in the fusion art.

Dr. Farnsworth measured neutrons as an indicator of the fusion reaction occurring within the sphere. With deuterium gas in the Mark II-Model 2 Fusor a count exceeding 50 Mega neutrons per second was recorded. This device eventually produced 1.3 Giga neutrons per second in a sustained reaction for more than one minute. These reactions were stable, completely under the operator's control, and could be duplicated on command.

On October 5, 1965 the Fusor Mark II-Model 6 was tested. A reconfigured, high precision ion gun arrangement produced 1 Giga neutrons per second, a world-record in the art. On December 28, 1965 tritium was admitted into the test chamber, producing 2.6 Giga neutrons per second. Higher voltages produced greater neutron counts. With a mixture of tritium and deuterium Dr. Farnsworth's team measured and recorded 6.2 Giga neutrons per second.

The Mark III Fusor produced startling high records in quick succession. By the end of 1965 the team was routinely measuring 15.5 Giga neutrons per second. It must be remembered that this Fusor was yet the size of a softball. A Fusor having a diameter of just one meter would permit greater ignition power for a smaller time period, while multiplying its output power volumetrically. A Fusor could be built to any size as power was required. Dr. Farnsworth reported that his team achieved a self-sustaining reaction on several occasions, and could repeat the effect. The thunderous vibrations of the Fusor are well reported by those who worked with Farnsworth. Many laboratory workers saw the brilliant white light of the Fusor in its early test-runs...right through the metallic shielding!

Dr. Farnsworth once invited a few individuals to watch a test-run of this feat one evening. As power was applied to the Fusor the neutron-reading meter achieved a steady threshold and there remained. Only a slight additional increment of power was applied. Then the needle went off the scale and stayed fixed. The room thundered. The light released behind the shield would have instantly and permanently blinded anyone. Although the ignition power was completely removed, the needle remained off-scale in excess of thirty seconds as the fusion reaction sustained itself. Controlled self-sustaining nuclear fusion of tritium nuclei was historically achieved in 1965.

Success had come. The patent record shows that Farnsworth finally achieved that goal in 1965. The upscaled Mark IV would have completely cornered the electric utility market for ITT. Large Fusor systems could be set up everywhere. The Fusor System proved successful throughout its forgotten seven year research history. The establishment of Fusor power stations would have been more than cost-effective for ITT. It would have made them trillions the world over. Fusor reactors were simple to build, maintain, and operate. Dr. Farnsworth and his team had computed each company cost to the penny! Therefore, who called ITT to stop production?

FUSION CONTROLLED

With the announcement of these final achievements, Farnsworth was met by a totally unexpected and contradictory turn of events. ITT had been gradually absorbing the entire Fusor project throughout the few record-making years. All related patents were assigned to ITT even as Dr. Farnsworth's achievements arrived in successive steps. Suddenly ITT was "not interested" in the Fusor System.

It is both curious and contradictory that, while steady progress was being achieved at minimal cost, ITT was already planning to drop the Fusor project completely. Influenced by powerful professionally hired "lobbyists", executive board members were urging the eradication of the project. During this strange time, certain Wall Street analysts were publishing their "concerns" for ITT and its absorption of the Farnsworth subsidiary as a "terrible mistake". Farnsworth himself was made the direct focus of every corporate death-word thereafter. Hired to assassinate the project and the project leader by yet unknown outside agencies, ITT folded up like wet cardboard under the pressure.

This complete contradiction is all too conspicuous, a familiar pattern in American technology. Outlandish accusations against ITT remain in the indelible historic record. Newspapers from the time period, journals, and other publications show the campaign. Nevertheless, and equally indelible, are the patents and periodical records which Farnsworth has left to us on controlled fusion. Who had "spoken" to ITT, dissuading them from further development of the Fusor reactor? The AEC was mounting the uranium fission race and the "anti-fusion" race simultaneously, using every tactic to achieve total dominance of the energy field.

A large reception at the Waldorf was astir with executive unrest concerning the Farnsworth research project. While dressing for the dinner that evening, Farnsworth suffered a stroke. He was thereafter suddenly "relieved" of his research project, now on the basis of his "now failing health". Furthermore, ITT formally and publicly announced that the Fusor project was "a failure...a dead-end". Dr. Farnsworth suffered another stroke on a plane ride back home.

COOL DOWN

During his long recuperative period Farnsworth decided that the Fusor should be privately developed to its complete perfection. After all, the Fusor was Farnsworth's own creation, why should he not pursue the course alone? Dr. Farnsworth tried to obtain his patents back from ITT. Considering their public announcement of his "dead-end" he believed they would be more than happy to sell him back the "wasted" patents. Since the Fusor was "a miserable failure" it would be he (not they) who be taking the loss by buying back what was considered worthless on the academic market.

He therefore contacted ITT and honestly announced his hopeful intentions. The answer came with quick, cold, and ready calculation, negative and impersonal. Under no circumstances would they ever release to him the right to pursue the Fusor project. Moreover, ITT legally warned Farnsworth that it would dominate all of his own private Fusor research forever, despite its "infeasibility". ITT cut all formal financial ties with Farnsworth and left him virtually bankrupt in 1966. Several Farnsworth patents yet maintain the entire ITT operation to this very day. In quick, methodically accurate legal moves, ITT asserted its complete ownership of all Fusor applications for the future. This curious response surrounds a device which is declared a "dead-end"!

In July 1969 Farnsworth built a small Fusor lab in a Brigham Young University cellar room. With privately purchased equipment he continued his research with generous University support. Suddenly, however, creditors began crowding him on every side. Furthermore, it was impossible to obtain the necessary fuel materials. Deuterium and tritium gases were already regulated by legal means, and he was barred from purchase. During this time an offer came to him from SONY. By now he was unable to continue. Physically ill and emotionally scarred from his dealings with both RCA and ITT, he died in 1971. ITT sent nothing to his poor widow.

LIGHTS OUT

There are those noble individuals in whose hearts ride the dreams of whole societies and futures. Most younger academicians will not even recall Farnsworth's project. These individuals will usually protest that such a claim is not "scientifically possible". They arrogantly base their confident refutation on the vacuum of the critical piece of information: the Patent Record itself. Such scientifically biased refutations are patterned personality reactions, based on incomplete knowledge. Only suspicion best explains the cavalier manner in which the term "pseudo-scientific" has been flung about in the academic-industrial world these days. It is indeed marvelous that the phrase makes its appearance, after, always after a threatening scientific achievement has been scored by private researchers.

There are deeply entwined reasons why few have ever heard of Dr. Farnsworth's Fusor System. These reasons exceed the modern academic censure of this possibility. ITT now holds the Farnsworth patents, and bears the social debt of responsibility for suppressing Fusor Technology. ITT will not release them to public domain for licensing. ITT maintains this stance despite the twenty year statute of limitations normally granted to United States Patents. Since 1982, the patents should have been fair game. ITT yet conspicuously withholds the rights of all privateers from formally developing and marketing the Fusor device. This is indeed an awkward poise for a device which is "a miserable failure".

What threatens traditional financial dynasties more than a new discovery? Regulating what discoveries are "permissible" precedes what regulators decide is "allowable knowledge". In other words, the control of discovery precedes the control of knowledge. The control of knowledge precedes the control of awareness. And the control of awareness prevents new discovery on behalf of those whose financial interests are potentially threatened at any moment.

Fundamental natural discovery is the force around which corporations scurry, fearing the often violent social and economic changes which have historically followed the appearance of new discovery. To be ignorant of fundamental scientific discoveries is to be ignorant of both the present world-condition and the future world-direction. Fundamental revolutionary technology represents a complete elevation of society into a new consciousness and world-condition. Fundamental revolutionary technologies such as Dr. Farnsworth's Fusor have been deliberately suppressed. We do not know with certainty the actual depth and extent by which the commands of suppression are dispatched. We do not know how far government agencies are involved in this process of suppressions. What we do know for certain is that a degenerate technology, a distorted and synthetic fragment of lost science, now guides the course of world history.

Technological revolution is real revolution. It is that which the dynasties most dread. Deferring potential technological revolutions infers control at the fundamental level: at the patent Office, at the market place, in the very courts of government. Deferred technology maintains the financial stability of a few "old families" at the expense of humanity at large. Science, the servant of providence and humanity, has lost its first love. It has lost its way because its ways were deranged by excessive and unwarranted financial involvements.

But, where is the knowledge of lost technology? Where does this knowledge reside? How does the knowledge surface? Who are the ones through whom the lost information is socially regained, proliferated, and acquired? Look in the patent archives. You will find them all safely, and legally, preserved. Hot fusion was achieved, scrutinized, judged, condemned, assassinated, buried, and censured. It is an episode which is now "forbidden" to mention. It is remarkable that individuals in the fusion research teams across the world are even aware that their goal was realistically attained in 1965. Thirty years ago.

Among the incredibly prolific patents of Dr. Farnsworth remain two working designs for achieving practical nuclear fusion: patents 3.258.402 and 3.386.883 as found in the Registry. The patents themselves are textbook lessons in the fusion art. The designs of the device which attained sustained fusion is elegantly simple and can be examined. Such technological options, as global property, must be cultivated among disadvantaged nations. New technologies must never again die on Puritanic shores before visiting the other shores which lie beyond these national gates.

One remarkable property of natural discoveries is the incessant manner in which they appear. Discoveries appear before social crisis requires their development. Discoveries represent providential manifestations of grace. They must be honored as such, comprehended as messages against some desperate future hour. Discoveries are not restricted to specific locales. No single nation rules the flow and dispersion of natural discoveries. Those who seek the eradication of discoveries throughout the world will be destroyed. The consequences of eradicating technologies have a mysterious way of finding us out, by their deadly absence. In the hour of need, they keep their silence.

In truth, the vision which is carved in stone cannot be ruined. There it remains. Let those who study and devise their twisting way plot. Plan, scheme, bend, and turn. None will stop the day. The bright sapphire, a pure night vision, remains. Starpoints in the great radiant blackness, from which come all things. The dream seas surge, asking no permissions and giving gifts liberally. In a land not far off, where dreams and dreamers walk as one, there the love reigns and waits.

CHAPTER 10

MIND, FIRE, AND THERMODYNAMICS

FIRE FALL

The discovery of fire has been repeatedly applauded as the single greatest civilizing catalyst, chief among the consciousness expansive agents of humanity. Though such superficial examinations are found in almost every anthropological science fiction, the true negative significance of fire and its discovery has rarely been addressed. The legacy of fire and its damaging effect on individual world perception must be engaged before the possibility of realistic new technology can ever be approached.

Less the archaic benefactor of society, and more the cunning conqueror, changes wrought by fire became permanent features of the human conscious repertoire. Deeply permeating effects were introduced to the dream-well of consciousness through the discovery and implementation of fire, effects which were both violent and damaging to the basic nature of our being.

Before fire, conscious experience was entirely flooded with water symbols: gushing rain, sweet showers, rushing streams, babbling falls, bubbling springs, forested lakes. The ancient terms describing stellar space was always related to water. If Nature was said to melt, it was melting from blue white ice to trickling water, while generating the watery green life of springtime. But the "watery conscious" experience evaporated away when the crackling red flame was found.

The external, material vision of fire was one having no prior equal. From the moment in which those isolated adventurers grasped the famed "flaming branch" and ran back to their respective tribes, a series of new and permanent awarenesses flickered into the mind of humanity. For those who facilitated the dramatic retrieval there were only honors. Those who witnessed these events were the enthralled co-discoverers. The entire cluster of humanity was part of the focussing effect which fire wrought. Fire was the fascinator.

The early novelty of fire was not applauded by the elders, who saw in fire an enemy. Ignoring their wisdom, younger leaders clave to the fire. The first steps of the fire learning process seemed to be only mind-elevating ones. Fire gripped and thrilled the senses of those who could not cease looking into and through its flames. All other activities stopped before the fire. The discovery of fire placed a sudden halt on the daily ritual activities. The discovery of fire was like a religious event out of context, one in which a natural force had seemingly summoned one cluster of "special people" together. Societies which discovered and trapped fire believed themselves alone in the event.

Typical of the total surrender which it commanded, each awestruck archaic population imagined that fire was a secret which they alone possessed. That quick and easy first surrender would soon become devotion and allegiance. Gazing into the magickal effluence, fire seemed a supernatural agency. Fire evidenced innumerable strange aspects which had only been seen among the celestial lights. Fire was a visitor from the sun then, The obvious first of its wonders was the continuous and seemingly independent production of light. The light of fire was different and separate from sunlight or moonlight. Fire light seemed to be a special and "independent" light. Firelight seemed to be an autonomous, self-generating entity whose radiance was eternal. Fire seemed to be in everything they saw now.

Terrestrial manifestations were projected into celestial realities. Now, everything in the skies were made of fire too. Lightning was fire. The sun was fire. The stars were fires. Only the moon was a cold man crying. Why was he crying? He had no fire. Fascinated only with the radiance which they saw, firelight seemingly "went on and on". It was the same with the heat produced by fire. Those who beheld fire imagined themselves in possession of the "sun body". Fire was viewed as a mediator between humans and Natural Power. Fire was believed to be a token, a manifestation of favor. When these properties of fire were discovered, humanity lingered in a promised-filled dawn of wonders.

FIRE IS ALL

The initial fascination with fire phenomena, for a time prevented humanity from realizing its negative aspects. At first, fire seemed a wonderful gift. A blessing, an agent of good, a gift from the sun. Fire was magick. It was the power which coursed through the sky, and they alone possessed it!

Fire fascinated and captivated. Nothing could be seen except the flames and the power the flames gave. Social rank began to emerge and solidify. There were those who were privileged to stand nearest the fire. Others fed the fire, tending its delicate and deadly body. Others were banished from the fire.

All cherished the flame. Children gathered the best branches from the forest, delivering them as offerings to the fire. Younger rulers owned it. Specified persons carried it. Elders watched and learned as the watery green offerings of sprigs and saplings were rejected. The rejection of green by red fire was a symbol which humanity especially internalized during these early parts of their experience with fire. It said that fire rejected the old ways, the old images, the old song-myths, the green watery ways. Only dry, dead branches were to be delivered. Dead branches to the transparent, impassive flames of red.

As great fires were made, new discoveries were made by accident. Plants writhed in the fire-heated pots, giving up their green blood. Forests now fed families and cured the infirm. The secrets of plants were tortured into the pot by fire. Stones, used to surround the fire, trickled liquids under the magick flames. In the morning, shiny flattened trails were found flowing out. In time, these were sharpened and wielded. Tin blades were like razors. Fire the invincible! Fire consciousness gave humanity a false sense of world pre-eminence. With the new found ally of fire, humanity assumed a dominant world poise over Nature.

Torches, children of the fire, were carried up into the deep caves. The once frightened rulers now bravely ventured into the mouths of the mountains, bearing fire before them into caverns were bears once ruled. There, plant blood was splattered over hands and feet, marking the innumerable secret places which archaeologists now thrill to find. Deep in cavern recesses, the song-myths were enacted. Against the flickering red, where shadows of "the tellers" went up and down, children learned the story of the world.

A change had come. The young rulers were pleased. Some remembered the fearful time before fire. They told that the world was a great darkness, a haunt for frightful creatures. Before the coming of fire, there were monsters in the dark. No one remembered the water songs any longer. A few elders knew the secret which would restore the former time, the watery songs of lush green and blossoming life. But the young tribal rulers would kill them if it were now told that water could destroy fire. They knew it, but kept the secret deeply hidden. Fire was "invincible".

FIRE FOCUS

Fire completely captivated each young mind, each body, and each tribal group which discovered it. The first and most fundamental fire-provoked change occurred in the mind of humanity. Streaming forth as a new and powerful sense stimulator, fire fascinated human attentions out into the external. Fire coaxed consciousness away from its true and central inner focus. The change was subtle, but definite. With the image of fire came a comparison of externals and internals. What fire made was solid. Food, tin, copper, spearheads, blades, pots. What dreams made, vanished. Songs. Pictures. Talk. Therefore fire was thought "true and real" while consciousness, the very wellsprings of being, was considered "the lie".

The once continuous flow of imagery between Nature and individuals was now interrupted, replaced with transparent streams of fire lacking image and content. Fire was a vacuous fascination. There were no images, no feelings, no memories, and no message in the flame. Fire was the one item in Nature which had no soul, and could give no life. Fire took life. Fire was the antithesis of life, and humanity desired it for what it promised. Survival. Protection from fierce beasts, light in the darkness, and warmth against the deadening cold of winter. For this, humanity innocently traded its inner fortune.

Having no previous associations or images other than itself, fire captivated and permeated the minds of those who beheld its brilliant dance. Individuals who were once in total fusion with the radiance of their innermost thought life were now externally captivated by the dancing display of lights. Individual attention was completely externalized on the central fire, challenging the eternal underground streams of consciousness. This is the period of fire-learning during which a strange and damaging transmutation of consciousness occurred.

INVASIVE FIRE

The generations-old treasury of dreams, symbols, images, associations, and themes were now invaded by the fire god. Conscious associations were once freely shared, a wash of imagery and meanings flowing between individuals and every aspect of the Natural environment. The water myths. This formed the conscious integrity of individuals with their world. Nature suffused individual consciousness with its own animistic supra-vitality, the mindstate very evidently undamaged in young children.

Until the advent of fire, the otherwise freely expressed mind projections, the dream symbols and dream images, once successfully propelled humanity toward natural and empirical discovery. The old world was the watery time, the time of water songs. Fire sent its red tongues, like so many invasive rootlets, deep into the wells of human consciousness. Its ruthless seizure and incorporation of every symbol, imagination, and theme destroyed the old watery world view and its experience.

Fire destroyed the old world, piercing, appropriating, and re-associating all the song-myths of water. When fire was found, it alone focussed all human attentions. All consciousness, with its symbolic floods and portions focussed on fire. Burned in the fire. The result was an imageless, restructured consciousness in which fire itself played its own unifying role. Fire, the very image of self. False autonomy. Deceptive self-reliance. Fatal self-help.

This fire-fused amalgamation no longer permitted the oceanic consciousness which society once freely enjoyed. Now consciousness was focussed into a constricted singularity which effectively excluded each one who experienced its appearance. Fire had nothing to offer the soul, nothing with which to fill the inner heart. It glimmered and fascinated, a consumptive being unto itself "out there". No feelings, no pictures, no stories, no songs. Fire blocked and sealed all the former imaginal stores.

Having thus literally threaded through the store of images and symbols, opportunistic fire threshed its hurtful path through the thesaurus of mind, ripping the conscious stratum wherever it went. Societies which possessed fire developed new and unnatural mythologies. Fire myths. The external interwoven connectivity of mind and Nature, the extensive vinework of conscious experience, was totally damaged. Rent asunder from within and reconnected in fiery streams. All naturally derived experience, thought, imagination, and myth was now smelted into the fire theme. The invasion had new and unexpected consequences, every component of consciousness being "thrown out" into the fiery focus. The first curse of fire was its invasion of consciousness and its enslavement of conscious process, a fusion which has permeated every mind to this very day.

What did fire do to our minds? Fire blocked us from realizing, with the inner urges which drive our being, the whole of natural provision and the plethora of survival-available energies. Fire took humanity out of the cavern of mind, the safe inner place from which all symbols, images, and other survival urges flow. This dangerous and vulnerable poise intensified our outer dependency on fire, while separating, alienating us from our true ground. Contact with our true selves was lost.

CAPTIVATOR

Not only because of its captivating presence, but because of the power it gave, all minds were pulled into the fire. The water myths represent a whole way of being, not just a cycle of imaginations. In the ancient water songs and water myths are found the very wellsprings of consciousness, the experiential mindstate in which all things natural blend with all things imaginal.

These song-myths did not envision the death of light. Water myths see the sun reborn each dawn, nothing required of humanity. After rain, after dew, after the night, water comes. The sun comes, all flows together and solidifies. The dreams flow at night, the sun going down. All flows in the waters of forever. The dreams, the visions, the knowledge, the past, the future. All flows like waters flowing. The dreams crystallizing into the morning. Found in the forest are the dreams and the dream gems, sparkling. In the water songs, light is eternal. Endless light. Fire placed a halt on these old dreams, on their symbols, on their flowing images, and on the survivalistic fusions with which inner symbol guides outer experience. Watery inner imagery ceased, and fire flickered on. The focus now outside. The old stories of the world and its creation from the water were heard no longer.

Fire divided the mind of society, driving attentions away from the true and radiant conscious source. Out of the mind cavern. Fire de-focussed consciousness and its images of wonder, pulling all attention to itself. It was cruel and demanding. What was the first fuel which the fire demanded and received? It was the whole former world and its song-myths, plucked, burned from the minds of those who yielded to the flames.

FIREWALL

All other symbols and potentials were drawn into the flames, serving them alone. Crafts, taken to the flames, became transformed in nature and intent. Metallurgy, defense, medicine, art, all such were the fire drawn crafts. Fire focussed all attentions outside. Fire demanded attention into itself alone. To serve the flame meant to survive. Fire was the consumptive supplier, an image of death. Fire became synonymous with survival. Humanity was now always only a flicker away from the close-wandering predators. Fire was the wall. Predators without, humans within. Fire was a cage, an entrapment. Fire produced a mindstate which eventually projected outward to become generalized as an apparent worldstate.

Fire provoked a defined schism in perception and thought process. The consequences to this schism, this alienation from the inner world, was devastating. Fire and society were bound together in an unhealthy, unnecessary union which successfully masked humanity's deepest visions. Humanity was now alienated from the life of its own conscious expressions by a technological agency whose presence has remained. Fire lent a false independence from Nature's provisions. Imagining themselves independent of natural dictates. Humanity now believed that Nature was a subordinate entity. Humanity was alienated from Nature's dreams and dream-symbols. Fire burned human consciousness, teaching it a lie. Fire-autonomy brought unnatural separations. The inflation of self, derived from savage use of fire craft, produced a defined schism between humans and the natural world upon which they truly relied.

With fire craft came a new outrage. The "torture of Nature" by deliberate human intent. Those who used fire no longer perceived themselves as passive recipients of natural benefits. They now took whatever provisions they demanded by force. Nature recoiled from the high wielded torch. Plants and animals alike, whole forests and herds all instinctively withdrew and perished by the flame. This second curse of fire successfully alienated humanity from Nature. At war with the Nature which freely and fluidly gave her bounty, humanity wielded frail fire as a hard weapon. Because of the projective will by which Nature could be "subdued", many societies lost the continuity between their own dream projections and Nature.

All aspects of life were gauged against fire. Fire became the social reference, the ultimate expression of dominion. New song-myths even dared to tell that Nature was really "fire born". Fire focussed humanity on fire. Nature was seen through fire. The fire of Nature was sought. Wherever naturally found, fire was worshipped. Fire was strong. Water was weak. Fire burned water. Fire repelled water. All of Nature was seen through the flames.

Mountains seemed to be frozen flames of red and violet, the green trees resembling "green flames".

FIRE SLAVES

The absolute survival need for fire was revealed as an insidious new dependence. Fire began enunciating laws to its willing servants. Empirically learned, these demands were simple equations. No fuel, no fire. No fire, no survival. No fuel...no survival. The simple idea came with a swift and terrifying finality. The third fire curse was the enslavement of human action to demands made by flame, determinations which had now become absolute survival necessities. Should the fire vanish, survival would be totally imperiled.

Humanity learned, and never forgot the lesson. This fear travelled deep into the mind. Humanity has been branded by its mark. The lesson equates the death of fire in the absence of labor. Work makes light. Work makes warmth. Work makes good. All good comes from work. Work makes life. Work makes survival. It equates survival with fire, fire with labor, and light with death. Remarkably modern. The theme re-emerges in the arts periodically.

FIRE PRIDE

Fearful humanity entered into league with the demands which fire enunciated, serving those demands. Now unable to envision an alternative power equalling fire, survival was determined exclusively by one obsession: the acquisition of fuel. Tribal addiction to fire. Fire taught humanity that there were only two realities on which humanity's survival would forever depend: fire and fuel. Servitude to fire taught humans that they were the secondary beings. Fire alone was the supreme Natural expression, the apex by which light was gained against the night, and heat was gained against the death of winter.

The loss of fire was now the greatly feared event. Fear of the cold. Fear of the dark. Fear of what lay waiting in the dark. The pride of fire led humanity out into frozen wastelands where, formerly in times of the water-myths, none would dare venture. Torches high, the adventurers led their tribes onward, ever onward. Elders died in the snows, watching their bodies melting the ice. The water songs sung as they died. The women complained, and children shrieked in the cold. The pride of fire forced the tribal leader to drive them all northward to new lands.

Not heeding the cry of their needy, cries from the sea of consciousness, the proud forged onward. Fire would keep them. Fire would protect them. Fire would drive the ice back. Now amid the icy steppes and glaciers, humanity found itself in a new nakedness. Vulnerability by seduction. Fire had led them forth into dependency zones, where their lives were in new and unfamiliar perils. Too late to recall sense and reason, many tribes were destroyed by fire's ploy.

In northern mountains of blue ice there was no fuel. Many tribes became extinct when fire was lost. Did they stoop to burn each other for the fire? Frozen bodies seated on stone terraces in circles, thick brown hair blowing in the frigid wind. Lost fire. Lost life. All of northland survival was now desperately focussed on fire. Fire, the once wonderful token, began unmasking its true personality. Fire taught humanity that, in the absence of its presence, there would be no survival. Fire was audacious, existing at the expense of human life. Fire supplied certain needs while threatening the lives of those who cared for it. Its chief threat was extinction. If the fire went out, humanity would extinguish.

FIRE EATER

Fire light and fire warmth were both identified with labor, loss, and impermanence; three deadly killers of the human soul. The death of light in the absence of fuel translates directly to the mind in the death of hope. It was a lie which could have been first undone by simply looking into the sun, or gazing at the stars. The old water songs. Humanity's fire promise of freedom and survival became gross enslavement. Fire was more like a beast than magick now.

Fire devoured food of its own. And when devoured, fire vanished with its quarry. Fire would require the whole world for its food, the very ones who tended the flame being consumed. Fire would soon devour everything if so allowed. It was only in this place of betrayals that the transparency of fire was thoroughly understood. Fire was a betrayer of humanity. Fire was the malicious and cunning one. New fire songs became fearful, telling of the fiery fate awaiting the world.

Ultimately all of Nature would be drawn into the flame as its food. Fire would eat the world. Fire hissed in red...a dragon waiting to be fed. Fire suddenly became a hideous and malicious being of evil intent in song-myths. Worse than the predators which roamed at night, those from which fire gave immunity. Fire was the ultimate predator, king of the man eaters, and he had brought into the camp with honors! Now historically viewed and mythically pictured as a deceiver, fire became a fire bird, a fire breathing dragon, a captivating power from which humanity could not escape. The sting of fire in the all-too sensitive heart of humanity left a defined wound. An impression, a lie, a false reality.

And then, fire taught the deadly lie which has singed minds and pursued technologies to this very moment. The single lesson which fire burned into the mind of humanity was one from which all laws of thermodynamics are directly drawn. Fire taught that its light, all light, would come at a price. Fire needs more than labor. Fire needs death. Fire devours the dead branch Fire devours all branches. When the forest is eaten, fire would devour another.

Fire became the eater of dead things. The devourer of once-living things for the gift of its light. A simple exchange. A deal which humanity struck with its new master. Fire light came from the souls of trees, from the death of the watery green forest. Light came from the death of Nature. Monstrous fire required a price, a life, before it would give its light. The equation was a simple one. Fire gives light, fuel gives fire, and death gives fuel. This equation of logic defines the permitted bounds of fire survival, away from which humanity may not stray. In engineering terms, the equation sums the laws of "thermodynamics".

We are yet living in the boundary which fire had set for society. That boundary is a world-model which has been rigidly consolidated for several centuries by countless academicians. The fire lesson so burned itself into the human psyche that it now becomes impossible for them to envision any other than the thermodynamic world model. Today, the hard social demand for machinery is predicated on the survival demand. More light, more heat, more cold, more motion, more food, more media. Fire taught humanity that fixed natural rules would limited their every escape from its dominion.

FIRE MASK

Fire was the technological first wonder, and the first technological failure. Fire produced a paradigm of loss, of consumptions, of exchanges, of conversions, of thermodynamic boundaries. The archaic disappointment produced a distorted view of Nature, and a warped quantitative science. There is no escape today for those whose insistence on the fire paradigm has bound their vision into its cruel service. Fire defined the engineering concepts of the "impossible".

Fire blocked contact with the inner water world. Fire placed a mask, a template over the mind of humanity. To this day, consciousness is likened to a "crucible" where life's fire brings our thoughts to "fusion". Thermodynamic concepts, proceeding from the historical use of fire, have so rooted themselves in human thought structure that all energy forms are comprehended in terms of heat. Energy, all energies, are always each referenced against their "heat equivalent". Thermodynamic laws actually represent the archaic template through which all energies, all technologies are referenced against...fire. Thermodynamic law is the obvious artifact of fire, the template of which we speak. It totally binds our vision, fixating our minds. We see no other way. We see nothing else. It is a world of entropy, of lost fire. Ever losing the fire. The fear dominates. Fire masked the mind.

The thrumming machine algorithm is the yet persistent desire for fire, a desire based on a false premise early learned by those who first used fire craft. The undying demand for new machinery is framed in a lie which promises new survival, but which is only a cunning instantaneous demand for fuel. If there is no fuel, then there is no heat flow. No heat flow means no motion. No motion means no machinery. No machinery means no survival. If there is no fire, there is no society. The lie persists, the boundaries policed by academicians and made secure by financiers. It claims that without heat flow, without fire, there is neither machinery nor human survival.

Fire craft, as the unfortunate first technology, produced negative templates in human consciousness which forever biased human technological expectations. Present technology embraces the fatal thermodynamic theme, embodying the "boundary" which fire craft taught humanity. It is the expected limit in which a machine, exchanging an energetic flow, will perform. In actuality, the thermodynamic boundary is that perimeter which fire has jealously set for humanity, binding attentions so close to its side that we have forgotten the whole of Nature and its thesaurus of energetic wonders.

We design machines exclusively with the fiery "working substance" in mind. If a machine has no working substance, it cannot really work. This explains the sharp separations existing between works of art and works of technology. Art, in the archaic world, functioned in the watery world. Visions moved through and around artworks. Art worked. Now, if the art does not move, get hot, ring or light, it "does not work". It is not machine. It is art. Now it is said that artwork is no work at all. Art is fancy. Machine is real. The fire lie again.

WATER SONGS

The harsh lessons which fire craft brought archaic humanity might never have found so secure a place in the human psyche had the more fundamental energy been technologically explored and socially developed. What fire did to humanity profoundly distorted and divided the human perception of Nature, forever fixating both our world view and rigidifying technology within the so-called "thermodynamic limitations".

Grave damage has been done to consciousness, permanently filtering our vision and thought life. The damage has poisoned our technological dreams. Our desperate need for more heat, more light, more motion, is incapable of discovering any other means for attaining those manifestations without fire. Machineworks without fire are deemed "impossible". Few engineers consider the possibility that there may exist more fundamental natural energies than fire or heat. Those energies run eternally. Those energies are everywhere, not the conventional natural manifestations of sea and sun, stars and wind alone. There are other energies as there are visions and dreams. In all our thoughts concerning survival, we have not been willing or able to escape the fire paradigm.

Despite this sad mindstate from which many have not escaped, the horrid vision-binding template of firecraft does not represent a complete world view.

Fire is not the only energy which Nature has to show us. It is time again to consider what the watery song-myths of the old green forest world yet sing. There is fire, and it kills. But Nature is a garden, a thesaurus of dynamics.

The working substance of the water world fundamentally is visions, symbols, archetypes. And there are technologies by which these visionary streams are summoned and magnified. But, these visionary currents do find their materializations in the natural world. Children know this. Empirical inventors seek those materializations. And those materializations, ever present and scarcely seen, are periodically and "accidentally" discovered.

It is only when the visionary currents so suffuse individuals that the fire mask, the template, is lifted for a rare instant. It is then that "anomalous discoveries" are made. In truth the "anomaly" is our own mind block. The lifting of that template requires enormous watery suffusions of dream and faith in the minds of those who seek. This is why those who discover "anomalous" energies and who build "anomalous" technologies are rarely part of the academic fold, where the fire template is harsh and killing. Where the dreams are shunned, hated, feared, eradicated, and buried. In truth, the watery suffusions of symbol, image, and dream destroy the fire template. The elders knew the secret. Water quenches the fire. Water break fire.

WATER WAY

It is both astonishing and conspicuous that thermodynamics, a development of great minds, had absolutely no regard for the consciousness which spawned its mathematical phrasing. Consciousness should be the very first consideration when studying any science. Consciousness is the energy by which we experience and by which we know our being.

Had discoveries of the naturally prolific energy, preceded the discovery of fire, then the human psyche and its various technological expressions would not have focussed on fire and its "thermodynamic limits". There exist fundamental energies whose nature is to grow from negative states, contrary to theoretical limits. The naturally prolific energy is the energy of dreams. The real materializations of dream currents are those "anomalous" energies, everywhere potential. They are real. They are not ephemeral. They have been found and used by numerous inventors, and they are embodied in patents which work.

Unlike fire, plants do not exist to consume and then vanish. Living things everywhere seen defy the thermodynamic rule. Concerning the growth process and the tendency toward living organization, fire can teach nothing. Survival, machines, fire, and fuel are each links in a heavy chain which has been cast around the neck of humanity far too long.

Fire. Fuel. Fire. Fuel. The ring of captivity from which neither minds, souls, nor bodies have escaped. The ring tightens, enslaving human vision along its closed loop. It is the very image of the snake, devouring itself and choking humanity to death in the process. Fire split the mind of humanity, fixating all vision out along its winding smokey paths. No one can even envision another way to obtain energy. It falls to those sensitives and dreamers who are fortunate recipients of accidental observations and rare phenomena to find the way out.

The magick which breaks the circular spell is beyond those who have forgotten how to dream. The images we require have always been speaking to us. They are accessible. They do not require arduous exercise or meditations. The images are forever flowing. They are the deepest radiance. It is we who have forgotten how to recognize them apart from the background of thoughts which occupy us.

Those whom fire has seared cannot and will not accept these ideas, but there exists an energy stratum in which both inner dream and outer Nature are in fusion. The art of observing imagery and deeper meanings is the common experience of those who love rich poetry. It is a sensitive art, requiring a willingness to enter the written script by which poets lead us.

Unlike poetry, life has no pre-written "image script". Nevertheless, our daily activities are replete with spontaneous emergence of images, impressions, and essential meanings. This deep radiance is cherished by those who recognize their manifestations against the background of the ordinary. Distinguished, as it were, among the world-visual spaces which surround us.

WATERY REFLECTIONS

Though energies are both employed and liberated in organic life cycles, the growth of living things is not primarily driven for energetic reasons. Energetic conversions, flow, and exchange may be traced throughout organisms. Nevertheless, we are innately aware that these thermodynamic maps do not in any way represent the primary organismic function or reason for being. One may cite numerous references which indicate that a primary sensitivity and consciousness is shared by vegetation (Bose, Backster, Lawrence). Despite these marvels of discovery, we must look to the laboratory of our own experience for deepest understanding.

But, some will argue, consciousness and matter are "completely different" realms! Nevertheless, one will notice that specific imaginal sequences persistently occur in specific locations. They come unbidden, though forgotten for years. It is remarkable how they come, and with their attendant moods and sense. Exact synaesthesial similacra. Conscious energy flows through material forms, being discharged into percipients who wander near them. Consciousness alone is the energetic flow which discerns, enjoys, and participates in world reality. The fundamental world reality is a sea of consciousness, a watery flood of prolific conscious energy.

Symbols and forms of thing unknown, emerging from the dream sea, once exclusively guided human actions. Inspirations, revelations, and visions spoke as nature spoke, driving the sailed ships of civilization toward the sun. There was no difference, no separation between dream and Nature. Consciousness, conducted through our being as images and meaningful symbols, can never be measured in thermodynamic terms. The energy of consciousness can never be gauged according to the ridiculous limitations set by the fire paradigm. Consciousness cannot be measured directly. Despite the quantitative inability to measure this most fundamental energy form, consciousness persists.

Dreams and dream-symbols are evidence of the natural radiance. Images, symbols, and dreams come, unbidden messengers up from the deepest world. The great minds have each attested that this is their true and only source of creative potential. The numerous mythologies clearly picture the magick of radiance, of endless light, throughout their lovely rhymes. The old watery green gabled world of mountains and forests. The deepest radiance is one which is first experienced as consciousness itself. The evergreen tree, watery forest symbol of endless life, would have taught humanity everything required without fire.

After centuries of languishing in the fire paradigm, ancient natural philosophers refocused their vision on the fundamental world of dream and mind. There they learned that subjective fusion with both dream and Nature is fundamental world reality. In this ground of being they began making astonishing discoveries. Emergent symbols and images formed wonderfully sinuous paths into the external world. The symbols envisioned, radiant and undying, were sought and found by visionary seekers. The arts, born of these realizations, were called Alchymy and Geomancy.

The world is a garden of conscious symbols which has materialized. Like crystals, half in and half out of their ground fully formed, portions of the dream symbols were found in Nature. Metadimensional visitors, bridges between the worlds, the radiant natural correspondences were located and retold. All the dim shadows of the radiant world.

The desire for the magickal and the wonderful does not die. Desires exist, it has been said, because the objects of desires exist. Though yet unseen, the sweet thorn of an unrequited desire cries out for its fulfillment. Desires are messages to us, illuminated visions of realities which can be touched. Dreams, images, and symbols of eternally radiant lamps, of endless light, represent a world which lives within that world which appears. It is the very soul of the apparent world which projects these symbols. The sensitive few who quest for these realities are never disappointed.

MY MANY THANKS

Mr. John LaCorte, Theresa Cappici, and Anthony DeNonno
for bibliographic materials on Antonio Meucci

The warm and dear people of Murray, Kentucky
Mrs. Dortha Bailey and Mr. E.R. Bailey,
Mrs. Baker, Ms. Alexander for Nathan Stubblefield files

My Many, Many Thanks to the entire Farnsworth Family

William Lehr for all your remarkable revelations and fellowship

Dan Winter for your friendship and wonderful studies!

Preston Nichols for your rare and encouraging Anecdotes
on Tesla and Moray

Mr. D. Crnosiya for personal experience with Tesla

We All Thank You Mr. John Crane

Bob Nelson of REX RESEARCH: JEAN, NEVADA.

BORDERLAND SCIENCES RESEARCH FOUNDATION

BIBLIOGRAPHY

BOOKS

Abrams, A. *New Concepts in Diagnosis and Treatment*, San Francisco, 1916, Reprinted by Borderland Sciences Research Foundation.

Cheney. *Tesla: Man Out Of Time*, 1981.

Corliss, W. *Unusual Natural Phenomena*, Arlington House, New York, 1986.

Corliss, W. *Lightning, Auroras, Nocturnal Lights*, Sourcebook Project, Maryland, 1982.

Corliss, W. *Earthquakes, Tides, Unidentified Sounds*, Sourcebook Project, Maryland 1983.

Devereaux. *Earthlights*, Turnstone Press, UK, 1982.

Dollard, E. *Theory of Wireless Power*, Borderland Sciences Research Foundation, 1986.

Drown, R. *The Drown Homo-Vibra Ray and Radiovision Instruments*, Borderland Sciences Research Foundation, 1951.

Farnsworth, E. *Distant Vision*. Pemberley Kent Publishers, Salt Lake City, Utah. 1989.

Hackh's Chemical Dictionary, Grant, McGraw-Hill, New York, 1944.

Kilner, W. *The Human Atmosphere*, London, 1911.

LeBon, G. *Evolution of Matter*, Walter Scott, New York, 1907.

LeBon, G. *Evolution of Forces*, Dryden House, London, 1908.

Lodge, O. *The Ether of Space*, Harper and Brothers, London, 1909.

Martin. *Lectures of Nikola Tesla*, New York, 1893.

Mendeleev, D. *A Chemical Conception of The Ether*, New York, 1904. Reprinted by Rex Research, Jean, NV.

Moray, T.H. *Radiant Energy*, Los Angeles, 1928, Reprinted by Borderland Sciences Research Foundation.

Moray, T.H. *The Sea of Energy*, Salt Lake City Press, 1930.

Moray, T.H. *Beyond The Light Rays*, Res. Institute, Salt Lake, 1931.

O'Neill, J. *Prodigal Genius: The Life of Nikola Tesla*. Brotherhood of Life, 1996.

Reichenbach, K. *Dynamics of The Vital Force*, 1851, Borderland Sciences Research Foundation.

Reichenbach, K. *Letters on Od And Magnetism*, 1852, Borderland Sciences

Research Foundation.

Reichenbach, K. *Somnambulism And Cramp,* 1860, Borderland Sciences Research Foundation.

The Stubblefield Papers, Pogue Special Collections Library, Murray State University, Kentucky.

Tesla, N. *Colorado Springs Diary* (1899), Nolit, Beograd, 1978.

Tesla, N. *Supreme Court Transcripts*, Suffolk County, N.Y.S., 1915.

Tomas. *We Are Not The First*, Bantam Books, 1971.

Vassilatos, G. *Vril Compendium Volume 2 (Vril Telegraphy)*, Borderland Sciences Research Foundation, 1992.

Vassilatos, G. *Vril Compendium Volume 4 (Vril Archeforms)*, Borderland Sciences Research Foundation, 1992.

Vassilatos, G. *Vril Compendium Volume 5 (Vril Connection)*, Borderland Sciences Research Foundation, 1992.

Vassilatos, G. *Vril Compendium Volume 6 (Vril Telephony),* Borderland Sciences Research Foundation, 1992.

Vassilatos, G. *Vril Compendium Volume 7 (Vril Dendritic Ground Systems),* Borderland Sciences Research Foundation, 1992.

White, G.S. *The Finer Forces of Nature,* 1929, Reprinted by Borderland Sciences Research Foundation.

PATENTS

Ainsworth, 1.145.735 “Electric Wave Detector”, 1915.

Blackmore, 806.052 “Receiver For Wireless Telegraphy”, 1905.

Brown, 300.311 (UK) “Apparatus For Producing Force”, 1928.

Brown, 1.974.483 “Electrostatic Motor”, 1930.

Brown, 3.018.394 “Electrokinetic Transducer”, 1957.

Brown, 3.022.430 “Electrokinetic Generator”, 1962.

Brown, 3.187.206 “Electrokinetic Apparatus”, 1965.

Burke, 3.409.820 “Electric Power Apparatus”, 1968.

Gavreau, Saul 131.551 (Fr) “Infrasonic Generator”.

Gavreau, Saul 437.460 (Fr) “Infrasonic Generator”.

Gavreau, Saul 1.536.289 (Fr) “Infrasonic Shield”.

Farnsworth, 2.071.515 “Electron Multiplying Device”.

Farnsworth, 2.071.516 “Oscillation Generator”.

Farnsworth, 2.143.262 “Means Of Electron Multiplication”.
Farnsworth, 2,141,837 “Multi-Stage Multipactor”.
Farnsworth, 2,184,910 “Cold Cathode Electron Discharge Tube”.
Farnsworth, 2,217,860 “Split Cathode Multiplier”.
Farnsworth, 2,141,838 “Split Cathode Multiplier Tube”.
Farnsworth, 3,258,402 “Producing Interaction Between Nuclei”.
Farnsworth, 3,386,883 “Producing Nuclear Fusion Reactions”.
McElrath, 2.032.545 “Electron Tube”, 1931.
Meucci Caveat “Galvanic Battery” 1859.
Meucci 36.619 “Mineral Oil in Paint” 1862.
Meucci 38.714 “Hydrocarbons in Paint” 1863.
Meucci Caveat “Sound Telegraph” 1871.
Moray, 2.460.707 “Electrotherapeutic Apparatus”, 1943.
Tesla, 381.968 “Electromagnetic Motor”, 1888.
Tesla, 381.969 “Electromagnetic Motor”, 1888.
Tesla, 381.970 “System of Electrical Distribution”, 1888.
Tesla, 382.280 “Electrical Transmission of Power”, 1888.
Tesla, 405.858 “Electromagnetic Motor”, 1889.
Tesla, 447.921 “High Frequency Alternator”, 1890.
Tesla, 462.418 “High Frequency Currents”, 1891.
Tesla, 577.670 “High Frequency Electric Currents”, 1897.
Tesla, 583.953 “High Frequency Electric Currents”, 1897.
Tesla, 609.245 “Electric Circuit Controller”, 1898.
Tesla, 645.576 “Electrical Energy Without Wires”, 1899.
Tesla, 645.576 “Transmission of Electric Current”, 1900.
Tesla, 787.412 “Electrical Energy Through Natural Mediums”, 1902.
Tesla, 1.119.732 “Transmitting Electrical Energy”, 1914.
Wallace, 3.626.605 “Generating Gravitational Force Fields”, 1971.
Wallace, 3.626.606 “Generating A Dynamic Force Field”, 1971.
Winkelmann, 1.650.921 “Vacuum Tube”, 1923.

ARTICLES

Brown, T.T. “Gravitation”. *Science And Invention*, August 1929.
Brown, T.T. “Optical Frequency Gravitational Radiation”, *Notebooks*, Au-

gust 1976.

(Farnsworth), "New Amplifier Amazes Radio Engineers", San Francisco *Engineer* (March 5, 1936).

Fawcett, W. "Wireless Telephony", *Scientific American*, May 1902.

Fehr, "Infrasound From Artificial and Natural Sources", *Journal*

Geophysical Research, May 1967.

Gavreau V. "Infra-Sons", *Acustica*, Vol.17, 1966.

Gavreau, V. "The Silent Sound That Kills", *Science and Mechanics*, January 1968.

Halloran, "Farnsworth's Cold-Cathode Electron Multiplier Tube", *Radio,* October 1932.

Hoffer, Thomas. "Nathan B. Stubblefield and His Wireless Telephone", *Journal of Broadcasting*, 1971.

Houston, "Ether Density", *Electrical Engineer*, March 1894.

Markovitch, "Apparatus For Transmitting Electrical Energy", *Rex Research, Jean, NV.*

Mohr, "Effects of Low Frequency Noise on Man", *Aerospace Medicine*, Sept. 1965.

Moos, "Fohn Weather/Accidents", *Aerospace Medicine*, July 1964.

Moray, T.H. "Gamma Rays", *Research Institute*, 1946.

Moray, T.H. "Recovery of Minerals", *Research Institute*, 1964.

Moulton, "Gravity Conquered At Last?" *Technical World Magazine,* November 1911.

Nipher, "Gravitation And Electrical Action" *Transactions Academy of Sciences St. Louis*, February 1916.

Nipher, "Can Electricity Destroy Gravitation?", *Electrical Experimenter*, March 1918.

Piggot, "Overcoming Gravitation", *Electrical Experimenter*, July 1920.

Payne, B. "An Apparatus For Detecting Emanations From Planets", p. 7, *Journal of Borderland Research*, November-December 1990.

RIFE [Collection Obtained Through Borderland], "Contrast Methods in Microscopy", Olympus Corporation, 1987.

"Filterable Bodies Seen Through The Rife Microscope", *Science*, 1928, vol.74.

"Filterable Germ Forms Seen With New Supermicroscope", *Science,* December 1931,

"Observations on Bacillus Typhonsus", *California and Western Medical Jour-*

nal, Dec. 1931.

"Filtration of Bacteria", *Science*, March 1932.

"Poliomyelitus Virus Seen Under Electron and Light Microscopes", *Science,* Proc. Staff Meeting, Mayo Clinic, Feb 1942.

"The New Microscopes", *Smithsonian Institute Report*, 1944.

"The New Microscopes", *Journal of The Franklin Institute,* Feb 1944.

"Observations With The Rife Microscope", *Science*, Aug 1932, Sagnac.

"Matter On Ether", *Comptes Rendu*, November 1899.

Smith, Gene, "ITT Hopeful On Experiments To Harness H-Bomb Power", *N.Y.Times*, (January 4, 1961).

Tesla, "Transmission of Electrical Energy Without Wires", *Electrical World and Engineer*, March 1904.

Williams, Ernest "Farnsworth's A-Power Unit Near Reality", *Fort Wayne News Sentinel,* (January 3, 1961).